MALIGNANT HYPERTHERMIA

MALIGNANT HYPERTHERMIA

edited by

Beverley A. Britt

Martinus Nijhoff Publishing
a member of the Kluwer Academic Publishers Group
Boston Dordrecht Lancaster

Distributors

for the United States and Canada: Kluwer Academic Publishers,
101 Philip Drive, Assinippi Park, Norwell, Massachusetts 02061, USA

for the UK and Ireland: Kluwer Academic Publishers, MTP Press Limited,
Falcon House, Queen Square, Lancaster LA1 1RN, UNITED KINGDOM

for all other countries: Kluwer Academic Publishers Group, Distribution
Centre, P.O. Box 322, 3300 AH Dordrecht, THE NETHERLANDS

Library of Congress Cataloging-in-Publication Data

Malignant hyperthermia.

Includes bibliographies and index.
1. Malignant hyperthermia. I. Britt, Beverley A.
[DNLM: 1. Anesthesia—adverse effects. 2. Malignant
Hyperthermia. WO 245 M2505]
RD82.7.M3M33 1987 617′.96 87-20241
ISBN 0-89838-960-7

Printed in the United States of America

CONTENTS

CONTRIBUTING AUTHORS

BEVERLEY A. BRITT
Associate Professor, Departments of Anaesthesia and Pharmacology, University of Toronto; Senior Staff Anaesthetist, Toronto General Hospital, Toronto, Ontario M5G 2C4

A.K.W. BROWNELL
Professor of Clinical Neurosciences and Medicine, The University of Calgary; The Malignant Hyperthermic Detection Program, Foothills Hospital, Calgary, Alberta T2N 2T9

H.M. CARLIN
Assistant Professor, Department of Anesthesiology, Uniformed Services University of the Health Sciences, Bethesda, Maryland 20814

K.S. CHEAH
Agricultural and Food Research Council, Institute of Food Research, Bristol Laboratory, Langford, Bristol BS18 7DY, U.K.

D.G.F. HARRIMAN
Reader in Neuropathology, The University of Leeds, Neuropathology Laboratory, Department of Pathology, University of Leeds, Leeds LS2 9JT, U.K.

G.G. HARRISON
Professor and Head, Department of Anaesthetics, University of Cape Town and Groote Schuur Hospital, Cape Town, South Africa 7925

D.E. JONES
Instructor in Anesthesia, Baylor College of Medicine and University of Texas; Staff Anesthesiologist, Heights Hospital, Houston, Texas 77008

W. KALOW
Professor, Department of Pharmacology, University of Toronto, Medical Sciences Building, Toronto, Ontario M5S 1A8

H. KIKUCHI
Associate Professor, Department of Anesthesiology, Hiroshima University School of Medicine, Hiroshima City 734, Japan

L. KLEIN
Associate Professor of Anesthesia, University of Pennsylvania, School of Veterinary Medicine, Kennett Square, Pa. 19348

J. LERMAN
Assistant Professor, Department of Anaesthesia, University of Toronto; Staff Anaesthetist, Hospital for Sick Children, Toronto, Ontario M5G IX8

K. MATSUI
Instructor, Department of Anesthesiology, Hiroshima University School of Medicine, Hiroshima City 734, Japan

M. MORIO
Professor and Chairman, Department of Anesthesiology, Hiroshima University School of Medicine, Hiroshima City 734, Japan

S.M. MULDOON
Professor and Chairman, Department of Anesthesiology, Uniformed Services University of the Health Sciences, Bethesda, Maryland 20814

T.E. NELSON
Professor, Department of Anesthesiology, The University of Texas Medical Branch, Galveston, Texas 77550

G.A.R. O'CONNOR
Assistant Clinical Professor, Department of Anaesthesia, University of British Columbia; Staff Anaesthetist, British Columbia Children's Hospital, Vancouver, British Columbia V6H 3V4

H. ORDING
The Malignant Hyperthermia Register, Department of Anaesthesia, Rigshospitalet, University Hospital, DK-2100, Copenhagen, Denmark

J.E.S. RELTON
Associate Professor, Department of Anaesthesia, University of Toronto; Senior Staff Anaesthetist, Hospital for Sick Children, Toronto, Ontario M5G IX8

H. ROSENBERG
Professor and Chairman, Department of Anesthesiology, Hahnemann University Hospital, Philadelphia, Pa. 19102

J.F. RYAN
Associate Professor of Anesthesia, Harvard Medical School; Director, Pediatric Anesthesia, Massachusetts General Hospital, Boston, Massachusetts

D.J. STEWARD
Professor, Department of Anaesthesia, University of British Columbia; Chief of Anaesthesia, British Columbia Children's Hospital, Vancouver, British Columbia V6H 3V4

PREFACE

A HISTORY OF MALIGNANT HYPERTHERMIA

Malignant hyperthermia (MH) is a hereditary disorder of muscle. Undoubtedly, individuals have possessed this trait since time immemorial. However, because the trait is usually only unmasked in the presence of potent inhalational anaesthetic agents or non-depolarizing skeletal muscle relaxants, the existence of malignant hyperthermia was not suspected until well after the dawn of the modern anaesthetic era.

In the early years of ether and chloroform anaesthesia, monitoring was minimal. Body temperature was never measured. A finger on the pulse, and observation of respirations and skin colour were the most that could be expected. Death was not infrequent and usually unexplained (1). By the beginning of the twentieth century, reports of fulminant fever and tachycardia (rapid heart rate) during or immediately after anaesthesia often ending in death, were being described with increasing frequency in the medical literature (2-6). As a number of cases from New York had occurred during summer months, they were initially thought to be a form of heat stroke due to overly hot operating theatres (2-6). However, one enterprising anaesthetist (5) checked the weather reports for the days on which some of these so called "heat strokes" had occurred. He found that on the days in question the ambient temperature had never been in excess of 72°F. Environmental heat, therefore, could not have been a cause of at least some of these reactions.

Guedel, author of a famous anaesthetic textbook published in 1937 (7), reported that during 20 years of practice he had seen six cases of post-operative hyperthermia which he thought might have been due to diethyl ether. In all instances the temperature rose within a few hours to between 108°F and 110°F and all patients died. Necroscopy showed only cerebral oedema. Lee, in his 1953 edition of "A Synopsis of Anaesthesia" (8) noted that "ether convulsions" associated with excessive rises in temperature, had been reported with increasing frequency since 1926 and that many of the "convulsions" occurred in children and young adults. Hewer, in the sixth edition (1948) of "Recent Advances in Anaesthesia and Analgesia" (9), discussed a series of 22 patients afflicted with "late ether convulsions" which were associated with intense muscle activity and followed by high temperatures. Cullen, in the third edition of "Anaesthesia in General Practice" (10) described "ether convulsions" and high fevers in adolescents and robust young adults. He listed as contributory factors: high endogenous or exogenous temperatures, retention of carbon dioxide and hypoxia.

The first case report in which the genetic nature of per anaesthetic fever was recognized, was published by Denborough and Lovell (11) first in the July 2nd, 1960 issue of Lancet, and later, in more detail, in the June, 1962 issue of the British Journal of Anaesthesia (12). Denborough described a successfully treated case of fever triggered by anaesthesia in a young, previously healthy male who had been very apprehensive prior to induction because ten of his relatives had died during anaesthesia. The attending anaesthetist reassured the young man that there was no need to worry because he was going to be put to sleep with a new inhalational agent called halothane. However, during the anaesthetic, the patient rapidly developed pallor, tachycardia, cyanosis, and a hot and sweaty skin. He remained comatose for 30 minutes following the end of the operation. After a stormy post-operative course, he finally made a complete recovery. Denborough, an internist, was then asked to investigate this patient's family. He found that ten relatives had indeed died during anaesthesia. For all ten, the anaesthetic agents used were ethyl chloride and diethyl ether. In the three best documented cases the deaths were preceded by convulsions and fulminant fevers. The pattern of inheritance was compatible with a dominant gene or genes.

In the early and mid 1960s, a flurry of similar cases began to occur across North America triggered not only by halothane, but also by other new and potent inhalational anaesthetic agents and by the recently discovered depolarizing skeletal muscle relaxant, succinylcholine. Dr. R.A. Gordon, Professor of Anaesthesia of the University of Toronto, was gravely concerned about these unexplained deaths from fulminant hyperthermia. He, therefore, organized a symposium in Toronto which was attended by thirteen Canadian anaesthetists, each of whom had personally experienced one or more of these cases. The details of the reactions described at this symposium were subsequently published in 1966 in the Canadian Anaesthetists' Society Journal (13). In his editorial preceding these articles, Dr. Gordon coined the name "Malignant Hyperthermia (MH)" to describe this hitherto unnamed syndrome. This publication roused great interest in the anaesthetic world and stimulated reporting of other cases and much speculation as to the cause of malignant hyperthermia.

In 1969, a family was described, living in northern Wisconsin, containing 138 individuals at risk, 21 of whom had had MH reactions during anaesthesia (14). Eight of these had died during or shortly after completion of the surgery. Commencing in 1971 an even larger French-Canadian kindred was investigated (15) who lived mainly in Essex and Kent counties in southwestern Ontario and around Lake Temiskaming in northeastern Ontario and adjacent parts of northwestern Quebec.

Early research into MH was directed toward trying to find a defect in the central nervous system (16). These efforts were fruitless. Furthermore, post-mortem examinations never detected any central nervous system lesion which could account for the anaesthetic or stress-induced fevers. Attempts were then made to artificially create suitable animal models by infusing into rats and dogs catabolic agents such as dinitrophenol (17). These substances increased heat production but decreased useful work performed in the body's cells. Unfortunately, subsequent administration of inhalational anaesthetics such as halothane did not greatly aggravate the pre-existing fever or produce other typical signs of MH.

Epidemiological surveys of MH patients, commenced in the 1960s (18), revealed clinical features which suggested that the primary defect lay in the skeletal muscle. For example, common pre-anaesthetic findings included skeletal muscle abnormalities such as cramps, kyphoscoliosis, club foot, hernias, strabismus and ptosis; per anaesthetic extreme rigor-mortis-like skeletal muscle rigidity; and, massive elevation in the blood of potassium, myoglobin, muscle enzymes and of acid waste products of metabolism such as carbon dioxide and lactic acid. This suspicion was confirmed by the observation of a MH reaction in a patient, to whose arm a tourniquet had been applied. Rigidity occurred in all muscles except those in the arm distal to the tourniquet. Direct contact of the anaesthetic with the skeletal muscle was, therefore, shown to be essential for the reaction to be triggered, thus ruling out a central nervous system aetiology (19).

These early research efforts were hampered by the lack of an experimental animal model that possessed the genetic MH trait. This very serious problem was corrected one evening at a Royal Society banquet attended by Dr. Gordon, and by Dr. Neville Woolfe, an experimental animal pathologist from Cambridge. Dr. Woolfe related to Dr. Gordon the unfortunate premature end of a recent pharmacological experiment utilizing young Landrace pigs. A pig, one of a litter of three, had been anaesthetized with halothane and succinylcholine. Instead of the expected smooth and relaxed anaesthesia, the animal became intensely rigid, hyperthermic, cyanotic, tachycardic and arrhythmic. In a short time, the animal expired. A litter mate was then anaesthetized and also died. A third litter mate was then similarly anaesthetized but again met the same sorry fate. As tea time had arrived, experimental efforts for the day were terminated. Dr. Gordon, on hearing this story, immediately realized that these animals had undoubtedly suffered MH reactions. Shortly thereafter, Drs. M.C. Berman and G.G. Harrison from the University of Cape Town reported malignant hyperthermia in South African Landrace pigs (20). Then Drs. E.W. Jones and T.E. Nelson discovered malignant hyperthermia in Poland-China pigs in Oklahoma (21). The trait was also observed in Pietrain pigs in Belgium and Minnesota (22,23). Soon it was realized that the MH trait was widespread in pig herds throughout the world.

Dr. Nelson, in 1971 at the First International Symposium on Malignant Hyperthermia in Toronto (24), and shortly thereafter Dr. G.G. Harrison (25) suggested that porcine MH and Pork Stress Syndrome (PSS) were probably identical. PSS had long been a scourge of farmers (26-34). Pigs, bred to be lean and meaty to satisfy consumers tastes, showed a distressing propensity to develop reactions similar to those of MH, when upset or over-exercised. For instance, exposure to a very high ambient temperature, fighting for status in the pens, or being forced into trucks to go to market were likely to end with several pigs rigid, blue, hyperthermic and in extremis. Death was the usual outcome (25). Subsequent experience showed that these syndromes were in fact identical, the same animals being susceptible to both PSS and MH.

In the late 1960s it was hypothesized that the defect of MH lay in the membrane of the sarcoplasmic reticulum (SR) (35) so that in the presence of drugs known to trigger MH reactions too little calcium was taken up by the SR leaving too much calcium in the muscle cytoplasm (36,37). Later, work by Kim using much more sophisticated equipment showed that the defect in MHS SR was due to release of excessive amounts of calcium rather than to failure of uptake of calcium into the SR (38). Nelson, in a series of very elegant experiments has now shown that this extraordinarily high rate of release of calcium from the SR also required the presence of extra calcium on the outside of the SR--a phenomenon known as "calcium induced calcium release" (39).

Until the latter part of the 1960s no method existed of diagnosing with a certainty the MH trait prior to anaesthetic challenge. Then in 1970, Dr. W. Kalow and his team described in Lancet (40), a reliable test to diagnose MH. These workers found that the dose of caffeine required to induce a contracture in a small isometrically mounted strip of muscle was less in MHS than in normal muscle. Additionally, it was later observed that in very severely afflicted MHS muscle, halothane caused contractures which it never did in normal muscle (41). This test, now called the "caffeine halothane contracture test (CHCT)", became and has remained the standard method of diagnosing MH (40,41).

A very recent pertinent observation was that of Drs. Fletcher and Rosenberg who found that the sequence in which halothane and succinylcholine were given was an important determinant of the severity of the reaction (42). Thus reactions triggered by a halothane succinylcholine sequence were more severe than were those induced by a succinylcholine halothane sequence. They also observed that contractures in isolated muscle strips were larger if the muscle was exposed first to halothane and then to caffeine than if the muscle had been bathed in first succinylcholine and then in halothane. Using the former sequence they were able to improve the diagnostic discrimination of the halothane contracture test (42).

Between 1965 and 1975 mortality from MH reactions slowly declined because of increasing awareness of the condition by anaesthetists and because of improved patient monitoring (particularly body temperature and electrocardiograms) during anaesthesia. Nevertheless, the majority of patients continued to die or to survive, only permanently brain damaged. Treatment of acute MH crises remained difficult and often unsuccessful. Return to normal levels of temperature, heart rate, muscle tone, carbon dioxide and oxygen tensions, electrolytes and enzymes often did not bring about survival of the patient. Rather, brain death ensued (43,44).

However, in 1975, a small supply of a drug called dantrolene sodium in the powder form was given to Dr. Harrison of Groote Schuur Hospital at the University of Cape Town in South Africa, to determine if it would reverse MH reactions in his susceptible Landrace pigs (45). Oral dantrolene capsules had been used for several years previously to treat muscle spasms and cramps associated with various neuromuscular conditions (46-48). Dantrolene sodium was thought to exert its relaxant effect on skeletal muscle by inhibiting release of calcium from the SR, although the exact mechanism by which it did so was, and still is, not well understood (49). No intravenous form of dantrolene was in use at that time because of its extreme insolubility in water.

Nevertheless, Dr. Harrison found that by mixing dantrolene sodium powder with mannitol and sodium hydroxide, it was possible, after about three laborious hours of vigorous heating and stirring, to dissolve about half a milligram of dantrolene in each millilitre of solution. For his first experiment, he prepared a small volume of the dantrolene-mannitol-sodium hydroxide mixture. A pig known to be susceptible to MH was then challenged with halothane. In due course, a typical MH reaction--with rigidity, tachycardia, ventricular arrhythmias, cyanosis, fever and acidosis--developed. Dantrolene infusion was immediately begun. Within a few minutes the animal began to recover. The fever was subsiding, the heart rate was slowing and becoming more normal in rhythm, the skin colour was improving, the muscles were relaxing and the metabolic acidosis was disappearing. However, the precious supply of dantrolene was rapidly disappearing. Soon it was all gone! Rapidly the reaction recrudesced. Dr. Harrison tried to reconstitute more dantrolene but before he could do so the pig died (45).

Fortunately, Dr. Harrison still had several more malignant hyperthermic susceptible (MHS) pigs in his holding pens. He made up a much larger supply of intravenous dantrolene and then anaesthetized each animal with halothane as before and when the MH reaction commenced again infused dantrolene. For these pigs the larger amount of dantrolene available proved to be adequate and all survived (45). A new era in the treatment of MH had dawned.

When the anaesthetic world heard of this magic new drug, many anaesthetists attempted, with varying degrees of success, to make an intravenous formulation by dissolving dantrolene capsules. In most instances the blood levels of dantrolene actually achieved by these early preparations did contain significant amounts of available dantrolene. Thanks to these later formulations the survival rate from acute MH reactions began to dramatically improve--at a much more rapid rate than previously (50). In 1979 a commercial formulation was approved for use in Canada and then in the United States, consisting of a lyophillized dantrolene-mannitol-NaOH mixture which could be reconstituted with water in three minutes instead of three hours.

Recently considerable efforts have been made to find a non-invasive test to diagnose MH in humans and pigs, particularly in human relatives of known MH individuals. While some of the tests so far reported have some value as screening guides, none are reliable diagnostic tools since all are plagued by a high incidence of false negative and false positive results (50–57). However, several performed in concert can certainly raise one's index of suspicion that the patient may possess the MH trait. Furthermore, these non-invasive tests have considerably enhanced our knowledge of the aetiology of MH in that they have revealed that MH is not a defect confined to the skeletal muscle, but rather is a widespread membrane defect involving many types of connective tissue throughout the body--for instance, bone, nerves, heart muscle, red blood cells and lymphocytes (50–57).

In the 1960's, the mortality rate from MH reactions was 84% (15). As the years have passed the mortality rate has steadily declined, slowly at first, and much more rapidly recently to about 7% at the present time (58). In the 1970s declines in mortality were mainly due to a better understanding by anaesthetists of the clinical signs and pathophysiology of MH and to progressive improvements of per-anaesthetic monitoring of patients. In the 1980s, reductions in the death rate have been primarily due to the widespread early use of intravenous dantrolene during MH reactions as well as to the general availability of more sophisticated monitoring devices such as pulse oximeters and end tidal CO_2 monitors. It should prove possible in the future to reduce the mortality rate from inadvertent MH reactions to zero. Better still would be the development of a reliable, simple, cheap and non-invasive test which could be performed on every patient presenting for anaesthesia. This type of test would virtually prevent untoward reactions from ever occurring.

Until such a test is developed it is hoped that this book will assist the reader in: detection of MHS individuals before surgery; providing safe anaesthetic techniques for those who require elective operations; making early diagnoses of acute MH crises during anaesthesia and other stressful situations; and in managing MH reactions effectively. To achieve these goals, an understanding of the pattern of inheritance and of the biochemical aetiology and pathophysiology is essential. Detailed discussions of these aspects of the syndrome are, therefore, included. I hope that this issue will contribute to a reduction in the occurrence of MH and to an improved survival of those patients who continue to sustain inadvertent reactions.

I wish to extend my deepest gratitude to Mrs. Marilyn Berg, whose expert organization and ever present cooperation in preparation of this volume is most sincerely appreciated. As well, many thanks to my technicians, Mrs. Elizabeth Scott and Mrs. Wanda Frodis for their continued and valued assistance in our laboratory. My special thanks goes not only to the contributors to this issue, but also to the surgeons, technicians, nurses and secretaries, without whom the research described on the following pages would not have been possible.

Beverley A. Britt

REFERENCES

1. Sykes, W.S. Essays on the First Hundred Years of Anaesthesia. 1972. Huntington, NY.: R.E. Krieger Publishing Company.

2. Gibson, C.L. JAMA 1900;35:1685.

3. Johnson, A.B. JAMA 1900;35:1685.

4. Brewer, G.E. JAMA 1900;35:1685.

5. Tuttle, J.P. JAMA 1900;35:1685.

6. Moschcowitz, A.V. JAMA 1900;35:1685.

7. Guedel, A. Inhalation Anaesthesia. A Fundamental Guide. 1937. New York: The MacMillan Company, p. 133.

8. Lee, J.A. A Synopsis of Anaesthesia. 3rd Edition. 1953. Bristol: J. Wright and SOns.

9. Hewer, C.L. Recent Advances in Anaesthesia and Analgesia (Including Oxygen Therapy). 7th Edition. 1948. London: J.A. Churchill Ltd.

10. Cullen, S.C. Anesthesia in General Practice. 3rd Edition. 1951. Chicago: The Year Book Publishers, p. 88.

11. Denborough, M.A. and Lovell, R.R.H. Lancet 1960;1:45.

12. Denborough, M.A., Forster, J.F.A., Lovell, R.R.H., Maplestone, P.A. and Villiers, J.D. Brit. J. Anaesth. 1962;34:395.

13. Gordon, R.A. Can. Anaesth. Soc. J. 1966;13.

14. Britt, B.A., Locher, W.G. and Kalow, W. Can. Anaesth. Soc. J. 1969;16:89.

15. Britt, B.A. In: Milton, A.S. (ed.). Handbook of Experimental Pharmacology, Pyretics and Antipyretics. 1982: Heidelberg: Springer-Verlag, p. 547.

16. Summers, R.J. In: Gordon, R.A., Britt, B.A., Kalow, W. (eds.). International Symposium on Malignant Hyperthermia. 1973. Springfield: Charles C. Thomas, p. 363.

17. Wilson, R.D., Nichols, R.J., Dent, T.E. and Allen, C.R. Anesthesiology 1966;26:232.

18. Britt, B.A. and Kalow, W. Can. Anaesth. Soc. J. 1970;17:293.

19. Satnick, J.H. Anesthesiology 1969;30:472.

20. Berman, M.C., Harrison, G.G., Bull, A.B. and Kench, J.E. Nature 1970;225:653.

21. Jones, E.W., Nelson, T.E., Anderson, I.L., Kerr, D.D. and Burnap, T.K. Anesthesiology 1972;36:42.

22. Van den Hende, C. and Lister, D. Brit. J. Anaesth. 1976;48:821.

23. Elizondo, G., Addis, P.B., Rempel, W.E., Madero, C., Martin, F.B., Anderson, D.B. and Marple, D.N. J. Animal Sci. 1976;43:1004.

24. Nelson, T.E. In: Gordon, R.A., Britt, B.A. and Kalow, W. (eds.). International Symposium on Malignant Hyperthermia. 1973. Springfield: Charles C. Thomas, p. 191.

25. Harrison, G.G. J. S. Afr. Vet. Assn. 1972;43:57.

26. Allen, W.M. and Bell, J.C. Third Symposium on Neuromuscular Disorders. 1973;1:.

27. Anonymous. Vet. Med. 1977;361.

28. Ball, R.A., Annis, C.L., Topel, E.G. and Christian, L.L. Veterinary Medicine/Small Animal Clinician 1973;1:1156.

29. Ball, R.A., Topel, D.G., Annis, C.L. and Christian, L.L. Proc. 76th Ann. Meeting U.S. Animal Health Association 1972;1:517.

30. Briskey, E.J. Adv. Food Res. 1964;13:89.

31. Fee, R.J. Successful Farming 1973;18.

32. Williams, C.H. Perspectives in Biol. Med. 1976;20:120.

33. Williams, C.H. In: Henschel, E.O. (ed.). Malignant Hyperthermia--Current Concepts. 1977. New York: Appleton-Century-Crofts.

34. Williams, C.H., Houchins, C. and Shanklin, M.D. Brit. Med. J. 1975;3:411.

35. Hasselback, W. Prog. Biophys. Mol. Biol. 1964;14:167.

36. Britt, B.A., Kalow, W., Gordon, A., Humphrey, J.G. and Rewcastle, N.B. Can. Anaesth. Soc. J. 1973;20:431.

37. Britt, B.A., Endrenyi, L., Cadman, D.L., Ho, M.F. and Fung, H.Y-K. Anesthesiology 1975;42:292.

38. Kim, D.H., Sreter, F.A., Ohnishi, S.T. et al. Biochim. Biophys. Acta 1984;775:320.

39. Nelson, T.E. J. Clin. Invest. 1983;72:862.

40. Kalow, W., Britt, B.A., Terreau, M.E. and Haist, C. Lancet 1970;ii:895.

41. Ellis, F.R., Harriman, D.G.F., Keaney, N.P., Kyei-Mensah, K. and Tyrell, J.H. Brit. J. Anaesth. 1971;43:721.

42. Fletcher, J.E. and Rosenberg, H. Anesthesiology 1985;

43. Britt, B.A., Kwong, F.H-F. and Endrenyi, L. In: Henschel, E.O. (ed.). Malignant Hyperthermia, Current Concepts. 1977. New York: Appleton-Century-Crofts.

44. Britt, B.A., Kwong, F.H-F. and Endrenyi, L. In: Henschel, E.O. (ed.). Malignant Hyperthermia, Current Concepts. 1977. New York: Appleton-Century-Crofts.

45. Harrison, G.G. In: Huelsz, et al. (eds.). Anaesthesiology 1977. 1977. Amsterdam: Excerpta Medica, p. 452.

46. Gelenberg, A.J. and Poskanzer, D.C. Neurology 1973;23:1313.

47. Chyatte, S.B., Birdsong, J.H. and Robertson, D.L. Arch. Phys. Med. Rehab. 1973;54:365.

48. Aguilar-Bernal, O.R., Bender, M.A. and Lacy, M.E. J. Roy. Soc. Med. 1986;79:277.

49. Ward, A., Chaffman, M.O. and Sorkin, E.M. Drugs 1986;32:130.

50. Britt, B.A., McComas, A.J., Endrenyi, L. and Kalow, W. Anesthesiology 1977;47:490.

51. Britt, B.A., Harrison, J.E. and McNeill, K.G. Can. Anaesth. Soc. J. 1979;26:117.

52. Kelstrup, J., Reske-Nielsen, E., Haase, J. and Jorni, J. Acta. Anaesth. Scand. 1974;18:58.

53. Harrison, G.G. and Verburg, C. Brit. J. Anaesth. 1973;45:131.

54. Huckell, V.F., Staniloff, H.M., McLaughlin, P.R., Britt, B.A. and Morch, J. In: Aldrete, J.A. and Britt, B.A. (eds.). Second International Symposium on Malignant Hyperthermia. 1978. New York: Grune and Stratton, p. 373.

55. Huckell, V.F., Staniloff, H.M., Britt, B.A. et al. Presented at the American College of Cardiology, March 6-9, 1978, Anaheim, California.

56. Mambo, N.C., Silver, M.D., Huckell, V.F. et al. Human Pathology 1980;11:381.

57. Klip, A., Britt, B.A., Elliott, M.E., Walker, D., Ramlal, T. and Pegg, W. Biochem. Cell Biol. 1986;64:1181.

58. Britt, B.A. Hotline (publication of the Malignant Hyperthermia Association), Volume 2, Number 2, Spring, 1984.

MALIGNANT HYPERTHERMIA

1

MALIGNANT HYPERTHERMIA - THE ACUTE CRISIS

D.J. STEWARD AND G.A.R. O'CONNOR

Department of Anaesthesia, University of British Columbia; British Columbia Children's Hospital, Vancouver, British Columbia V6H 3V4

ABSTRACT

A knowledge of the clinical manifestations and biochemical changes of the acute crisis stage of malignant hyperthermia will enable the physician to recognize the syndrome at an early stage, to anticipate the possible progression of the disease, and to immediately apply appropriate therapy.

INTRODUCTION

Early recognition of the acute malignant hyperthermia (MH) reaction is vital if the patient is to be saved from death or permanent injury. Rapid diagnosis of the disease must be followed by immediate withdrawal of triggering agents and institution of specific therapy if morbidity and mortality is to be minimized.

Many MH susceptible (MHS) patients lack obvious pre-operative diagnostic stigmata of the disease and, therefore, may be given the commonly used triggering agents. In addition, some patients who are recognized to have the MH trait may develop the syndrome despite all precautions (1). To ensure an early diagnosis every anaesthesiologist must be thoroughly familiar with the initial signs of an acute MH reaction. If these appear, surgery must be rapidly terminated and appropriate action taken. A knowledge of the later clinical course of the acute crisis is necessary in order that appropriate continuing therapy can be administered.

PRESENTING SIGNS OF THE ACUTE CRISIS

The initial presentation of an acute MH crisis is determined by the triggering agent(s) (particularly succinylcholine) which have been used. The severity of the disease and the prior use of modifying drugs (eg. Dantrolene) will also influence the course of the crisis.

Muscle Rigidity

The first indication of a MH crisis may be rigidity rather than relaxation, following administration of a standard dose of succinylcholine. This rigidity which is often first noted in the jaw muscles may later be evident in all skeletal muscle. The rigidity is often, but not always, preceded by massive fasciculations. The muscles are in a state of contracture as opposed to contraction. The muscle bellies feel hard, are unindentable and the joints become unmovable.

Rigidity of the jaw muscles may make intubation difficult or even impossible, but in most cases adequate ventilation can be maintained by means of manual ventilation with bag and mask. The rigidity of MH will not be relieved by a second dose of succinylcholine, or administration of a non-depolarizing relaxant drug.

Fortunately, not every patient with masseter spasm following succinylcholine has the MH trait. Masseter spasm has been described in association with muscle disease, such as myotonia congenita (2), myotonia dystrophica and polymyositis. In addition, a form of masseter spasm may occur quite often (in up to 1% of patients) (3) when succinylcholine is administered after induction of anaesthesia with halothane. This latter observation has raised some question as to the true diagnostic importance of masseter spasm as a predictor of the MH trait (4). It is, however, our continued belief that any unexpected appearance of masseter spasm should be treated as a possible early sign of malignant hyperthermia (5). The anaesthetic should be abandoned, if possible and the patient re-evaluated for other stigmata of MH trait (clinical or biochemical). The rigidity will usually resolve over ten to twenty minutes and recovery is usually uneventful (3). The patient should be carefully monitored for twenty-four hours in case further signs of MH develop. If the surgery must proceed, anaesthesia should be continued with a suitable technique which excludes known triggering agents and the patient very carefully monitored for other early signs of the MH syndrome.

The development of muscle rigidity varies with the severity of the syndrome, the anaesthetic agents used and the presence of other clinical or subclinical muscle disease. It has been suggested that prior administration of atropine may increase the incidence of muscle rigidity (6). If succinylcholine is not administered at induction, muscle rigidity may develop insidiously as the MH crisis progresses. Though subtle at first this rigidity may become intense leading to a state of opisthotonus. Occasionally muscle rigidity is absent (6).

Tachycardia and Other Cardiac Arrhythmias

Tachycardia is a very consistent early sign of malignant hyperthermia, but is often ignored as it is assumed to be due to light anaesthesia. If other causes of a rapid heart rate, such as anticholinergic drugs, hypovolaemia, hypercarbia, or light anaesthesia have been eliminated, blood gases should be taken to further the diagnosis. If still in doubt the patient should be treated as a potential case of malignant hyperthermia. Other arrhythmias commonly seen early in the crisis include bigeminy, frequent ventricular extrasystoles and ventricular tachycardia. The electrocardiogram may show tall peaked T waves characteristic of hyperkalaemia. Ventricular fibrillation may occur as a terminal event.

Blood Pressure Changes

Hypertension may occur during the early stages of the acute MH crisis and an instability of the blood pressure has often been noted. Later in the disease, as cardiac action becomes progressively impaired, arrhythmias appear and hypotension occurs.

Ventilation

The early metabolic changes in malignant hyperthermia result in a marked increase in carbon dioxide production and lead to tachypnoea if the patient is breathing spontaneously. The patient who is on a mechanical ventilator will begin to "fight the ventilator" if neuromuscular block is incomplete. Levels of end-tidal carbon dioxide increase despite the rise in ventilation and the end tidal CO_2 monitor is a valuable adjunct to diagnosis (7). In a patient with MH the end tidal carbon dioxide has been reported to increase from 4.5 mmol/L to 9.3 mmol/L within thirty minutes despite a four fold increase in ventilation (7). If a carbon dioxide absorber is in use, the temperature of the canister may rise due to the greatly accelerated chemical reaction within the soda-lime. Pulmonary oedema may occur as the disease progresses and cardiac failure becomes apparent.

Skin Changes

An erythematous rash which rapidly progresses to a mottled cyanosis may be observed over the head, neck and upper chest. Simultaneously the surgeon may comment upon the appearance of dark blood at the site of the skin incision. Unexplained cyanotic mottling of the skin associated with tachycardia in a previously healthy, well oxygenated and ventilated patient is almost pathognomenic of the MH syndrome.

LATER SIGNS OF THE ACUTE MH CRISIS

Pyrexia

Although pyrexia is the hallmark of malignant hyperpyrexia it is a later sign and the objective of the anesthesiologist must be to diagnose the condition and commence therapy before the temperature has risen markedly. The time of onset of pyrexia is influenced by the anaesthetic agents used. Pyrexia occurs earlier and rises more rapidly when succinylcholine is given in conjunction with a potent volatile inhalation agent such as halothane (8). Once the body core temperature starts to rise it does so rapidly, sometimes by $1^{o}C$ every five minutes (8). Final temperatures in excess of $46^{o}C$ ($115^{o}F$) have been recorded.

A rapidly rising body temperature should always prompt an immediate search for other signs of an MH crisis, which will certainly be evident at this stage of the disease.

Pyrexia is a result of and not the cause of the early biochemical derangements. Later pyrexia itself may contribute to irreversible damage to organ systems.

The excessive heat production occurs primarily in the skeletal muscle and secondarily in the liver. Variations in body surface temperature from a muscular to a less muscular site may be noted.

Left Ventricular Failure

Acute pulmonary oedema as a result of left ventricular failure commonly occurs in the terminal stages of an acute MH crisis. Coarse rales are heard on auscultation of the chest and frothy, blood stained fluid is obtained when the endotracheal tube is suctioned. Later, large volumes of pulmonary oedema fluid may issue from the endotracheal tube into the anaesthetic circuit.

Coagulopathy

During the MH crisis excessive bleeding may become apparent (9). Blood may ooze from previously dry edges of the surgical incision and venipuncture sites may begin to bleed again. Later, extensive bruising may become apparent.

BIOCHEMICAL CHANGES DURING THE MALIGNANT HYPERTHERMIC CRISIS

Laboratory investigations should be used to confirm or exclude the diagnosis of acute malignant hyperthermia and to guide continuing therapy, if any of the aforementioned early signs of the disease appear.

Blood Gas and Acid-Base Changes

Early in the crisis the carbon dioxide tension in the blood rises as the rate of carbon dioxide production in skeletal muscle exceeds the capacity for CO_2 excretion via the lungs. In the very early stage of the crisis mixed venous carbon dioxide tension may be significantly elevated while the arterial tension remains normal.

As the disease progresses a major increase in tissue oxygen demand develops. In the early stage this leads to a fall in mixed venous oxygen saturation. Later arterial oxygen tension falls probably as a result of extreme venous desaturation and the development of intrapulmonary shunting. As the oxygen demand of the hypermetabolic tissues outstrips oxygen availability anaerobic metabolism occurs with increasing lactate production and a state of metabolic acidosis develops.

The classic blood gas and acid-base status is, therefore, of a mixed respiratory and metabolic acidosis. It has been suggested that a $PaCO_2$ of 60 mm Hg or more with a base deficit of -5 or more is diagnostic of MH in a patient who has the signs of the disease listed previously (8). Initial $PaCO_2$ may, in fact, commonly be as high as 100-200 mm Hg and the initial blood pH may range as low as 7.15-6.8.

Serum Electrolyte Levels

The acute crisis of malignant hyperthermia is accompanied by major fluctuations in serum electrolyte levels (10). The levels of individual electrolytes alter as the disease progresses and many variations have been described. Usually, however, the serum potassium level is elevated early in the crisis. This may contribute to the production of cardiac arrhythmias. Later in the crisis, a marked and prolonged fall in the serum potassium level occurs that may be difficult to correct.

Hypercalcaemia is reported to occur during the initial stages of the crisis, but an early hypocalcaemia may be seen in some patients (11). During the late course of the crisis serum calcium levels may again fall as calcium moves into the muscle cells. Serum phosphate levels as high as 4.4 mMol/L (normal 0.6-1.6 mMol/L) have been reported, but as with other electrolytes this high level is followed later by abnormally low serum values.

Serum Enzyme Levels

Creatine kinase (CK) levels in the blood increase markedly during the crisis, but in survivors, reach their highest levels 24-48 hours later. CK levels may reach as high as 1,000,000 units.

Serum concentrations of lactate dehydrogenase (LDH) and aspartate amino transferase are also found to be very high and increase over the first 72 hours in survivors (10).

Haematological Studies

Haemolysis may occur during the crisis and may be detected if a blood sample is centrifuged or urine is examined. Myoglobinaemia and myoglobinuria are commonly present.

Thrombocytopenia occurs in many patients during the MH crisis and this, together with lowered levels of factor VIII and fibrinogen results in impaired coagulation. A coagulation screen should be obtained when the disease is first suspected and be repeated periodically during therapy.

The cause of the coagulopathy may be a pre-existing platelet defect or may be due to diffuse intravascular coagulation and secondary factor depletion triggered by the crisis.

AFTER THE CRISIS

Muscle Pains

Patients who survive an acute malignant hyperthermia crisis with rigidity usually complain of severe muscle pains which may last for days or weeks. Muscles, especially those of the thighs, are usually swollen and tender. Later as the swelling disappears, muscle weakness and wasting may become apparent and may persist for a period of months, in association with generalized fatigue.

Central Nervous System Damage

Unfortunately some patients who survive a MH crisis are left with permanent neurological defects. The incidence of serious neurological damage is highest in patients suffering the most fulminant reactions. The degree of damage varies from patient to patient. Some never regain consciousness and proceed to neurological death. Continuing coma, convulsions, fixed dilated pupils and absent tendon reflexes are ominous signs of brain death. Paralysis, speech defects, blindness, deafness and impairment of intelligence have occurred in some survivors.

The causes of this cerebral damage may include the extremely high fever, hypoxemia and other biochemical derangements which accompany this condition. It is also possible that a primary defect of nerve tissue is present and contributes to the problem.

Electrolyte Disturbances

Electrolyte disturbances occurring during the acute phase have already been described, but it is important to remember that further derangements may follow. These are the result of continuing impairment of cell membrane function, poor renal function and the effects of drug therapy. Electrolyte imbalance following the MH crisis is difficult to treat, since the response to infused solutions is quite unpredictable. Dangerous fluctuations in the serum potassium level are particularly likely to occur during therapy. Successful treatment demands very careful continuous biochemical monitoring.

Renal Failure

In patients who have survived the initial crisis there is a considerable danger of renal failure. In such cases myoglobinuria is followed by oliguria and an increase in the blood urea nitrogen. Renal damage is thought to result primarily from the direct effects of myoglobin on the renal tubules.

Recurrence of the Syndrome

Some patients who have apparently survived an acute MH crisis have suffered a fatal recurrence of the disease a few hours later, so it is of vital importance that all patients be very carefully monitored following the crisis. Rarely, the initial signs of MH may be delayed until the post-operative period (12). Hence, signs of MH in any patient in the post anaesthesia room should be considered as a possible onset of the acute crisis.

THE NEUROLEPTIC MALIGNANT SYNDROME

The Neuroleptic Malignant Syndrome (NMS) is a potentially fatal syndrome associated with the use of neuroleptic medications in psychiatric disorders (13). Clinical similarities between this syndrome and malignant hyperthermia suggest the possibility of a common pathophysiology especially as they both respond to dantrolene (14). However, the two diseases also have many features which argue against a close link (15): the triggering agents are quite different; neuroleptic agents do not trigger MH but do trigger NMS; and succinylcholine and volatile anaesthetic agents have been given to patients with NMS with no adverse effects. Non-depolarizing neuromuscular blocking drugs will not relax the rigidity of MH but will produce flaccid paralysis in patients with NMS.

Clinical Characteristics

The onset of NMS is more insidious than occurs with the MH syndrome, with signs of the disease developing over 24-72 hours. A generalized muscle rigidity with a "lead-pipe" character develops, which is accompanied or followed by elevation of the body temperature to as high as 41-42°C. Autonomic dysfunction is usually present, evidenced by labile heart rate and blood pressure, excessive sweating, and incontinence. Extrapyramidal signs are often present.

Laboratory findings are inconsistent, but usually a leucocytosis is present. Liver function tests show elevated serum transaminases, lactate dehydrogenase, and alkaline phosphatase. CK levels may be as high as 16,000 IU/L and myoglobinuria may be present due to rhabdomyolysis.

REFERENCES

1. Fitzgibbons, D.C. Anesthesiology 1981;54:73.

2. Ellis, F.R. and Halsall, P.J. Brit. J. Anaesth. 1984;56:381.

3. Schwartz, L., Rockoff, M.A. and Koka, B.V. Anesthesiology 1984;61:772.

4. Badgwell, J.M. and Heavner, J.E. Anesthesiology 1984;61:230.

5. Donlon, J.V., Newfield, P., Sreter, F. and Ryan, J.F. Anesthesiology 1978;49:298.

6. Britt, B.A. and Kalow, W. Can. Anaesth. Soc. J. 1970;17:293.

7. Baudendistel, L., Goudsouzian, N., Cote, C. and Strafford, M. Anaesthesia 1984;39:1000.

8. Gronert, G.A. Anesthesiology 1980;53:395.

9. Daniels, J.C., Polayes, I.M. and Villar, R. Anesth. Analg. 1969;48:877.

10. Rutberg, H., Henriksson, K.G., Jorfeldt, L., Larsson, J., Martensson, J. and Schildt, B. Brit. J. Anaesth. 1983;461.

11. Stanec, A., Spiro, A.J. and Lent, R.W. In: Aldrete, J.A. and Britt, B.A. (eds.). Second International Symposium on Malignant Hyperthermia. New York: Grune and Stratton, 1978.

12. Schulte-Sasse, V. and Eberlein, H.J. Can. Anaesth. Soc. J. 1983;30:635.

13. Caroff, S.N. J. Clin. Psychiatry 1980;41:79.

14. Weinberg, S. and Twersky, R.S. Anesth. Analg. 1983;62:848.

15. Allsop, P. and Twigley, A.J. Anaesthesia 1987;42:49.

2

AETIOLOGY AND PATHOPHYSIOLOGY OF MALIGNANT HYPERTHERMIA

B.A. BRITT

Departments of Anaesthesia and Pharmacology, University of Toronto, c/o Toronto General Hospital, Toronto, Ontario M5G 2C4

ABSTRACT

Malignant hyperthermia reactions are initiated by a sudden rise in the concentration of calcium in the muscle cytoplasm. This elevation of myoplasmic calcium appears to be secondary to an increase in calcium induced calcium release from the sarcoplasmic reticulum (SR). This in turn may be a result of either or both: heightened sensitivity of the calcium channels of the SR to normal amounts of "trigger" calcium; or to increased amounts of "trigger" calcium. Excessive sensitivity of the SR calcium channels to "trigger" calcium may be induced by elevated levels of long chain fatty acids. Rises in cytoplasm long chain fatty acid levels in turn may be brought about by heightened phospholipase A_2 activity in the mitochondria. Elevated levels of "trigger" calcium may originate in the mitochondria or the sarcolemma. Accelerated release of calcium from the mitochondria may be precipitated by the increased phospholipase A_2 activity. Excessive amounts of calcium entering the cytoplasm from the sarcolemma may be due to: accelerated release of "trigger" calcium from the sarcolemma, or to increased transport of calcium from the extracellular fluid into the interior of the cell. The rate of calcium release from the SR may also be indirectly stimulated via heightened sensitivity of the skeletal muscle catecholamine receptor to endogenous adrenaline or noradrenaline.

The elevated cytoplasmic calcium produces contracture of the myofibrils, increased heat, lactic acid and carbon dioxide production and accelerated oxygen consumption. ATP production falls while ATP consumption rises. Without ATP there is no substrate for active pumping of ions and molecules against concentration gradients across cellular and intracellular membranes. Ions and molecules, therefore, simply seep across these membranes in the direction of their natural concentration gradients. This in turn leads to further rises in cytoplasmic calcium. Excessive amounts of sodium and water also enter the muscle cell. In the reverse direction abnormally large amounts of potassium, magnesium, phosphate and at a later time of larger molecules such as myoglobin and enzymes are lost from the muscle cell to the extracellular fluid and from thence to the serum.

AETIOLOGY

The aetiology of malignant hyperthermia (MH) has not yet been entirely clarified in spite of a number of years of intensive investigation. Nevertheless, the most likely immediate cause of the acute hypercatabolism of MH appears to be a sudden rise in the concentration of myoplasmic calcium (1-4) (Figure I). This

FIGURE I

IMMEDIATE CAUSE OF A MH REACTION

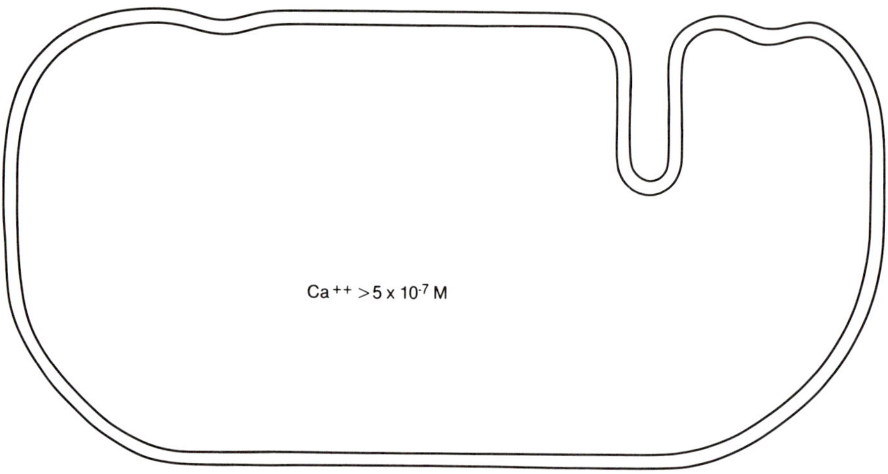

hypothesis is supported by the observation that drugs which raise myoplasmic calcium, for example, lidocaine, cardiac glycosides, caffeine, and calcium salts, worsen the prognosis of in vivo MH reactions (5,6). Lower doses of caffeine in vitro are required to induce contractures in malignant hyperthermia susceptible (MHS) than in normal skeletal muscle fascicles (3,4,7-10). These contractures are further enhanced by calcium (9,10). On the other hand, drugs which lower myoplasmic calcium, for instance, dantrolene (11-13) and procaine amide (13-17) improve survival from in vivo crises (6). In vitro these agents, along with diltiazem and verapamil (calcium channel blockers that also lower myoplasmic calcium) (18-20), and calcium free Krebs Ringer solution attenuate caffeine-induced contractures of MHS skeletal muscle fascicles (21). Firmer evidence in favour of elevated cytoplasmic calcium being responsible for MH has been the recent work of Lopez et al. (22). These workers have, by means of intracellular calcium electrodes, found that porcine myoplasmic calcium does indeed rise excessively during MH reactions.

The defect that could account for the sudden and rapid rise of myoplasmic calcium during a MH reaction has yet to be firmly established. Some reasonable possibilities based on known muscle physiology are abnormalities in one or more of the following:

1. mitochondria;
2. abnormal sarcoplasmic reticulum (SR);
3. excitation contraction coupling;
4. calmodulin;
5. sarcolemma
6. adrenergic innervation of the muscle

These possibilities are not necessarily mutually exclusive since MH may well be a widespread membrane disease involving not only membranes of different cell types, but also different membranes with a given cell type. Thus a defect that permits excessive release of calcium from the SR might also simultaneously allow increased release of calcium from the mitochondria and the sarcolemma.

ABNORMAL SARCOPLASMIC RETICULUM

The SR normally takes up and binds calcium during muscle relaxation (Figure 2) and then releases calcium during muscle

FIGURE 2

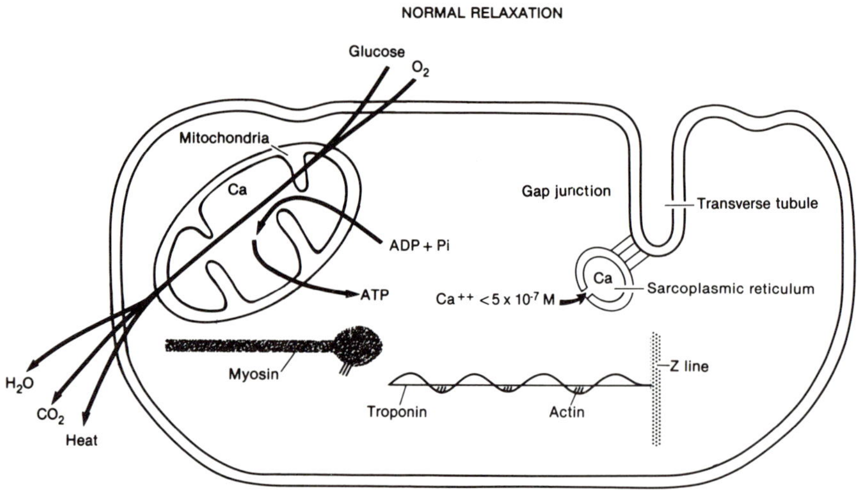

contraction (Figure 3) (23-25). The uptake of calcium is against a concentration gradient, the energy for which is obtained through coupling of calcium uptake to hydrolysis of ATP by SR Ca^{++} ATPase. Calcium is bound to calsequestrin, a protein component of the SR membrane (26,27) (Figure 4). Release of calcium from the SR, being with the concentration gradient, is a passive phenomenon. Evidence for a primary defect in MHS muscle (presented in detail by Nelson in Chapter 3 of this volume) in one or both of these functions is considerable (2,28-40) although it is not entirely without challenge, particularly evidence for uptake (41-46). The great variation in reported results is probably due to a combination of: different experimental techniques and use of different species (humans vs. pigs) and different breeds of pigs (Pietrain, Landrace and Poland China).

FIGURE 3

NORMAL CONTRACTION

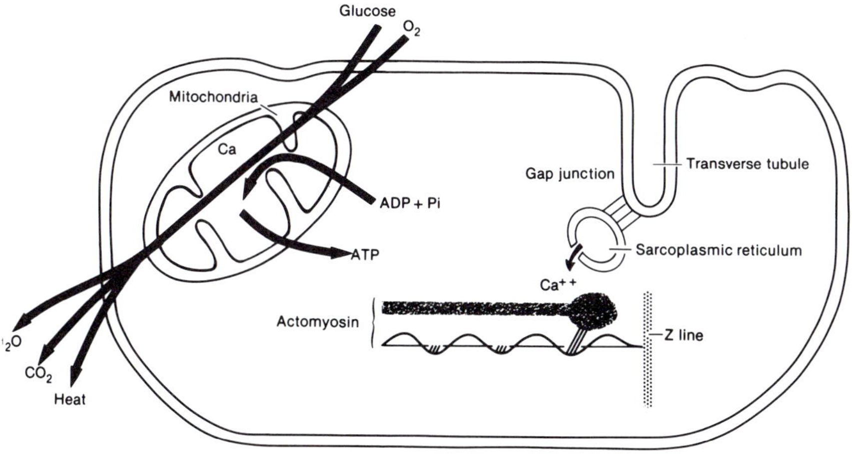

FIGURE 4

STRUCTURE OF SARCOPLASMIC RETICULUM MEMBRANE

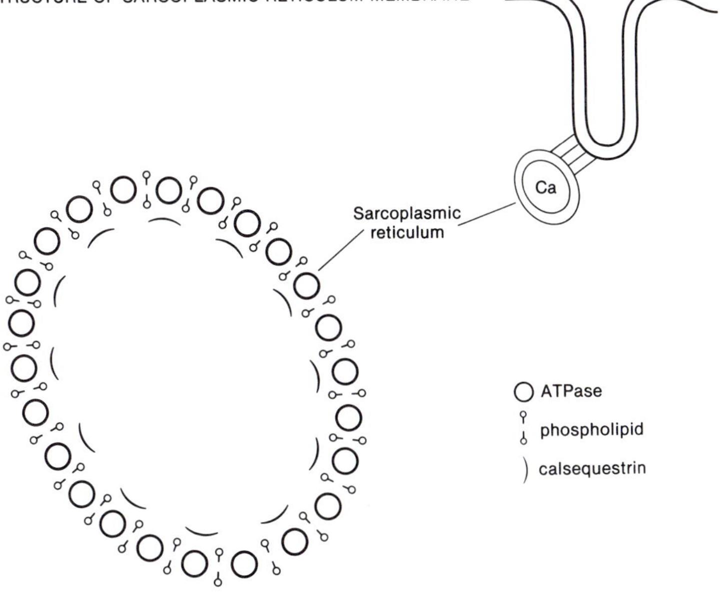

Recently, Nelson (25) and Ohnishi (47), employing isolated SR, and Horiuti et al (48), using skinned fibres, have shown that less than usual amounts of exogenous calcium are needed to trigger release of calcium from isolated MHS SR (42) (Figure 5) and that

FIGURE 5

PRIMARY DEFECT OF MH POSTULATED TO BE INCREASED CALCIUM-INDUCED CALCIUM RELEASE

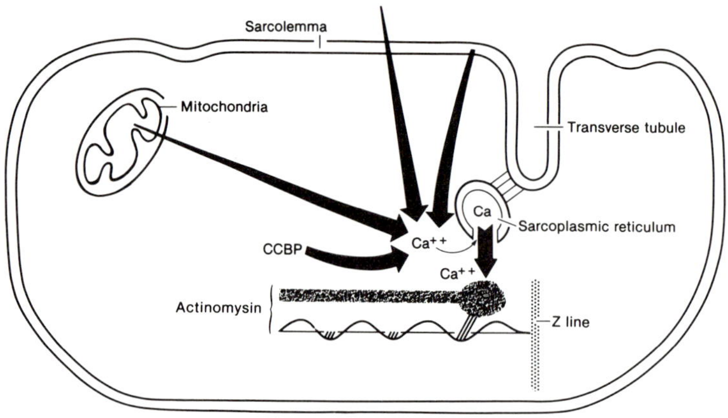

CCBP = Cytoplasmic Calcium Binding Proteins

MH CONTRACTURE TRIGGERED BY CALCIUM-INDUCED CALCIUM RELEASE

this difference is increased by either halothane or caffeine (25,49,50). Furthermore the amount of 55 kDa phosphoprotein (which probably represents an autophosphorylated calmodulin-dependent substance) is 30% less in MHS SR than in normal SR (51).

ABNORMAL MITOCHONDRIA

A primary defect in the skeletal muscle mitochondria of MHS patients has been postulated (51-71). The mitochondria in addition to being the site of cell respiration are also storage vesicles for calcium. They function as long-term repositories for excess calcium which is beyond the capacity of the SR to take up during each relaxation cycle. The evidence for a primary mitochondrial abnormality being responsible for abnormal rises in the concentration of myoplasmic calcium is considerable (51-71) and is presented in detail by K.S. Cheah in Chapter 4 of this volume (60). For example, Cheah (57-59,72) has shown that porcine MHS mitochondria contain overly active phospholipase A_2 (Figure 6). This leads to calcium induced uncoupling of oxidative phosphorylation from electron transport, swelling of the mitochondria, excessive release of calcium from the mitochondria, and hydrolysis of phospholipids in the mitochondrial membranes resulting in the liberation of long chain unsaturated fatty acids (Figures 5 and 6). Cheah postulates that the calcium and/or the long chain unsaturated fatty acids released from MHS mitochondria may in turn trigger the release of much larger amounts of calcium from the SR or may prevent uptake of calcium into the SR (Figure 6) (56).

FIGURE 6

PRIMARY DEFECT OF MH POSTULATED TO BE INCREASED ACTIVITY OF PHOSPHOLIPASE A_2

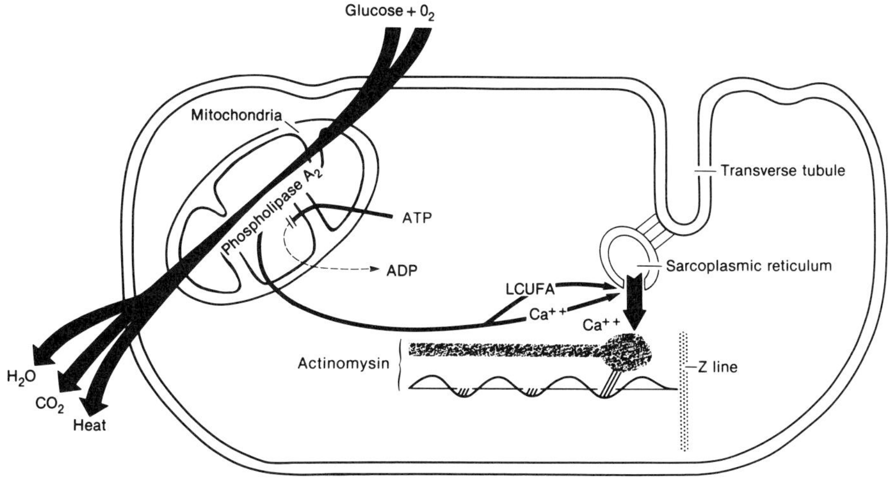

LCUFA = Long Chain Unsaturated Fatty Acids

The work presented by Cheah (58,60,72) and that by Nelson (25), Ohnishi (47) and Horiuti (48) are not necessarily contradictory. Long term exposure of the SR to enhanced levels of long chain unsaturated fatty acids may cause permanent damage to their calcium channels with resultant increased permeability of these channels to calcium. Even normal levels of cytoplasmic calcium would, therefore, be expected to induce release of increased amounts of calcium from the SR. If the mitochondria are in fact releasing larger than usual amounts of calcium as suggested by Cheah (61) then the calcium-induced calcium release from the SR would be all the greater.

ABNORMAL EXCITATION-CONTRACTION COUPLING

Because of difficulties of direct investigation of the connecting links between the transverse tubules (TT) and the SR, little research has been done on the possibility that the defect of MH lies in the excitation-contraction coupling step between the TT and the SR (Figure 7). Some interesting indirect investigation has recently been reported by Sullivan et al (73). These workers have observed that 8-(N,N-diethylamino)-octyl-3,4,5-trimethoxybenzoate (TMB-8), a calcium antagonist, produces, in MHS but not in normal muscle, substantial contractures (greater than those produced by caffeine), potentiates contractures induced by halothane, caffeine, succinylcholine and inhibits contractures induced by KCl. These effects are abolished by prior excitation-contraction uncoupling with either glycerination of the muscle or dantrolene administration. They have also found that TMB-8 is unable to alter calcium levels in MHS SR, under conditions where halothane is able to induce extensive release. Furthermore, lack of a highly charged functional group on the TMB-8 molecule suggests that a direct interaction with calcium is unlikely. Sullivan and his co-workers (73) believe, therefore, that these results indicate that in MH muscle the basic molecular abnormality occurs at the level of excitation-contraction coupling (Figure 7). Sullivan's thesis is further supported by the prolonged electro mechanical coupling times in MHS pigs which have been reported by Nelson and his team (74).

FIGURE 7

PRIMARY DEFECT OF MH POSTULATED TO BE ABNORMAL EXCITATION-CONTRACTION (EC) COUPLING

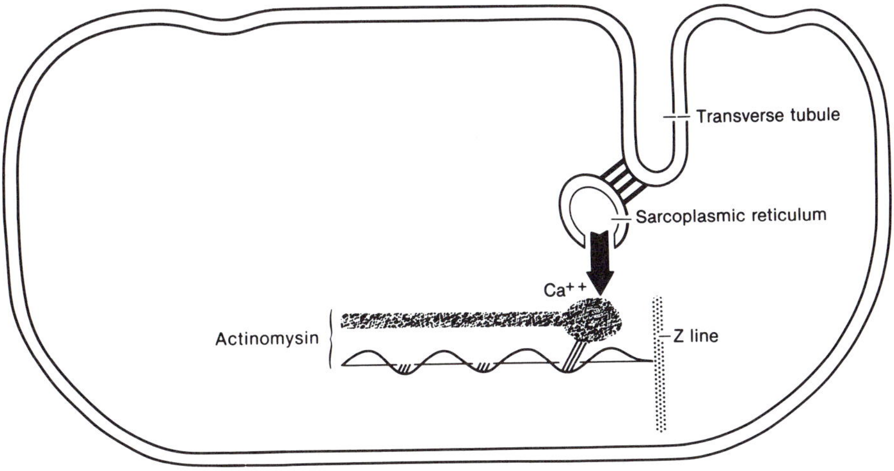

MH CONTRACTURE TRIGGERED BY ABNORMAL EC COUPLING

ABNORMAL CALMODULIN

Calmodulin is the main intracellular receptor for calcium (75) and stimulates calcium-dependent phosphodiesterases (76,77). Marjanen et al (78) have found that the ability of the calmodulin inhibitor trifluoperazine to reduce calmodulin-stimulated phosphodiesterase activity in MHS pig muscle homogenates is not different from normal pig muscle homogenates. These results suggest, but do not prove, that there is no difference between calmodulin molecules from normal and MHS muscle. On the other hand, Cheah et al (79) have recorded higher than normal levels of calmodulin in adult MHS landrace pigs. They hypothesize that these elevated calmodulin levels may be responsible for the enhanced phospholipase A_2 activity that they have detailed in these pigs (79).

ABNORMAL SARCOLEMMA

What evidence exists for a defect in the ability of the sarcolemma to keep calcium out of the myoplasm? Niebroj-Dobosz et al. (45,80) have reported that the phospholipid and cholesterol content of MHS porcine sarcolemma is less than normal while McIntosh and Berman (81) have shown that the phospholipids of MHS porcine sarcolemma and SR contain fatty acids which are more unsaturated than normal. They and others have postulated that this abnormality may result in an increased permeability of the sarcolemma to extracellular fluid calcium (Figure 8). Since the normal concentration of calcium in the ECF is about 2.5×10^{-3}M, while that in the myoplasm is only about 5×10^{-7} M, calcium would, when the sarcolemma is incompetent, be expected to flow from the ECF into the myoplasm. Even prior to anaesthesia MHS sarcolemma may be somewhat leakier than normal since in many susceptible human (82-87) and porcine (84,88-91) subjects serum CK concentrations are elevated--the elevations apparently being due to

FIGURE 8

PRIMARY DEFECT OF MH POSTULATED TO BE ABNORMAL SARCOLEMMA

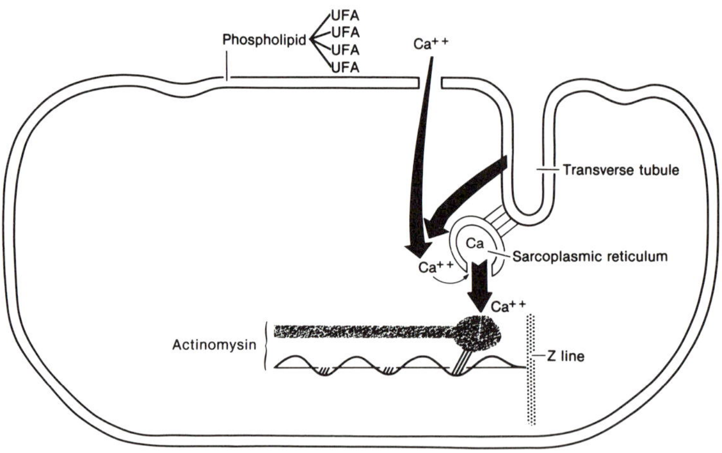

UFA = Unsaturated Fatty Acids

MH CONTRACTURE TRIGGERED BY INCREASED UNSATURATED FATTY ACIDS IN SARCOLEMMA

outward seepage of the enzyme across the super-fragile sarcolemma. The hypothesis of the sarcolemma being overly permeable to extracellular fluid calcium is supported by the observation that MH is more common in cold-acclimated than in warm-acclimated humans. However, Mickelson has reported that calcium transport in MHS sarcolemma is normal (92).

The sarcolemma, as well as controlling calcium flux, binds calcium. During contraction some of the bound calcium is released into the myoplasm in the vicinity of the SR where it acts as a trigger for the release of still more calcium from the SR (25,35,49,93-95). It may be that in MH patients reactions are initiated, at least in part, by the release of excessive amounts of "trigger" calcium from the sarcolemma or by increased sensitivity of the SR to trigger calcium emanating from the sarcolemma.

Other abnormalities observed (71) in MHS porcine sarcolemma are reduced activities of $Na^{+}K^{+}$-ATPase and Mg^{++}-ATPase. Furthermore while activities of these enzymes decrease further during halothane anaesthesia in MHS sarcolemma, the reverse is true in normal sarcolemma. These reports have, however, partially been contradicted by Mickelson who observed normal $Na^{+}K^{+}$-ATPase activity in MHS sarcolemma (92). Mickelson (92) has also described normal cholesterol, phospholipid, adenylcyclase, calmodulin and calcium transport in MHS sarcolemma (92). On the other hand, with increasing times of calcium transport MHS sarcolemma accumulate less calcium than normal sarcolemma (92). Furthermore, resting sarcolemmal voltage depolarization augments calcium induced calcium release of mechanically skinned fibres, especially of MHS fibres (96).

ABNORMAL CATECHOLAMINE INNERVATION OF THE MUSCLE

An abnormality of catecholamine innervation may play a contributory role in the precipitation of some MH reactions, especially stress-induced reactions (Figures 9 and 10). Depletion of SR calcium induced by stresses may be secondary to an exaggerated sensitivity of sarcolemmal adenyl cyclase to catecholamines (97-100). Ahern and his colleagues (101) have observed that treating

FIGURE 9

ADRENERGIC INNERVATION OF SKELETAL MUSCLE
(INACTIVE)

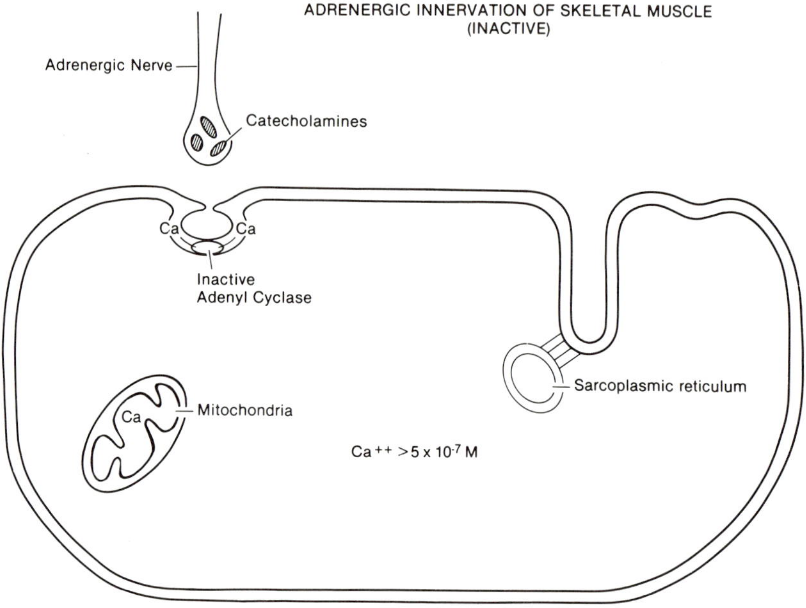

FIGURE 10

POSSIBLE SOURCE OF EXCESS MYOPLASMIC CALCIUM
DURING MH REACTION
(SUPER SENSITIVITY OF ADENYL CYLASE TO
CATECHOLAMINES)

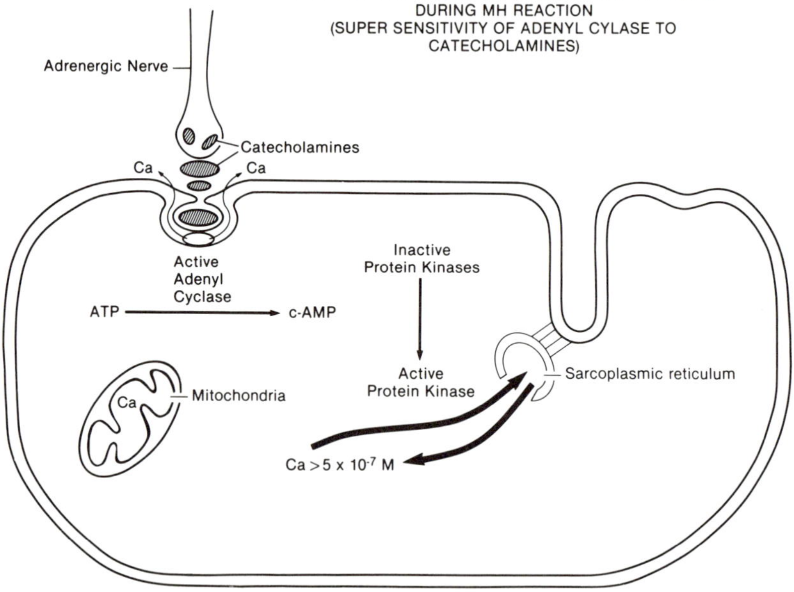

susceptible Pietrain pigs with haloperidol retards the onset of halothane-induced MH reactions by an unknown mechanism. Similarly, Hall et al (102), Lister et al. (103,104), and Short et al. (105,106) have pretreated MHS hogs with other drugs which reduce the availability of catecholamines to the skeletal muscle adrenergic receptor sites, for example, with reserpine and phentolamine. These animals when subsequently challenged with halothane have not developed MH crises. Lucke and his co-workers (107,108) have found that combined adrenalectomy and bretyllium infusion prevents the induction by halothane of MH reactors in susceptible pigs. As a corrollary of the above the same group have reported that noradrenaline plus propranolol or phenylephrine infusions induce fatal MH reactions in susceptible Pietrain pigs (109). The inability of other workers to reproduce these results in other strains of MHS swine, for instance in Poland China pigs (80), may be simply a reflection of the lesser severity of the MH defect in the latter breed of pigs.

Kerr et al. (110) have reported that epidural anaesthesia of MHS Poland China pigs prevents MH reactions in the blocked but not in the unblocked part of the body. Gronert and his team (111), however, note that tetracaine epidural anaesthesia of MHS Poland China pigs does not prevent MH induced by halothane and succinylcholine but does alleviate the expected increase in epinephrine and norepinephrine. They have also observed that pre-treatment of the pigs with dantrolene not only prevents the occurrence of MH during halothane challenge (112) but also inhibits rises in blood catecholamines due to stress, combined respiratory and metabolic acidosis, or haemorrhagic hypotension.

Gronert and Theye (113) have also noted that during halothane-induced MH reactions in pigs significant elevations of blood epinephrine and norepinephrine occur much later than do rises of other parameters, eg. lactate, potassium and oxygen consumption. Gronert et al. (111) claim, therefore, that increased release of catecholamines play only a secondary role in the pathophysiology of MH reactions. Their report does not, however, consider the possibility that the muscle catecholamine receptors (eg. sarcolemmal adenyl cyclase) or the cAMP-protein-kinase pathway might be abnormally sensitive to normal catecholamine concentrations. In another publication, Gronert and his colleagues (114) present data which does support this possibility. This latter data demonstrates that in MHS pigs the ability of halothane and succinylcholine to raise myocardial oxygen consumption is blocked by prior administration of propranolol, a beta adrenoceptor blocking drug.

In humans, hypersensitivity of the muscle catecholamine receptors does appear to be involved in the pathogenesis of MH reactions. For example, in humans, administration of catecholamine-like vasopressors, such as phenylephrine and methamphetamine, markedly increases mortality while administration of drugs which have some antisympathetic action, for example, chlorpromazine and propranolol, improves survival (115).

The stress-induced reactions which occur in both MHS pigs and in MHS humans may, therefore, be triggered by an uncontrolled activation of the adrenergic pathway within the muscle cell. The findings of Willner et al. (116) support this concept since they have found that in skeletal muscle excised from MHS humans adenyl cyclase is more active than in skeletal muscle removed from normal humans. They have also observed that in this same MHS skeletal muscle levels of cAMP are elevated. Additionally, Stanec and Stefano (117) have reported that following exercise plasma cAMP is higher in MHS than in normal humans. It has not been established whether the hyperactive adenyl cyclase is solely responsible for MH crises or whether it merely activates another defect in a more distal part in the excitation-contraction coupling system.

It is known that the skeletal muscle beta adrenergic receptor (118) consists of two components. An outer part facing the synaptic cleft is a key hole which receives the catecholamine key. The inner component facing the myoplasm is the enzyme adenyl cyclase. This enzyme, in the absence of the catecholamine transmitter, is kept inactive through the presence on it of two calcium ions. When the catecholamine combines with the receptor the two calcium ions are extruded--not to the myoplasm as might be expected, but rather to the extracellular fluid. The loss of the two calcium ions activates the adenyl cyclase so that ATP is converted to cAMP

(119). The cAMP in turn activates a number of protein kinases. These have the capacity to reinforce acetylcholine-induced muscle contraction and its several support systems. For example, a protein kinase increases the uptake of calcium into the SR, but since the extra calcium taken up is not bound within the SR, more calcium is available for release during the next cholinergic-induced action potential. The net effect, therefore, of this protein kinase is to increase release of calcium from the SR during each contraction (120). cAMP may also increase myocardial contractility by accelerating calcium influx through the calcium-conducting pores of the sarcolemma during the plateau phase of the action potential (121).

PATHOPHYSIOLOGY

The elevated myoplasmic calcium that occurs during a MH reaction, exerts a number of heat producing effects that markedly

FIGURE 11

EFFECT OF EXCESS MYOPLASMIC CALCIUM AND EXCESS CATECHOLAMINES ON GLYCOGEN CATABOLISM DURING MH REACTIONS

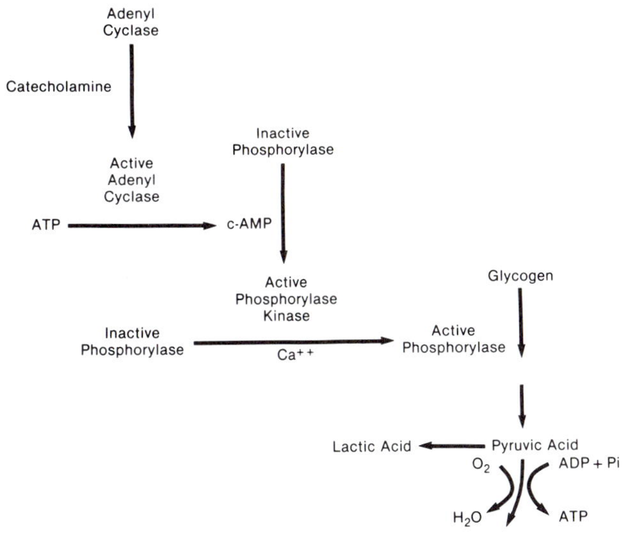

elevate the metabolic rate (122). Modest increases of myoplasmic calcium activate phosphorylase a (123), thereby increasing catabolism of glycogen to lactate, carbon dioxide and heat (124–127). The activation of phosphorylase a is further reinforced by a protein kinase, phosphorylase kinase (128–133) (Figure 11). In turn this protein kinase is activated by catecholamine stimulation (Figure 10). The activation of phosphorylase a drives electron transport in the mitochondria. Oxygen consumption is correspondingly increased. Excess heat production resulting from the activation of phosphorylase a is, therefore, both anaerobic and aerobic (134).

The excess lactate is transported to and taken up by the liver (135) where it is converted to glucose or catabolized to carbon dioxide and water. Di Marco et al. have shown that the incorporation of lactate into the livers of MHS pigs is less than half that of normal (136). Catabolism of lactate to $CO2$ and $H20$ is heat producing and probably accounts for the early and marked rise in liver temperature during MH crises. The inability of MHS livers to metabolize lactate adequately during halothane anaesthesia may be partially due to the reduction in hepatic blood flow (2), secondary to the influx of water into the muscle cells (see below) that occurs with the onset of hyperthermia.

Certainly no substantial evidence exists in favour of a primary defect in the liver. Liver function tests such as serum bilirubin and BSP retention remain normal or only slightly elevated during and after a MH crisis. Furthermore, measurements of temperatures, oxygen, carbon dioxide, and lactate levels in halothane-containing blood entering and leaving isolated MHS and normal pig livers reveal no differences in these parameters between the two categories of hogs (137).

The patient with a mild MH reaction, because of either an inherently moderate defect or because of anaesthesia with weak triggering agents (cyclopropane or d-tubocurarine), probably experiences no more than an increase in catabolism of muscle glycogen (Figure 11) and liver lactate. Clinically these events are expressed as fever and respiratory and metabolic acidosis. Skeletal muscle rigidity is not observed.

Still higher concentrations of myoplasmic calcium have several additional effects that further raise heat production and also induce muscle contracture (Figure 12). The elevated myoplasmic calcium combines with troponin, a filament of globular proteins conjoined with long and thread-like tropomyosin which lies within the helixes of actin. The calcium induces a conformational change and a rotation of the troponin, which in turn causes a movement of the tropomyosin in such a way as to reveal the receptor sites on the actin to the myosin cross-bridges. During stress another of the cAMP-activated protein kinases potentiates the formation of this calcium-troponin complex (115) (Figure 12).

FIGURE 12

EFFECT OF EXCESS MYOPLASMIC CALCIUM ON MYOFIBRILS DURING MH REACTION

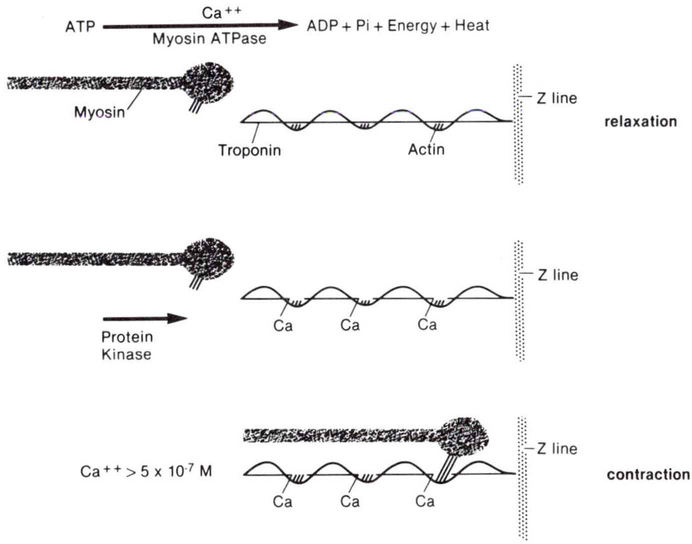

The elevated myoplasmic calcium also causes excessive activation of myosin ATPase during MH crises (138) (Figure 12). Hydrolysis by myosin ATPase of ATP to ADP plus phosphate and heat is, therefore, accelerated. The ATP hydrolysis induces formation of actomyosin (139) intermittently in normal muscle and permanently during reactions in MHS muscle. During relaxation the myosin cross-bridges are negatively charged because of the presence on them of ATP. They, therefore, project away from the shank of the myosin filament since it is also negatively charged. During contraction the addition of calcium to the ATP on the myosin cross-bridges (now combined with actin receptors) neutralizes their negativity. The bridges now bend inward toward the myosin shank pulling the myosin along the axis of the actin filament. The calcium-induced ATPase activity of the myosin head splits the ATP to ADP and phosphate, thus breaking the calcium-linked connection between the myosin bridge and actin receptor. Almost immediately a similar reaction sequence occurs at the next myosin cross-bridge so that the myosin is pulled along yet another step. In this way the myosin slides over the actin in a ratchet-like fashion forming short and rigid actomyosin (140,141).

As the muscle temperature rises the above events become self-sustaining since the increased temperature eliminates the calcium requirement for myosin-actin interaction (142). This thermal inactivation of the Ca^{2+}-regulating mechanism of actomyosin is potentiated by the decline in ATP levels. Hence two key events in the MH reaction, namely a rise in muscle temperature and a fall in muscle ATP, can perpetuate a muscle rigor independent of the intracellular calcium concentration (143).

In normal muscle, once the acetylcholine-induced muscle action potential ceases, calcium is taken back into the SR and the muscle returns to its resting relaxed state (Figures 2 and 12). During a MH crisis, however, the myoplasmic calcium remains permanently elevated so that muscle contracture, heat, carbon dioxide and lactic acid production and oxygen consumption continue unabated.

Some of the excess calcium may be absorbed by the mitochondria (Figure 13). Within the mitochondria, the resulting toxic calcium concentration uncouples oxidative phosphorylation from electron transport, thereby decreasing ATP production but further accelerating oxygen consumption and output of lactate (113,144), carbon dioxide and heat (Figure 13). This effect is aggravated by lipid soluble volatile anaesthetics (145).

Early in the reaction, ATP levels in the muscle are reasonably well maintained through conversion of creatine phosphate and ADP to creatine and ATP (144,146). Once, however, the muscle's stores of high energy creatine phosphate have been exhausted (143), muscle ATP declines rapidly (147) (Figure 13). The fall is particularly marked in red muscle. This ATP decline may possibly be exacerbated by the lack of complete availability of another source of ATP replacement, namely, adenylate kinase

$$2\ ADP \rightarrow ATP + AMP$$

This is because, for reasons not fully understood, MHS skeletal muscle has been reported to suffer from a deficiency of adenylate kinase (148), although the work of Marjanen and Denborough would appear to refute this claim (149).

FIGURE 13

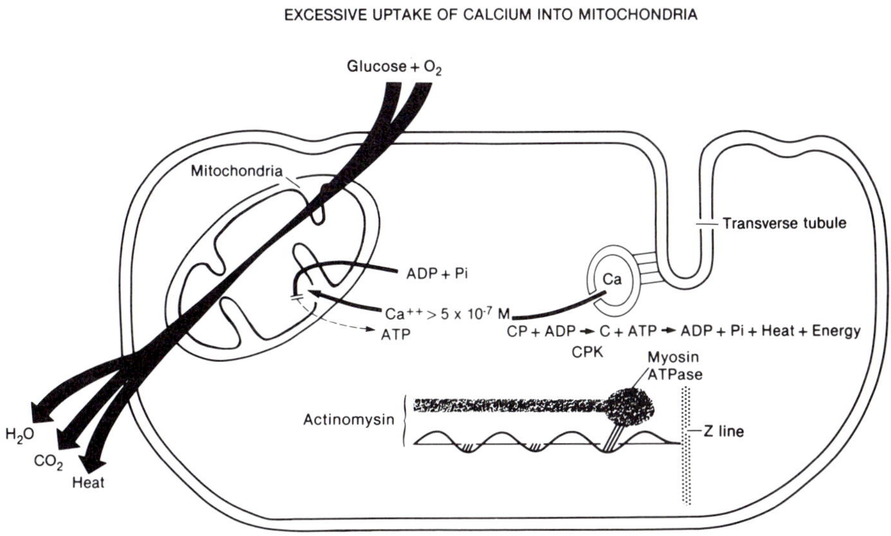

When depletion of ATP occurs the MHS crisis becomes irreversibly fatal (Figure 14). First, ATP (although not ATP hydrolysis) is necessary for muscle relaxation. There is, however, not sufficient ATP to separate actin from myosin. Rigor mortis, therefore, develops even before death has occurred (150). Second, not enough ATP remains to serve as substrate for SR ATPase. Calcium is, consequently, not retaken into the SR (Figure 14). This effect is independent of and in addition to any primary defect that might exist in the SR membrane. Third, the sarcolemmal ATPases (Na^+K^+-ATPase and Ca^{++}-ATPase) fail through lack of their substrate ATP (Figure 14). Ions and molecules, therefore, simply follow their natural concentration gradients. Thus, potassium, magnesium, phosphate, and, somewhat later, large molecules such as myoglobin and then the even larger enzyme such as CK leak outward.

The myoglobin obstructs the renal tubules, thereby inducing acute renal failure. Calcium, because its concentration gradient is in the reverse direction, flows into the myoplasm and so exacerbates the already existing calcium-dependent biochemical dearrangements (Figure 14).

FIGURE 14

CONSEQUENCES OF LACK OF SUBSTRATE ATP

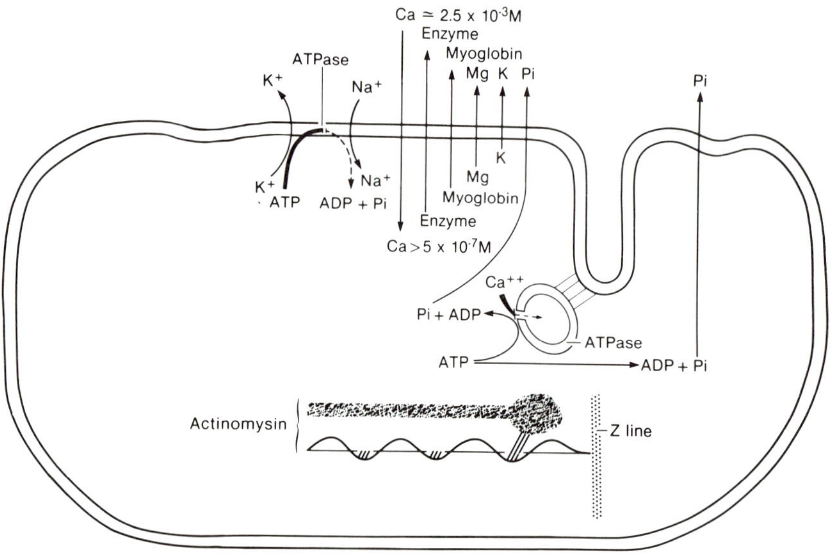

Another reason for the fulminant fever of MH may be decreased heat loss. Williams et al. (97) have observed that in MHS pigs during the early phases of a reaction, in spite of the increased temperature of the perfusing blood, peripheral blood vessels undergo intense spasm. Consequently, the exposed muscles at the operative site appear gelatinous and bloodless even though the driving pressure in the large central vessels is still well maintained (150). This spasm may be secondary to catecholamine stimulation or may be due to a primary MH defect of the vascular smooth muscle.

MALIGNANT HYPERTHERMIA—A WIDESPREAD MEMBRANE DEFECT

Malignant hyperthermia appears to be a widespread membrane defect, involving both different membranes within the same cell type, as indicated in the preceding pages, and membranes of different cell types. For example, in skeletal muscle there is evidence for calcium-related defects in sarcoplasmic reticulum, mitochondria and sarcolemma, in particular the catecholamine receptor area of the sarcolemma which may be a molecular abnormality common to all three membranes. The work of McIntosch and Berman (81) and Drabikowski et al. (151) suggests that the defect may lie in the structural lipids of these membranes. There is also some, although not irrefutable, evidence that similar calcium-related biochemical defects are present in other types of cells, for example, nerve (152,153) and brain cells, bone (154), platelets (155,156), red blood cells (157), lymphocytes (158), and cardiac muscle and hormone-producing and -secreting cells (159,160). Furthermore, concentrations of 3,4 dihydroxymandelic acid (DHMA) are higher in the hypothalamus and striatum of MHS than of normal pigs (161). Hence the severe neurological deterioration, consumption coagulopathy, haemolysis, bizarre arrhythmias, depletion of thyroxin and elevation of blood glucose that are characteristic features of fulminant MH reactions may be initiated by primary defects in tissues other than skeletal muscle rather than by the secondary effects of electrolyte, acid-base and temperature disturbances. It is uncertain whether the myopathy of MH is secondary to a primary neurogenic lesion or is a widespread abnormality involving independent but similar defects in membranes of many different cell types and in different membranes within given cell types. The work of Moulds (162) would suggest that MHS muscle is not denervated. He has shown that in denervated muscle electrical stimulation induces: a greatly increased twitch tension; a decreased frequency of stimulation required to produce a tetanus; and a tetanus which is folowed by an increased time to relaxation. These abnormalities of contraction and relaxation are all absent in MHS muscle. Harriman et al. (66) in a microscopic examination of 75 MHS families, never encountered the classical features of a denervated muscle, i.e. collateral reinnervation, type grouping or group atrophy have always been absent.

The role of brown fat and prostaglandins in heat generation during MH crises has yet to be investigated. Perfusion of isolated MHS and normal pig livers with halothane has revealed no differences in temperatures, blood gases, electrolytes or enzymes in the blood leaving the livers (137). This organ, therefore, is probably not abnormal in MHS pigs. This may be because it is of epithelial origin while muscle, brain, nerve, bone, red blood cells, and platelets are all of connective tissue origin.

MISCELLANEOUS OBSERVATIONS ON THE AETIOLOGY OF MH

Harriman et al. (163) have found that tissue cultures of human skeletal muscle cells show no significant differences between MHS and normal muscle with respect to rate of growth, progress of differentiation and cell morphology. Other characteristics of MHS skeletal muscle cells remain to be investigated, some of which may prove to be abnormal, particularly in the presence of triggering agents.

Okumura and Kuro (164) have not detected any differences between normal and MHS muscle protein separated by two dimensional SDS acrylamide electrophoresis. These results suggest that either the genetic defect does not involve the muscle fibre directly, that the defect is only revealed when susceptible muscle is exposed to anaesthetic agents or that the techniques employed have not been sufficiently discriminating. Evidence in support of the latter postulation is the more recent work of Oku and his colleagues

(165) on platelets. As with muscle they found that two-dimensional electrophoresis of MHS platelets revealed no consistent abnormalities. Treatment of the platelets with radioactive ATP prior to electrophoresis and with autoradiography after electrophoresis has demonstrated, however, that the phosphorylating proteins of the MHS platelets are markedly different from the normal platelets.

REFERENCES

1. Britt, B.A., Kalow, W., Gordon, A., Humphrey, J.G. and Rewcastle, N.B. Canad. Anaesth. Soc. J. 1973;20:431-67.

2. Britt, B.A. In: Milton, A.S. (ed.). Handbook of Experimental Pharmacology, Vol. 60. 1982. Springer-Verlag, Berlin Heidelberg, pp. 547-615.

3. Denborough, M.A. Int. Anesth. Clin. 1979;17:11.

4. Kalow, W., Britt, B.A., Terreau, M.E. and Haist, C. Lancet 1970;ii:895-898.

5. Britt, B.A., Kwong, F.H-F. and Endrenyi, L. In: Henschel, E.O. (ed.). Malignant Hyperthermia, Current Concepts. 1977. Appleton-Century-Crofts, New York, pp. 9-45.

6. Britt, B.A. In: Orkin, F.K. and Cooperman, L.H. (eds.). Complications in Anaesthesiology. 1982. Lippincott, Philadelphia, pp. 291-313.

7. Britt, B.A., Endrenyi, L., Kalow, W. and Peters, P.L. Can. Anaesth. Soc. J. 1976;23:624.

8. Ellis, F.R., Harriman, D.G.F., Currie, S., and Cain, P.A. In: Aldrete, J.A. and Britt, B.A. (eds.). Second International Symposium on Malignant Hyperthermia. 1978. Grune & Stratton, New York, p. 273.

9. Moulds, R.F.W. and Denborough, M.A. Br. Med. J. 1974;2:241.

10. Nelson, T.E. In: Aldrete, J.A. and Britt, B.A. (eds.). Second International Symposium on Malignant Hyperthermia. 1978. Grune & Stratton, New York, p. 23.

11. Ellis, K.O. and Bryant, S.H. Naunyn Schmiedebergs Arch. Pharmacol. 1972;274:107.

12. Ellis, K.O. and Carpenter, J.F. Naunyn Schmiedebergs Arch. Pharmacol. 1972:275;83.

13. Austin, K.L. and Denborough, M.A. Anaesthesia and Intensive Care 1977;5:207.

14. Harrison, G.G. In: Gordon, R.A., Britt, B.A. and Kalow, W. (eds.). International Symposium on Malignant Hyperthermia. 1973. Thomas, Springfield, p. 271.

15. Johnson, P.N. and Inesi, G. J. Pharmacol. Exp. Therap. 1969;169:308.

16. Moulds, R.F.W. and Denborough, M.A. Brit. Med. J. 1972;4:526.

17. Moulds, R.F.W. and Denborough, M.A. Clin. Exper. Pharmacol. Physiol. 1974;1:197.

18. Fleckenstein, A. In: Harris, P. and Opie, L. (eds.). Calcium and the Heart. 1971. Academic Press, London.

19. Nakajima, H., Hoshiyama, M., Yamashita, A. and Kiyomoto, A. Japan J. Pharmacol. 1976;26:571.

20. Uchida, Y. Jpn. Heart J. 1976;17:599.

21. Anderson, I.L. and Jones, E.W. In: Aldrete, J.A. and Britt, B.A. (eds.). Second International Symposium on Malignant Hyperthermia. 1978. Grune & Stratton, New York, p. 509.

22. Lopez, J.R., Alamo, L., Caputo, C., Wikinski, J. and Ledezma, D. Muscle & Nerve 1985.

23. Huxley, H.E. In: Paul, W.M., Daniel, E.E., Kay, C.M. and Monckton, G. (eds.). Muscle. 1965. Pergamon, Oxford, p. 3.

24. Sandow, A. Skeletal muscle. Ann. Rev. Physiol. 1970;32:1040.

25. Nelson, T.E. In: Britt, B.A. (ed.) Malignant Hyperthermia. Boston: Martinus Nijhoff. 1987.

26. Bowers, E. Br. Med. J. 1974;2:614.

27. MacLennan, D.H. In: Gordon, R.A., Britt, B.A. and Kalow, W. (eds.). International Symposium on Malignant Hyperthermia. 1973. Thomas, Springfield, p. 139.

28. Britt, B.A. Ann. Roy. Coll. Surg. Engl. 1971;48:73.

29. Britt, B.A. In: Gordon, R.A., Britt, B.A. and Kalow, W. (eds.). International Symposium on Malignant Hyperthermia. 1973. Thomas, Springfield, p. 451.

30. Britt, B.A., Endrenyi, L., Barclay, R.L. and Cadman, D.L. Br. J. Anaesth. 1975;47:647.

31. Heffron, J.J.A. and Mitchell, G. Proc. Int. Pig Vet. Soc., 1976 Congress, Ames, Iowa.

32. Kim, D.H., Sreter, F.A., Ohnishi, S.T., Ryan, J.F., Roberts, J., Allen, P.D., Meszaros, L.G., Antoniu, B. and Ikemoto, N. Biochim. Biophys. Acta 1984;775:320-327.

33. Nelson, T.E. and Bee, D.E. J. Clin. Invest. 1979;64:895-901.

34. Nelson, T.E. and Chausmer, A.B. J. Pharmacol. Exp. Ther. 1981;219:107-111.

35. Nelson, T.E. J. Clin. Invest. 1983;72:862-870.

36. Nelson, T.E. In: Britt, B.A. A Review of Malignant Hyperthermia. 1986. Martinus Nijhoff, Boston.

37. Takagi, A. Hiroshima J. Anesth. 1977;13:155.

38. Wood, D.S. Presented at the IVth International Congress on Neuromuscular Diseases, 1978, Montreal, Quebec.

39. Yagi, S., Ishizuka, T., Horiuti, K., Endo, M., Koga, Y. and Amaha, K. Hiroshima J. Anaesth. 1983;19:76.

40. Kim, D.H., Sreter, F.A. and Ikemoto, N. Presented at the 4th International Malignant Hyperpyrexia Workshop, York, September, 1986.

41. Bennett, D., Cain, P.A., Ellis, F.R., Louis, C.F. and Stanton, M. Br. J. Anaesth. 1977;49:979.

42. Blanck, T.J.J., Gruener, R., Suffecool, S.L. and Thompson, M. Anesth. Analg. 1981;60:492-498.

43. Britt, B.A., Endrenyi, L. and Cadman, D.L. Br. J. Anaesth. 1975;47:650.

44. Nelson, T.E., Jones, E.W., Venable, J.H. and Kerr, D.D. Anesthesiology 1972;36;52.

45. Niebroj-Dobosz, I. and Mayzner-Zawadzka, E. Br. J. Anaesth. 1982;54:885-890.

46. Steward, D.J. and Thomas, T.A. In: Gordon, R.A., Britt, B.A. and Kalow, W. (eds.). International Symposium on Malignant Hyperthermia. 1973. Thomas, Springfield, p. 409.

47. Ohnishi, S.T. Presented at the 4th International Malignant Hyperpyrexia Workshop, York, September, 1986.

48. Horiuti, K., Matsumura, N., Matsui, K., Iino, M. and Endo, M. Presented at the 4th International Malignant Hyperpyrexia Workshop, York, September, 1986.

49. Nelson, T.E. FEBS Letters 1984;167:123.

50. Nelson, T.E. Presented at the 4th International Malignant Hyperpyrexia Workshop, York, September, 1986.

51. Cohen, P.J. and Marshall, B.E. In: Fink, B.R. (ed.). Toxicity of anesthetics. 1968. Williams & Wilkins, Baltimore, p. 24.

52. Britt, B.A., Kalow, W. and Endrenyi, L. In: Gordon, R.A., Britt, B.A. and Kalow, W. (eds.). International Symposium on Malignant Hyperthermia. 1973. Thomas, Springfield, p. 387.

53. Britt, B.A., Endrenyi, L., Cadman, D.L., Ho, M.F. and Fung, H.Y-K. Anesthesiology 1975;42:292.

54. Hull, M.J., Webster, W.W. and Gatz, E. J. Oral Surg. 1971;29:640.

55. Wilson, R.D., Nichols, R.J., Dent, T.E. and Allen, C.R. Anesthesiology 1966;26:232.

56. Cheah, K.S. and Cheah, A.M. J. Sci. Food Agric. 1976;27:1137.

57. Cheah, K.S. and Cheah, A.M. FEBS Letters 1978;95:307.

58. Cheah, K.S. and Cheah, A.M. Biochim. Biophys. Acta 1981;638:40-49.

59. Cheah, K.S. and Cheah, A.M. Biochim. Biophys. Acta 1981;634:70-84.

60. Cheah, K.S. Biochem. Soc. Trans. 1984;12:358-360.

61. Cheah, K.S. In: Britt, B.A. (ed.) Malignant Hyperthermia. Boston: Martinus Nijhoff, 1987.

62. Denborough, M.A., Hird, F.J.R., King, J.O., Marginson, M.A., Mitchelson, K.R., Nayler, W.G., Rex, M.A., Zapf, P. and Condron, R.J. In: Gordon, R.A., Britt, B.A. and Kalow, W. (eds.). International Symposium on Malignant Hyperthermia. 1973. Thomas, Springfield, p. 229.

63. Gronert, G.A. and Heffron, J.J.A. Anesth. Analg. 1979;58:76.

64. Hall, G.M., Kirtland, S.J. and Baum, H. Br. J. Anaesth. 1973;45:1005.

65. Harris, R.A., Munroe, J., Farmer, B., Kim, K.C. and Jenkins, P. Arch. Biochem. Biophys. 1971;142:435.

66. Harriman, D.G.F., Ellis, F.R., Franks, A.J. and Sumner, D.W. In: Aldrete, J.A. and Britt, B.A. (eds.). Second International Symposium on Malignant Hyperthermia. 1978. Grune & Stratton, New York, p. 67.

67. Mitchelson, K.R. and Hird, F.J.R. Am. J. Physiol. 1973;225:1393.

68. Popinigis, J. Anaesthesiol. Resusc. 1973;1:63.

69. Ruitenbeek, W., Verburg, M.P., Janssen, A.J.M, Stadhouders, A.M. and Sengers, R.C.A. Acta Anaesthesiol. Scand. 1984;28:9.

70. Schafer, H.J. Stuttgart, New York: Gustav Fischer Verlag, 1979.

71. Stadhouders, A.M., Viering, W.A.L., Verburg, M.P., Ruitenbeek, W. and Sengers, R.C.A. Acta Anaesthesiol. Scand. 1984;28:14-26.

72. Cheah, K.S. Presented at the 4th International Malignant Hyperpyrexia Workshop, York, September, 1986.

73. Sullivan, J.S., Galloway, G.J. and Denborough, M.A. Clin. Exp. Pharmacol. Physiol. 1983;10:587-593.

74. Nelson, T.E., Flewellen, E.H. and Arnett, D.W. Muscle & Nerve 1983;6:263-268.

75. Walsh, M.P. Canad. Anaesth. Soc. J.

76. Cheung, W.Y. Science 1980;207:19.

77. Klee, C.B., Crouch, T.H. and Richman, P.G. Ann. Rev. Biochem. 1980;49:489.

78. Marjanen, L.A., Collins, S.P. and Denborough, M.A. Biochem. Med. 1984;32:283-287.

79. Cheah, K.W., Cheah, A.M. and Waring, J.C. Meat Science 1986;17:37.

80. Niebroj-Dobosz, I., Kwiatkowski, H. and Mayzner-Zawadzka, E. Med. Biol. 1984;62:250-254.

81. McIntosch, D. and Berman, M.C. S. Afr. Med. J. 1974;48:1221.

82. Barlow, M.B. and Isaacs, H. Br. J. Anaesth. 1970;42:1072.

83. Larard, D.G., Rice, C.P., Robinson, R.W., Spencer, R.W. and Westhead, R.A. Br. J. Anaesth. 1972;44:93.

84. Solomons, C.C. and Myers, D.N. In: Gordon, R.A., Britt, B.A. and Kalow, W. (eds.). International Symposium on Malignant Hyperthermia. 1973. Thomas, Springfield, p. 319.

85. Plotz, J. Anaesthesist 1983;32:158.

86. Auerbach, V.H., DiGeorge, A.M., Mayer, B.W., Hayden, M., Carpenter, G.G., Krumperman, L.W. and Truter, T.R. In: Gordon, R.A., Britt, B.A. and Kalow, W. (eds.). International Symposium on Malignant Hyperthermia. 1973. Springfield:Thomas, p. 30.

87. Denborough, M.A., Forster, J.F.A., Hudson, M.C., Carter, N.G. and Zapf, P. Presented at the Australasian Congress of Anaesthesiology, Melbourne, Australia, 1970.

88. Berman, M.C. and Kench, J.E. In: Gordon, R.A., Britt, B.A. and Kalow, W. (eds.). International Symposium on Malignant Hyperthermia. 1973. Thomas, Springfield, p. 287.

89. Woolf, N., Hall, L., Thorne, C. and Down, M. Br. Med. J. 1970;3:386.

90. Jones, E.W., Nelson, T.E., Anderson, I.L., Kerr, D.D. and Burnap, T.K. Anesthesiology 1972;36:42.

91. Gronert, G.A. and Theye, R.A. Anesthesiology 1976;44:36.

92. Mickelson, J.R. Presented at the 4th International Malignant Hyperpyrexia Workshop, York, September, 1986.

93 Endo, M., Tanaka, M. and Ogawa, Y. Nature 1970;228:34.

94. Thorens, S. and Endo, M. Proc. Jpn. Acad. 1975;51:473.

95. Thorpe, W. and Seeman, P. In: Gordon, R.A., Britt, B.A. and Kalow, W. (eds.). International Symposium on Malignant Hyperthermia. 1973. Thomas, Springfield, p. 152.

96. Gallant, E.M., Huetteman, D.A. and Donaldson, S.K. Presented at the 4th International Malignant Hyperpyrexia Workshop, York, September, 1986.

97. Williams, C.H., Shanklin, M.D. and Hedrick, H.B. In: Aldrete, J.A. and Britt, B.A. (eds.). Second International Symposium on Malignant Hyperthermia. 1978. Grune & Stratton, New York, p. 113.

98. Willner, J.H., Cerri, C.G. and Wood, D.S. J. Clin. Invest. 1981;68:1119-1124.

99. Wingard, D.W. In: Henschel, E.O. (ed.). Malignant Hyperthermia, Current Concepts. 1977. Appleton-Century-Crofts, New York, p. 79.

100. Wingard, D.W. In: Aldrete, J.A. and Britt, B.A. (eds.). Second International Symposium on Malignant Hyperthermia. 1978. Grune & Stratton, New York, p. 363.

101. Ahern, C.P., Somers, C.J., Wilson, P. and McLoughlin, J.V. Proceedings of the 3rd International Conference on Production Diseases in Farm Animals. Wageningen, 1976. PUDOC, Wageningen, p. 169.

102. Hall, G.M., Lucke, J.N. and Lister, D. Anaesthesia 1975;30:308.

103. Lister, D., Hall, G.M. and Lucke, J.N. Lancet 1975;1:519.

104. Lister, D., Hall, G.M. and Lucke, J.N. Br. J. Anaesth. 1976;48:831.

105. Short, C.E., Paddleford, R.R. and McGrath, C.J. Anesth. Analg. 1976;55:653.

106. Short, C.E. In: Aldrete, J.A. and Britt, B.A. (eds.). Second International Symposium on Malignant Hyperthermia. 1978. Grune & Stratton, New York, p. 175.

107. Lucke, J.N. and Denny, H.R. Res. Vet. Sci. 1977;23:372.

108. Lucke, J.N., Denny, H., Hall, G., Lovell, R. and Lister, D.D. Br. J. Anaesth. 1978;50:241.

109. Hall, G.M., Lucke, J.N. and Lister, D. Br. J. Anaesth. 1977;49:855.

110. Kerr, D.D., Wingard, D.W. and Gatz, E.E. Anesthesiology 1975;42:307.

111. Gronert, G.A., Milde, J.H. and Theye, R.A. Anesthesiology 1977;47:411.

112. Gronert, G.A., Milde, J.H. and Theye R.A. Anesthesiology 1976;44:488.

113. Gronert, G.A. and Theye, R.A. Anesthesiology 1976;44:36.

114. Gronert, G.A., Theye, R.A. and Milde, J.H. Anesthesiology 1978;49:330.

115. Britt, B.A. In: Hulsz, E. and Lunn, J.N. (eds.). Anaesthesiology. Proceedings of the VIth World Congress of Anaesthesiology. Mexico City, April 24-30, 1976. Excerpta Medica, Amsterdam, p. 439.

116. Willner, J.H., Cerri, C.J. and Wood, D.S. Neurology 1979;29:557.

117. Stanec, A. and Stefano, G. Br. J. Anaesth. 1984;56:1243.

118. Bowman, W.C. and Nott, M.W. Pharmacol. Rev. 1969;21:27.

119. Sutherland, E.W., Rall, R.W. and Menon, T. J. Biol. Chem. 1962;237:1220.

120. Rabinowitz, M., Desalles, L., Meisler, J. and Lorand, L. Biochim. Biophys. Acta 1965;97:29.

121. Opie, J.H. Jpn. Cir. J. 1978;42: 1223.

122. Williams, C.H., Houchins, C. and Shanklin, M.D. Br. Med. J. 1975;3:411.

123. Willner, J.H., Wood, D.S., Cerri, C. and Britt, B.A. New Engl. J. Med. 1980;303:138.

124. Gronert, G.A., Milde, J.H. and Theye, R.A. Anesthesiology 1976;44:124.

125. Jones, E.W., Nelson, T.E., Anderson, I.L., Kerr, D.D. and Burnap, T.K. Anesthesiology 1972;36:42.

126. Jones, E.W., Kerr, D.D. and Nelson, T.E. In: Gordon, R.A., Britt, B.A. and Kalow, W. (eds.). International Symposium on Malignant Hyperthermia. 1973. Thomas, Springfield, p. 198.

127. Lucke, J.N., Hall, G.M. and Lister, D. Br. J. Anaesth. 1976;48:297.

128. Drummond, G.I., Valdards, J.R.E. and Duncan, L. In: Paul, W.M., Daniel, E.E., Kay, C.M. and Monckton, G. (eds.). Muscle. 1965. Oxford: Pergamon Press, p. III.

129. Krebs, E.G., Graves, D.J. and Fischer, E.H. Fed. Proc. 1959:18;266.

130. Krebs, E.G., DeLange, R.J., Kempf, R.G. and Riley, W.D.B. Pharmacol. Rev. 1966;18:163.

131. Sutherland, E.W. and Rall, R.W. Pharmacol. Rev. 1960;12:265.

132. Ozawa, E. and Ebashi, S. J. Biochem. 1967;62:285.

133. Ozawa, E., Hosoi, K. and Ebashi, S. J. Biochem. 1967;61:531.

134. Hall, G.M., Bendall, J.R., Lucke, J.N. and Lister, D. Br. J. Anaesth. 1976;48:305.

135. Hall, G.M., Lucke, J.N., Lovell, R. and Lister, D. Biochem. Soc. Trans. 1978;6:587.

136. Dimarco, N.M., Beitz, D.C., Young, J.W., Topel, D.G. and Christian, L. J. Nutr. 1976;106:710.

137. Britt, B.A., Shandling, B.R., Endrenyi, L. and Kent, G.M. Can. Anaesth. Soc. J. 1978;25:373.

138. Krzywicki, K. J. Food Sci. 1971;36;791.

139. Perry, S.V. In: Paul, W.M. (ed.). Muscle. 1965. Pergamon Press, Oxford, p. 29.

140. Fuchs, F. Anesthesiology 1975;42:584.

141. Wakabayashi, T. and Ebashi, S. J. Biochem. (Tokyo) 1968;74:731.

142. Ebashi, S., Ebashi, F. and Kodama, A. J. Biochem. (Tokyo) 1967;62:137.

143. Waltemath, C.L. In: Gordon, R.A., Britt, B.A. and Kalow, W. (eds.). International Symposium on Malignant Hyperthermia. 1973. Thomas, Springfield, p. 16.

144. Verburg, M.P., Oerlemans, F.T.J.J., van Bennekom, C.A., Gielen, M.J.M., de Bruyn, C.H.M.M. and Crul, J.F. Acta Anaesthesiol. Scand. 1984;28:1-8.

145. Hall, G.M. Biochem. Soc. Trans. 1973;1:854.

146. Lohmann, K. Biochem. Z. 1934;271:264.

147. Nelson, T.E., Jones, E.W., Henrickson, R.L., Falk, S.N. and Kerr, D.D. Am. J. Vet. Res. 1974;35:347.

148. Schmitt, J., Schmidt, K. and Ritter, H. Humangenetik. 1974;24:253.

149. Marjanen, L.A. and Denborough, M.A. Clin. Chim. Acta 1982;122:225-230.

150. Britt, B.A. Fed. Am. Soc. Exp. Biol. 1979;38:44.

151. Drabikowski, W., Sarzala, M.G., Wroniszweska, A., Lagwinska, E. and Drzewieka, B. Biochim. Biophys. Acta 1972;274:158.

152. Britt, B.A., McComas, A.J., Endrenyi, L. and Kalow, W. Anesthesiology 1977;47:490.

153. LaCour, D., Juul-Jensen, P. and Reske-Nielsen, E. Acta Anaesthesiol. Scand. 1971;15:299.

154. Britt, B.A., Harrison, J.E. and McNeil, K.G. Can. Anaesth. Soc. J. 1979;26:117.

155. Solomons, C.C., Tan, S. and Aldrete, J.A. In: Aldrete, J.A. and Britt, B.A. (eds.). Second International Symposium on Malignant Hyperthermia. 1978. Grune & Stratton, New York, p. 221.

156. Britt, B.A. and Scott, E.A. Anesth. Analg. 1986;65:171.

157. Kelstrup, J., Hasse, J., Jorni, J., Reske-Nielsen, R. and Hanel, H.K. Acta Anaesthesiol. Scand. 1973;17:283.

158. Klip, A., Ramlal, T., Walker, D., Britt, B.A. and Elliott, E. Anesth. Analg. 1987 (in press).

159. Denborough, M.A., Warne, G.L., Moulds, R.F.W., Tse, P. and Martin, F.I.R. Br. Med. J. 1974;3:493.

160. Lister, D. Br. Med. J. 1973;1:208.

161. Bardsley, M.E., Whestley, A.M., Fowler, C.J., McCrodden, J.M., McLoughlin, J.V. and Tipton, K.F. Br. J. Anaesth. 1982;54:1313.

162. Moulds, F.R.W. In: Aldrete, J.A. and Britt, B.A. (eds.). Second International Symposium on Malignant Hyperthermia. 1978. Grune & Stratton, New York, p. 49.

163. Harriman, D.G.F., Ellis, F.R. and Spencer, J. In: Aldrete, J.A. and Britt, B.A. (eds). Second International Symposium on Malignant Hyperthermia. 1977. New York: Grune & Stratton, p. 37.

164. Okumura, F. and Kuro, M. Jpn. J. Anesth. 1975;24:525.

165. Oku, S., Ueyama, H. and Britt, B.A. Hiroshima J. Anesth. 1981;17:99.

3

SKELETAL MUSCLE SARCOPLASMIC RETICULUM IN THE MALIGNANT HYPERTHERMIA SYNDROME

T.E. NELSON

Department of Anesthesiology, The University of Texas Medical Branch, Galveston, Texas 77550

ABSTRACT

A general working hypothesis for the aetiology of malignant hyperthermia is that triggering anaesthetic agents produce an increased myoplasmic Ca^{2+} level that stimulates metabolic and contracture events. As long as the Ca^{2+} remains at a concentration capable of activating skeletal muscle heat, metabolic products and abnormal electrolyte balance will result, ultimately fulminating to the life-threatening malignant hyperthermia syndrome. Under a normal resting state, the Ca^{2+} concentration in myoplasm is 10^{-7} M or lower and at Ca^{2+} concentrations of 10^{-6} to 10^{-5} M or greater, the metabolic and contractile systems are activated. The primary regulatory system for Ca^{2+} in skeletal muscle is the sarcoplasmic reticulum which serves three main functions: Ca^{2+} storage, Ca^{2+} release and Ca^{2+} uptake. In this chapter, we discuss current understanding of the sarcoplasmic reticulum in performing these functions and review data from studies on sarcoplasmic reticulum function in malignant hyperthermia skeletal muscle. Studies on Ca^{2+} transport in sarcoplasmic reticulum from malignant hyperthermia human and pig muscle do not support a defect in the transport function of sarcoplasmic reticulum in malignant hyperthermia skeletal muscle. In addition to the Ca^{2+} pump protein, the sarcoplasmic reticulum membrane also contains Ca^{2+} channels through which the stored Ca^{2+} is released to initiate muscle contraction. Several studies have shown a defect in the Ca^{2+} channel that is receptor mediated with Ca^{2+} as an agonist.

These mechanisms related to electromechanical coupling in skeletal muscle are likely defective sites in malignant hyperthermia susceptible muscle.

INTRODUCTION

Malignant hyperthermia (MH) is a catabolic storm that represents a global loss in metabolic regulation seen in few other disease states. Unabated, it can very quickly, in a matter of minutes, produce life-threatening changes in normal body homeostasis. The clinical symptomatology of this dramatic syndrome is detailed elsewhere in this book. For a number of years, since a classic discovery reported in 1970 (1), several independent studies have provided different pieces of information that support a myogenic defect in Ca^{2+} regulation for the aetiology of the malignant hyperthermia syndrome. It is commonly thought that an abnormal sustained increase in myoplasmic free Ca^{2+} could be the primary aetiological event and that this triggers a cascade of metabolic processes that produce the syndrome. In this chapter, we review the evidence supporting the Ca^{2+} dogma, describe the physiology, biochemistry and pharmacology of the sarcoplasmic reticulum (SR) membrane system that regulates Ca^{2+}, and explore potential aetiological roles for this membrane in malignant hyperthermia.

CALCIUM--A UNIVERSAL CELLULAR MESSENGER

a) Cytoplasmic Versus Extracellular Concentrations of Free Ca^{2+}

From the most primitive single cell organism to the most complex cell of the human body, one factor remains constant; that being a relatively low (i.e. about 10^{-7}M) concentration of free Ca^{2+} ion (Ca^{2+}) in the cytoplasm. In most biological environments this produces a large Ca^{2+} concentration gradient across the plasma cell membrane. In man the total Ca^{2+} concentration in blood is about 2.5×10^{-3}M and about 50% of this total is unbound, ionized Ca^{2+}. Therefore, the concentration of Ca^{2+} in blood is about 1.25×10^{-3}M. Inside most resting cells the Ca^{2+} concentration is only 10^{-7} M or less and this produces a Ca^{2+} concentration gradient equal to

$1.25 \times 10^{-3}M/10^{-7}M = 12{,}500$. This enormous Ca^{2+} potential across the cell membrane is maintained by several systems including low permeability of the plasma membrane to Ca^{2+}, a metabolically active Ca^{2+} extrusion system, and Ca^{2+} binding and Ca^{2+} storage sinks inside the cell. Some cells, such as skeletal muscle, have specialized organelles, i.e., SR membranes, that serve to bind, transport, store and release Ca^{2+} within the cell and maintain a low resting Ca^{2+} concentration. Thus, if all the Ca^{2+} stored in the muscle, i.e., about 3 µmoles/g wet weight, were unbound, free Ca^{2+}, the Ca^{2+} concentration would be calculated to be about $3 \times 10^{-3}M$. Obviously, some systems must exist within the muscle cell to maintain the low Ca^{2+} concentration observed during the resting state.

b) Trigger Effects of Increasing Cytoplasmic Ca^{2+}

What happens when the intracellular Ca^{2+} concentration increases above levels found in the resting state? In practically every cell type an increase in Ca^{2+} results in activation of some process. A recent article reviews the messenger role of Ca^{2+} in cells (2). Starling's discovery of the importance of Ca^{2+} to contractility of the heart was only the beginning of our understanding of the basis on which cells utilize Ca^{2+} as a messenger to activate some event. Skeletal muscle is a convenient cell system for illustrating cellular mechanisms that utilize changes in Ca^{2+} concentration to trigger a variety of events from the opening of specialized Ca^{2+} channels to the activation of contractile elements. Some of these Ca^{2+}-regulated events may, as we discuss later, be involved in the aetiology of malignant hyperthermia.

c) Coupling Sarcolemma Depolarization to V_{O_2} and Contraction

The resting muscle cell has a Ca^{2+} concentration of $0.5 \times 10^{-7}M$ in contrast to the blood Ca^{2+} concentration of $1.25 \times 10^{-3}M$, representing a gradient for Ca^{2+} across the sarcolemma of 25,000 times greater on the outside. Inside the muscle cell most Ca^{2+} is stored in the SR, some is located in the mitochondria and a certain amount may be protein bound or superficially bound to the intracellular side of membrane structures such as the transverse tubule. Release of Ca^{2+} from the storage sites within the muscle cell is caused by depolarization of the sarcolemma membrane. In normal physiology, this depolarization is initiated by acetylcholine released from nerve cells at the myoneural junction. Under in vitro experimental conditions, it is possible to depolarize the sarcolemma in a stepwise manner by varying K^+ concentration outside the muscle cell. Under these experimental circumstances it has been shown that varying thresholds of membrane depolarization are required for stimulation of metabolic and contractile events (3). When the normal resting membrane potential of -90 mV is depolarized to -50 mV, there is marked increase in oxygen consumption without any activation of mechanical events (Figure I). Mechanical contraction occurs when the membrane is depolarized from -50 to -20 mV. When artificial depolarization of -90 to -50 mV is produced the amount of Ca^{2+} released from the SR is not sufficient to reach mechanical threshold. Thus, the Ca^{2+} concentration threshold for metabolic activation is considerably lower than that for activation of the contractile apparatus. On this basis it is theoretically possible to observe a "metabolic malignant hyperthermia" without the syndrome of skeletal muscle rigidity.

One consideration for the lower depolarization threshold for metabolism is that the rate of Ca^{2+} release from the SR is compensated for by the rate of active Ca^{2+} uptake back into the SR. Since the transport of Ca^{2+} back into the SR requires expenditure of energy (i.e., ATP and creatine phosphate), oxygen consumption may be increased in an attempt to restore the high energy compounds. Another possible explanation is that Ca^{2+} may be directly activating some enzyme systems (4). In contrast to stimulation of skeletal muscle oxygen consumption, activation of contraction requires a greater level of sarcolemma depolarization (i.e., 80 mM K^+ or about -20 mV). Thus, Ca^{2+} can activate oxygen consumption independently of its activation of muscle contraction. Since Ca^{2+} is important to skeletal muscle as an activator of metabolic and mechanical events, it is important to understand how

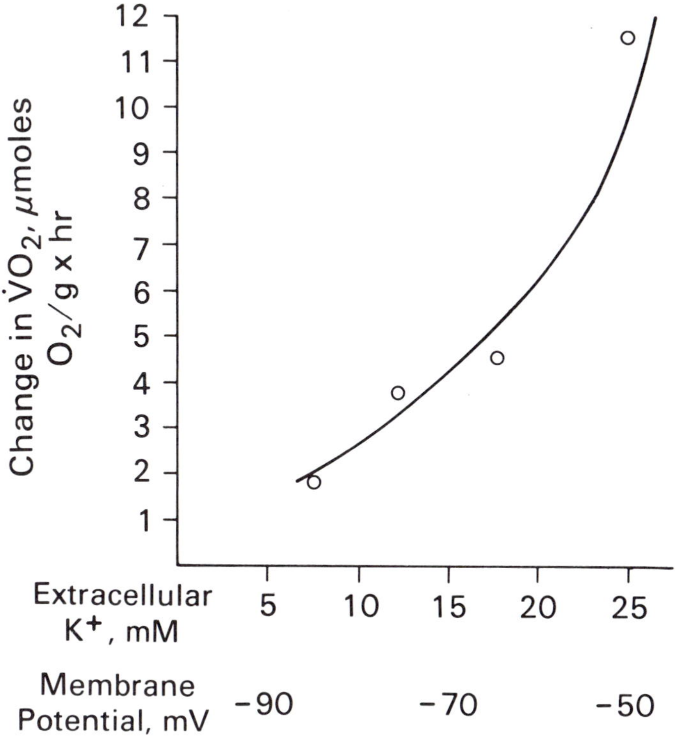

FIGURE 1. Relationship between extent of sarcolemma depolarization produced by increasing concentrations of extracellular K^+ on the change in oxygen consumption in skeletal muscle. (Data is plotted from reference #3 with permission from Dr. Bianchi.)

Ca^{2+} itself is regulated inside the muscle cell. It is generally accepted that the SR is the primary organelle for regulating Ca^{2+} concentration in skeletal muscle.

EVIDENCE FOR AN ABERRATION IN Ca^{2+} REGULATION IN MHS SKELETAL MUSCLE

a) Abnormal In Vitro Contracture Response of MHS Muscle

One of the more significant discoveries about MH was that by Kalow et al. (1) which showed that muscle from MH patients had abnormal in vitro contracture sensitivity to caffeine and to halothane. Prior to this observation it was already known that caffeine produced contracture in normal skeletal muscle due to its effects on the SR to produce a net release of Ca^{2+} into the myoplasm (5). The findings by Kalow et al. (1) not only provided significant information toward understanding the aetiology of MH, but even today these in vitro contracture responses are used as the most reliable diagnostic test for MH susceptibility. The abnormal contracture responses of MH muscle to caffeine and to halothane are illustrated in Figure 2.

b) **Caffeine and Halothane Effects on Contractility**

The ability of caffeine to alter skeletal muscle contractility and to produce contracture are concentration dependent. In normal human skeletal muscle lower caffeine concentrations alter contractility by potentiating the electrically evoked twitch response. This twitch potentiating effect is normally observed between 0.5 and 4 mM caffeine (Figure 2). The threshold for caffeine-induced contracture for normal human skeletal muscle is 4 mM and at caffeine concentrations greater than 4 mM, there is proportionately more contracture and less potentiation of the twitch. In muscle from phenotype H susceptible humans, concentrations of caffeine that only potentiate twitch in normal human muscle produce contracture in MH skeletal muscle. In a similar manner, the main effect of 1-3% halothane on isolated human muscle fascicles is to potentiate the twitch response whereas in MH skeletal muscle the same halothane concentrations produce contracture (Figure 2). Based on the interactions and effects of caffeine, halothane and dantrolene on normal human skeletal muscle, it has been postulated that more than one site or mechanism exists within the excitation-contraction (E-C) coupling system where these drugs could act to produce their effects (6). More significantly, it has been considered that low concentrations of caffeine act on a mechanism in E-C coupling with an effect that potentiates the normal twitch response. This suggests the possibility that an aberration at this site in MH muscle produces contracture effects in response to low dose caffeine instead of the normal twitch potentiation. Later we

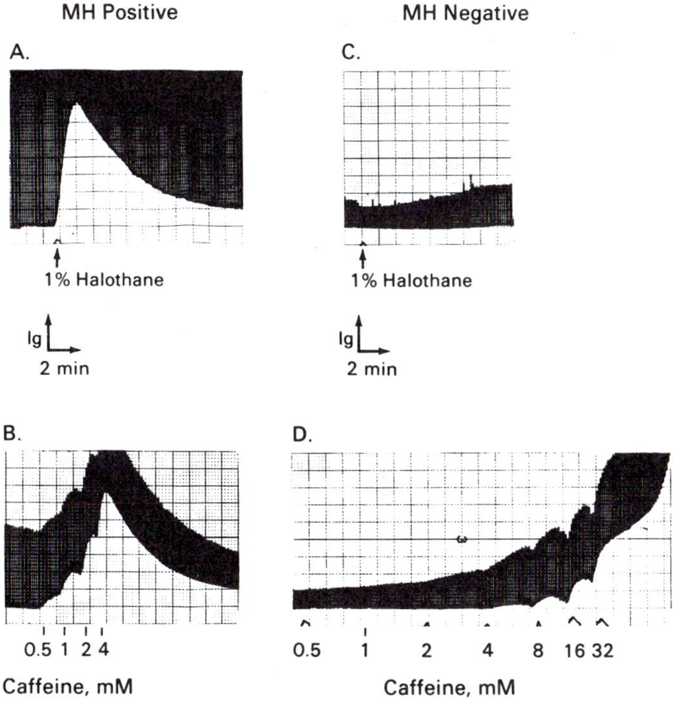

FIGURE 2. In vitro malignant hyperthermia diagnostic contracture responses of human skeletal muscle to halothane and caffeine. Halothane produces a large abnormal contracture in MH muscle (panel A) but ony potentiates the twitch in negatively diagnosed subject (panel C). MHS skeletal muscle has abnormal contracture sensitivity to caffeine with contractures initiated at 0.5 mM caffeine (panel B) compared to 4 mM caffeine in MH negative subject (panel D).

will draw on evidence to speculate that this E-C coupling site may be a Ca^{2+}-induced Ca^{2+} release pathway in the SR. Under any circumstance, it is difficult to deny that these abnormal contracture responses of MH muscle provide strong evidence that a loss in regulation of Ca^{2+} is produced by these drugs in MH skeletal muscle. That the unique muscle relaxant dantrolene completely blocks the abnormal halothane contracture of human MH muscle in vitro and is an effective prophylactic and therapeutic agent in man, is strong evidence for a primary defect in skeletal muscle. It should be noted that the same abnormal contracture responses occur in muscle from pigs genetically predisposed to MH (7,8).

c) **Clinical Evidence for Skeletal Muscle Defect**

The abnormal in vitro contracture responses of MH muscle and the theory of an aberration in Ca^{2+} regulation of MH skeletal muscle can be related to some clinical observations. In about 50% of families affected with MH there are individuals who have elevated resting CK levels in the blood (9). This suggests not only an underlying myopathy, but one that is active even in the absence of triggering anaesthetic agents. After exposure to triggering anaesthetic agents, especially halothane plus succinylcholine, a masseter muscle rigidity may occur and result in immediate cessation of anaesthetics and postponement of surgery. These cases are often associated with varying degrees of rhabdomyolysis as evidenced by elevated CK and myoglobinemia and/or myoglobinuria (10). Of those cases that are biopsied 50% test positive for MH (11). Finally, the frequent observation of skeletal muscle rigidity (sometimes global) in the fully developed MH syndrome indicates the involvement of the tissues in the syndrome. The prophylactic and therapeutic value of dantrolene for MH (12) is another piece of evidence for the aetiological role of skeletal muscle in MH.

Another abnormality observed in skeletal muscle from MH susceptible humans is that a lower than normal content of total Ca^{2+} occurs (13). This same observation was originally made in MH susceptible pig skeletal muscle (14) and was followed by a separate study comparing Ca^{2+} content and contractility in MHS muscle (15). Histologically, no lesion specific for or unique to MH skeletal muscle has been observed.

The most compelling evidence for abnormality of Ca^{2+} regulation in MH skeletal muscle is that discovered by Kalow and co-workers (1). The answers to MH aetiology can probably be found when we understand more exactly where and how drugs like caffeine and halothane act to produce abnormal contractures in isolated skeletal muscle. As we shall discuss later, some of these answers are being sought by studying isolated sarcoplasmic reticulum membranes.

d) Critical Ca^{2+} Concentration for Contraction: MHS Versus Normal

Some skeletal muscle systems have been studied in MH muscle and their aetiological role seems to have been eliminated. Utilizing chemically skinned single fibers, it has been determined that the amount of Ca^{2+} required to activate contractile elements does not differ between MH and normal human muscle (16,17). These observations suggest that no abnormality in the mechanisms activating and maintaining contraction occurs in the contractile apparatus of MH muscle. The same skinned fiber system has demonstrated abnormal sensitivity to caffeine induced contractures (16,17), consistent with results obtained using muscle fascicles in vitro.

EVIDENCE FOR AND AGAINST A DEFECT IN Ca^{2+} TRANSPORT IN SR FROM MH MUSCLE

a) Structure-Function Relationship of the SR

The functions observed in SR membranes isolated from skeletal muscle are determined by the anatomical origin in situ. Electron microscopy has provided data that allows construction of three-dimensional diagrams of the skeletal muscle fiber and this is depicted in Figure 3. The T system initiates as an invagination of the sarcolemma and traverses into the interior of the fiber. Associated with these transverse tubules are terminal sacs of the SR referred to as terminal cisternae. The terminal cisternae form on either side of the T-tubules and these three associated structures are referred to as a triad. The triad represents the structures that communicate the depolarization of the sarcolemma to the release of Ca^{2+} from the SR. This mechanism is referred to as E-C coupling. Electron dense material has been observed between the T-tubule and the terminal cisternae and these "feet" are thought to possibly carry signals for the E-C mechanism (Figure 3). When

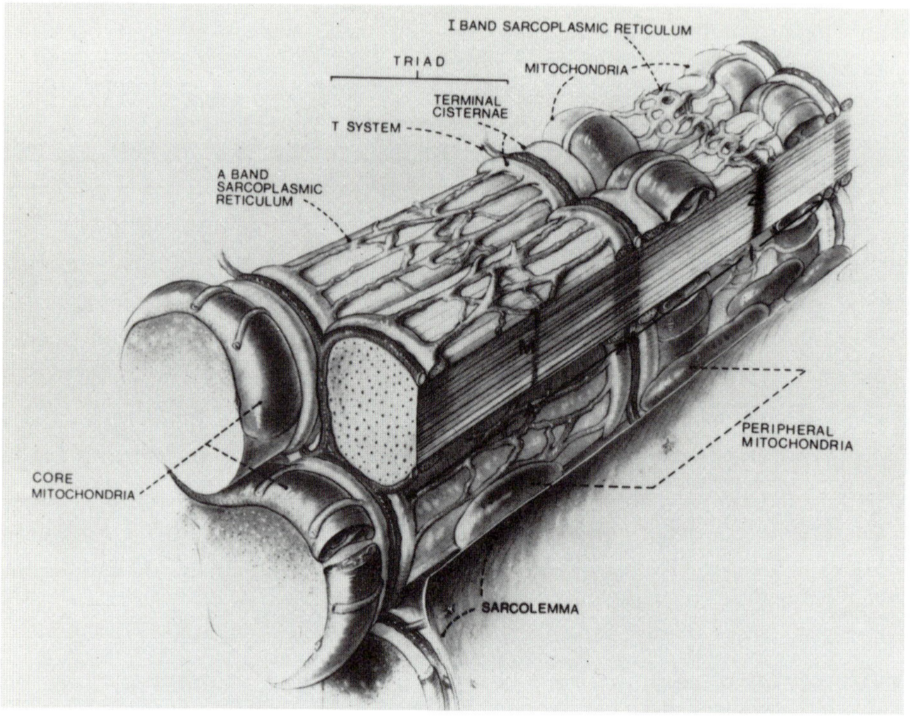

FIGURE 3. Schematic diagram of the sarcoplasmic reticulum as arranged in mammalian skeletal muscle. When muscle is homogenized, the SR membranes are broken and fragments seal to form membrane vesicles. Heavy SR membrane fragments may be comprised of terminal cisternae while light SR is usually derived from longitudinal elements. Special techniques of SR isolation can produce intact triad vesicles. (Photo from Eisenberg, B.R., Handbook of Physiology, Section 10, page 75, 1983, with permission of author and the American Physiological Society.)

muscle is homogenized, the SR membranes are sheared and these membrane pieces seal to form vesicles. These vesicles can then be separated by differential or density gradient centrifugation into different membrane fractions based on their densities. The terminal cisternae form the heavy vesicles and these are collected at low (i.e. 8,000 x g) centrifugal forces and commonly refered to as "heavy" SR fractions. Intermediate fractions can be collected at increasing centrifugal forces, or at lower density gradients, and lighter fractions at lower densities or higher centrifugal forces. Figure 3 illustrates how the SR branches off from the terminal cisternae to form narrow longitudinal elements. When muscle is homogenized, these longitudinal SR form vesicles that are low in density and referred to as "light" SR. At greater centrifugal forces or lower densities, lighter membrane fragments originating from the sarcolemma are obtained.

The different SR fractions vary in Ca^{2+} transport activity and in Ca^{2+} release activity through Ca^{2+} channels. This variation in function is determined, in part, by the protein composition of the SR.

b) **SR - Proteins**

The protein composition of the SR is also a function of its anatomical location. The major proteins of human skeletal muscle SR are shown in Figure 4 and are similar in distribution to that described for SR from other species. The longitudinal elements of the SR contain the largest percentage of the Ca^{2+}-ATPase or Ca^{2+} pump protein. This protein has a molecular weight of about 100,000 daltons and a considerable amount of detail is known about its structure-function relationship (18), including the scheme by which 1 mole of ATP transports 2 moles of Ca^{2+} across the SR membrane (19). In order to study Ca^{2+} transporting activity of SR, several factors need to be considered. Loss of Ca^{2+} from the SR during inward transport should be avoided and to do so an SR fraction with lowest amount of Ca^{2+} release channels should be isolated. Since Ca^{2+}-induced Ca^{2+} release channels have the greatest distribution in heavy SR fractions and whereas the Ca^{2+} pump protein is of higher percentage in light SR fractions, the light SR fractions are the most suited for studying Ca^{2+} pump activity. In Figure 4 it can be observed that the Ca^{2+} pump protein band, migrating between the 116,250 and 92,500 dalton standards, increases from the heavy to the light SR fractions. In the heavier (4-8,000 x g) SR fraction the Ca^{2+} pump protein is approximately 10% of the total SR protein whereas in the light SR fraction (i.e., 20-48,000 x g) it is as much as 88% of the total SR protein.

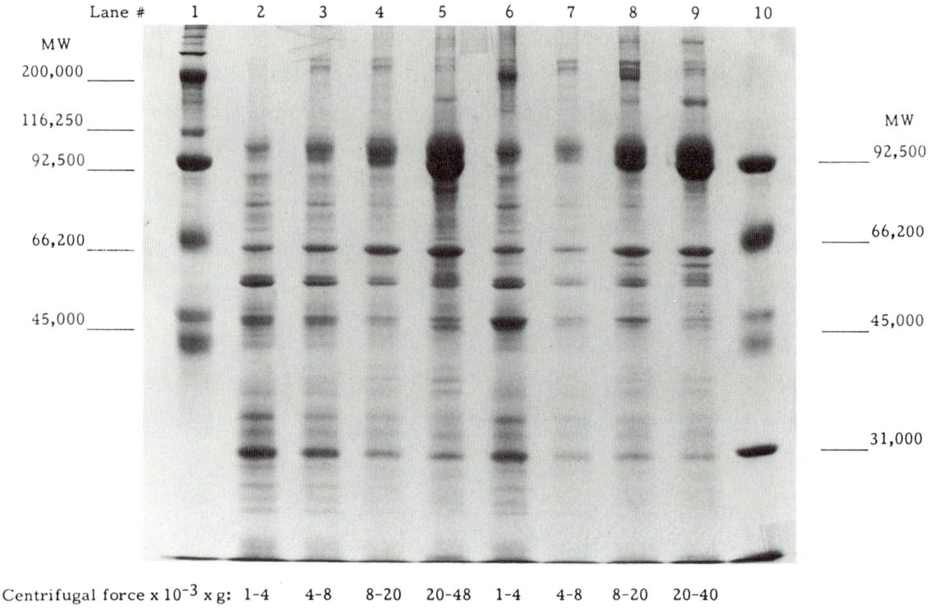

FIGURE 4. 10% Sodium dodecylsulfate-polyacrylamide gel electrophoresis of sarcoplasmic reticulum proteins from human vastus lateralis muscle. Lanes 1 and 10 are high and low molecular weight standard proteins respectively. Lanes 2-5 are SR fractions collected at 1-4,000, 4-8,000, 8-20,000 and 20-48,000 x g centrifugal forces from muscle of a MHS phenotype K diagnostic patient. Lanes 6-9 are SR fractions isolated as above but from muscle of an MHS phenotype H patient. Muscle is vastus lateralis and 30 μg of SR protein has been applied to each lane. Lanes 1 and 2 are high and low molecular weight standards respectively.

Another major protein in the SR is the Ca^{2+} binding protein or calsequestrin and in Figure 4 this protein band is near the 66,200 dalton protein standard. As a percent of the total SR protein, Ca^{2+} binding protein is greater in the intermediate (i.e., 8-20,000 x g) SR fraction where it may comprise as much as 25% of the total protein. Below the Ca^{2+} binding protein on the electrophoresis gel lies another major protein band of 55,000 dalton. The function of this glycoprotein is unknown and its distribution among the different SR fractions is similar to that for the Ca^{2+} binding protein (Figure 4).

c) **Ca^{2+} Transport**

There are two methods for studying Ca^{2+} transport, one in the absence of Ca^{2+} precipitating anions (i.e., oxalate or phosphate) and one in the presence of these anions. As the Ca^{2+} pump transfers Ca^{2+} across the membrane and into the SR lumen, the free Ca^{2+} concentration of the lumen increases and as it does, the release of Ca^{2+} from the ATPase enzyme is inhibited and ultimately Ca^{2+} transport is inhibited. The presence of anions such as oxalate form Ca^{2+} salts that actually precipitate in the SR lumen. The net effect is to prevent an increase in the free Ca^{2+} concentration and thus permit the Ca^{2+} pump to continue transferring Ca^{2+} across the membrane. In the presence of 5 mM oxalate, µ molar amounts of Ca^{2+} can be transported per mg SR protein. In the absence of oxalate, only nanomoles of Ca^{2+} per mg of SR are transported before the pump becomes inhibited. In summary, the ideal conditions for comparing Ca^{2+}-ATPase transport of Ca^{2+} across the SR membrane would be studies of a light (20-48,000 x g) fraction of SR in which Ca^{2+} transport is measured in the presence of 5 mM oxalate with all other substrates saturating.

d) **Ca^{2+} Transport in SR Isolated From MHS Human Skeletal Muscle**

Several studies have measured Ca^{2+} transporting activities of SR isolated from MHS human and pig skeletal muscle. The first reported investigation on SR isolated from human MHS skeletal muscle was that by Kalow et al. (1). In this study the only variable reported was that of total Ca^{2+} taken up by the SR in the presence of oxalate over a five minute time period. Effects of halothane, 2 and 5 volumes %, on Ca^{2+} uptake were also measured. Only two MH patients were studied and although the basal amount of Ca^{2+} uptake did not differ from controls, halothane had a much greater ability to reduce Ca^{2+} uptake in SR from the MH patient's muscle than it did for the control SR membranes (1). The small number of observations in this study precludes any definite conclusion about SR defect in MHS muscle. A subsequent study in the same laboratory compared SR membranes from five MHS patients to that of ten controls (20). The average amount of Ca^{2+} uptake in the absence of halothane was greater in controls compared to MH (20). Halothane, 0.5 - 2 volumes %, increased Ca^{2+} uptake into SR from control muscle while Ca^{2+} uptake was depressed by halothane effects on the SR from MH muscle. The SR preparations in this study were isolated at centrifugal forces of between 10,000 and 48,000 x g and probably contained mostly light SR membranes.

Independent studies on human SR in Isaacs' and Heffron's laboratories were performed on an SR fraction collected at between 1,500 and 200,000 x g (21). In addition to Ca^{2+} uptake, halothane effects on ATPase activity of the SR membranes were also measured. Lower amounts of Ca^{2+} uptake were observed for SR from three MH patients, consistent with the results obtained by Britt et al. (20). However, halothane increased rather than inhibited Ca^{2+} uptake in both MH and control SR membranes, a result at variance with previous studies. The basal ATPase enzyme activity was lower in SR from MHS muscle but halothane depressed ATPase activity equally in SR from control and MHS muscle (21). A subsequent study by Isaacs et al. on four asymptomatic suspected MH carriers produced results similar to their previous study (22). The SR membrane ATPase activity was lower than normal in two of three patients with muscle that produced abnormal in vitro muscle contracture response to halothane. Total Ca^{2+} uptake was lower than normal in all four MHS patients while the rate of Ca^{2+} uptake was below normal values in only one of the four subjects. Similar results were reported by Heffron and Isaacs in a subsequent report (23).

Yet another study looked at one patient who experienced an episode of succinylcholine-induced masseter muscle rigidity, was biopsied at a later date and isolated SR membranes were evaluated (24). Rate of Ca^{2+} uptake, total amount of Ca^{2+} uptake and ATPase activity of this patient's SR were lower than values for SR from normal human skeletal muscle (24).

Blank et al. (25) found that SR from five MHS family members did not differ in Ca^{2+} uptake or Ca^{2+} release properties. Further, they observed, in contrast to the results of Britt et al.

(20) and Isaacs and Heffron (21), that halothane inhibited Ca^{2+} uptake by the SR but that these halothane effects were similar for SR from MHS and normal skeletal muscle. In a subsequent study, Blanck et al. (26) compared Ca^{2+} uptake and release in muscle homogenates and in isolated SR membranes. Similar to their previous study (25) these authors found no differences in function between SR from MHS and normal human muscle. Halothane inhibition, pH and Ca^{2+} dependence for Ca^{2+} uptake was also the same for SR from normal and MHS muscle (26).

Dhalla et al. (27) isolated heavy (10,000-40,000 x g) and light (40,000-105,000 x g) SR fractions from six normal subjects and from a patient with MHS. Ca^{2+} uptake was greater in the heavy SR fraction but more than likely this fraction (40,000 x g x one hour) contained also a majority of the light SR protein while the fraction collected at 40,000-105,000 x g (x one hour) probably contained a high percentage of plasma membrane and very little Ca^{2+} pump protein. No difference was observed between SR from normal and MHS muscle and the rate and amount of Ca^{2+} transport were inhibited by halothane less in SR from the MHS muscle than that in the controls (27). Again, it is difficult to draw any conclusions from studies with only one MHS patient. The results of these studies on isolated SR membrane vesicles are summarized in Table I.

The studies on Ca^{2+} transport function of the Ca^{2+}-ATPase pump in SR membranes from MHS human muscle are variable in methods of isolation, and critical conditions such as temperature, pH, etc., can have substantial effects on the variables measured. For this reason it is difficult to compare values from one study to the next. The difference between total Ca^{2+} transport between normal and MHS muscle SR membrane tends to suggest lower amounts of Ca^{2+} taken up by SR from MHS muscle. However it should be stressed that only small numbers (i.e., maximum five in one study) of MHS subjects have been compared in single controlled experiments, making conclusions even more difficult.

TABLE 1. <u>Comparison of Ca^{2+} Transport Studies on Sarcoplasmic Reticulum Isolated from Human Malignant Hyperthermia Susceptible Skeletal Muscle.</u>

		No. Subjects		Total Ca^{2+} Uptake µmol/mg SR		Rate Ca^{2+} Uptake nmol/sec × mg		Halothane Effect on Total Ca^{2+} Uptake	
Ref	SR Fraction[a]	NORM	MHS	NORM	MHS	NORM	MHS	NORM	MHS
(1)	NA[b]	8	2	0.75–3.00	1.10–1.50	NA	NA	None	↓70%
(27)	10,000– 40,000 × g	6	1	1.40	1.30	NA	NA	↓50%	↓38%
(20)	10,000– 48,000 × g	10	4	0.76–3.10	0.87–1.40	NA	NA	0	↓52%
(24)	NA	NA	1	1.70–4.80	0.57	9.80–14.70	3.50	NA	NA
(21)	1,500–200,000 × g	5	3	1.80–2.20	1.40–1.70	NA	NA	↑12%	↑ 5%
(26)	8,700– 37,000 × g	1	2	0.12	0.09–0.10	NA	NA	NA	NA

[a] In each experiment the SR was collected by differential centrifugation. The values listed represent the gravitational forces between which the SR was collected. Centrifugal steps preceeding these varied among laboratories. Muscle used was vastus lateralis (refs. 1, 20, 26), abdominal (ref. 27) and not stated (refs. 21, 24).

[b] NA – data not available.

e) Ca^{2+} Transport in SR Isolated from MHS Pig Skeletal Muscle

Since there is good reason to consider that the lesion in MHS pig muscle may be similar if not identical to MHS humans, a comparison of SR function studies in pig muscle is warranted. No significant difference was found for Ca^{2+} transport between SR from MHS Poland China pig muscle and controls (28). In the presence or absence of oxalate, halothane stimulated Ca^{2+} binding in the SR from MHS pig muscle (28). Brucker et al. (29) measured the effect of halothane on net accumulation of Ca^{2+} in the presence of oxalate and found no effect on control SR whereas 5% halothane inhibited net Ca^{2+} uptake by 73% in MHS membranes. Also, at 5% halothane concentration, the ATPase activity of control and MHS membranes was stimulated more than two times (29). Berman and Kench (30) also studied SR isolated from control and MHS muscle and found that rates of Ca^{2+} uptake were greater in SR from MHS muscle and that halothane, 1-5%, had no effect on Ca^{2+} uptake by either MHS or control SR membranes. The studies of Steward and Thomas (31) on pig muscle SR produced results and conclusions very similar to those of Berman and Kench (30), namely that Ca^{2+} transport activity was greater in SR from MHS pig muscle and that halothane was without significant effect.

Britt et al. (32) studied SR isolated from MHS pig muscle before and after administering halothane to the MH susceptible pigs. Before halothane was administered to pigs, the SR isolates from MHS pig muscle had greater amounts of Ca^{2+} uptake (32). Halothane administered in vitro, increased Ca^{2+} uptake into SR from control and MHS muscle. After challenging MHS pigs in vivo with halothane, Ca^{2+} uptake was significantly changed in SR from MHS but not from control pig muscle (32).

Abnormalities were observed in SR membranes isolated from MHS pigs and from the offspring of MHS crossed to control pigs (33). In this study the effects of varying temperature on rate and amount of Ca^{2+} bound was compared among MHS, MHX (crossbred) and control pigs. The SR from both MHS and MHX pig muscle had a marked reduction in amount of Ca^{2+} bound when the temperature reached 35°C. This finding is consistent with an earlier study (34) which showed that the abnormal halothane contracture of MHS pig muscle occurred at 37°C but not at 25°C. This different temperature response of SR from MHS muscle suggests some structural difference in the lipid and/or protein structure that has a different temperature sensitivity than does SR from control pig muscle.

Studying a relatively light (15,000-45,000 x g) SR fraction from MHS pigs, Gronert et al. (35) found that rate of Ca^{2+} binding (without oxalate) and rate and amount of Ca^{2+} uptake were decreased while spontaneous Ca^{2+} release was greater in SR isolated from MHS pig muscle. The effects of various concentrations of halothane did not differ between SR from MHS and control muscle. These investigators concluded that the differences observed between SR from MHS pig muscle were not large enough to explain the aetiology of the MH syndrome. Studies of Ca^{2+} transport by SR isolated from MHS and control pig skeletal muscle are summarized in Table 2. It is evident from this table that results are variable from one laboratory to another. This is not surprising when one considers the variables that can affect the outcome of isolated SR membranes. First of all, it is important to consider whether or not and to what extent the MH syndrome has been triggered in the muscle prior to the isolation of SR membranes.

Studies by Britt et al. (32) show conclusively that SR function is altered in MHS muscle after exposure of the susceptible pig to halothane. Since stress alone can trigger MH in pigs, that stress occurring from restraint of animals for induction of anaesthesia to the stimulus directly to the muscle during surgical excision may produce intracellular changes in MHS muscle that could alter SR function. Secondly, if one assumes that muscle can be removed from MHS pigs without abnormally altering the muscle cell homeostasis, then the factors that prevent isolation of a homogeneous SR fraction must be considered. As shown in Table 2, a variety of muscles have been used for isolation of SR from MH pig muscle. Since the distribution of myofiber types in a muscle can influence SR function (36), it is difficult to compare SR isolated from diferent pig muscles. The efficiency of homogenization and chemicals (i.e., buffers and salts) used may affect the density of membranes formed and hence the variation in function of SR

TABLE 2. Comparison of Studies on Ca^{2+} Transport in SR from Control and MHS Pig Skeletal Muscle.

Ref	SR Fraction	Muscle	Temp °C	No. Pigs		Total Ca^{2+} Uptake µmol/mg SR		Rate Ca^{2+} Uptake nmol/sec × mg		Halothane Effect on Total Ca^{2+} Uptake	
				NORM	MHS	NORM	MHS	NORM	MHS	NORM	MHS
(28)	8-48,000 × g[a]	l. dorsi	25°	3	3	11.8	13.3	NA	NA	↑32%	↑ 9%
(29)	8-20,000 × g[b]	NA	40°	6	8	1.1	0.9	NA	NA	↑13%	↓67%
(30)	NA	NA	24°	NA	NA	1.1	1.2	NA	NA	None	None
(31)	NA	v. lateralis	25°	6	2	2.0	5.0	NA	NA	None	↑20%
(32)	10-48,000 × g	v. lateralis	25°	NA	NA	1.0	2.2	NA	NA	↑60%	↑60%
(35)	15-45,000 × g	semitendinous	37°	5	5	3.3	2.2	2.0	1.3	↓7.5%	↓ 5%

[a] In each experiment the SR was collected by differential centrifugation. The values listed represent the gravitational forces between which the SR was collected. Centrifugal steps preceeding these varied among laboratories.

[b] Differential centrifugation was followed by density gradient centrifugation in this experiment.

collected at a constant centrifugal force. How these types of factors influenced the studies on SR in malignant hyperthermia is unknown but could explain in part the discrepancies among results outlined in Table 2. One might expect that myofiber distribution would be more uniform in a given muscle of a strain of purebred pigs than it would be in humans, thus making this more of a factor in trying to investigate SR from human skeletal muscle.

TABLE 3. Comparison of Oxalate-Facilitated Ca^{2+} Transport in SR from Gracilis and Trapezius Muscle from MHS and Control Pigs

Pig Group[a]	SR	Ca^{2+} Transport, nmol/sec x mg Trapezius	Longissimus dorsi
MH	Light	24.5 ± 4[b]	16.5 + 5
Control	Light	31.4 ± 12	13.9 ± 8
MH	Heavy	25.3 ± 17	16.8 ± 7
Control	Heavy	21.9 ± 2	12.7 ± 1

[a]Trapezius and longissimus dorsi muscles were biopsied from 3 MHS and 3 control pigs anaesthetized with thiopental. Heavy SR = 8–12,000 x g. Light SR = 12–48,000 x g.

[b]Values are $\bar{X} \pm$ S.D. Ca^{2+} uptake rates were measured by dual wavelength (650–700 nm) spectrophotometry in a cuvette containing 1 ml volume of the following: SR 25–50 µg/ml, MgATP I mM, K^+-Oxalate 5 mM, HEPES (pH 6.8) 10 mM, NaN_3 5 mM, KCl 150 mM, arzenazo III 16 µM, a creatine phosphate-CPK system was present for regenerating ATP, and the reaction was initiated by addition of 75 µM Ca^{2+}.

f) Effect of Myofiber Type on Ca^{2+} Transport Studies

Recent studies in our laboratory have compared SR fractions isolated from different muscles in MHS and MH negative Poland China pigs and the Ca^{2+} transport activity has been compared between heavy and light SR fractions (Table 3). It has been found that the in vitro contracture response of MHS muscle is different in different muscles (37). The pig trapezius muscle contains 41% Type I myofibers and produces greater contracture response to caffeine or to halothane than does pig gracilis muscle which contains only 11% Type I myofibers. Differences in rate of Ca^{2+} transport are observed between SR isolated from trapezius versus longissimus dorsi muscle of MHS and control pigs (Table 3). No statistically significant differences in rate of Ca^{2+} transport have been observed between MHS and control pig muscle SR membranes (Table 3). Within muscle, rates of Ca^{2+} transport does not differ between light and heavy SR fractions from MHS and control pigs. Compared to SR from longissimus dorsi muscle, Ca^{2+} transport rates are significantly ($p < 0.05$) faster in SR from the trapezius muscle and this is true for both control and MHS muscles.

Although no differences have been observed between the heavy and light SR fractions shown in Table 3, a separate study which isolated SR at different centrifugal forces has shown a difference in distribution of SR protein among SR fractions from pig skeletal muscle (Table 4).

g) **Problems with SR Studies Related to MH**

It is difficult to draw any conclusions about a defect in SR from MHS human muscle because of large variations among studies and the small number of subjects in each study plus the opposite results obtained from one study to the next. Of the six investigations referenced in Table I, only one measured rate of Ca^{2+} transport. Since most of these studies incorporate oxalate in the transport assay, total amount of Ca^{2+} transported may be equal while rate of transport may differ. It is also possible that total amount of Ca^{2+} transported over a fixed time may vary if rates of Ca^{2+} transport differ between SR from control and MHS muscle. More studies are needed to determine if rate of Ca^{2+} transport is different. This can be done very simply today by the use of Ca^{2+}-sensitive dyes that absorb light when bound to Ca^{2+}. Examples are murexide, arsenazo III and antipyrylazo III (38). In addition to rate of Ca^{2+} uptake, efficiency of the transport system can also be measured. Efficiency of the Ca^{2+} transport system in SR is the ratio of moles Ca^{2+} transported to moles ATP utilized for the transport and in tightly coupled isolated SR membranes this ratio is

2.0 (39). If, for example, an isolated SR membrane is leaky to Ca^{2+} or if it is contaminated with other proteins, then the coupling ratio will be less than 2.0. A thorough investigation of rates and efficiency of Ca^{2+} transport needs to be carried out for uniformly isolated SR membranes from MHS skeletal muscle. We have been exploring both Ca^{2+} transport and Ca^{2+} release processes in SR isolated from human MH diagnostic muscle specimens.

TABLE 4. The Distribution of Ca^{2+} Transport and Ca^{2+}-Induced Ca^{2+} Release Activity Among Different SR Fractions Isolated from Pig Skeletal Muscle

Fraction[a]	Ca^{2+} Uptake nmol/sec x mg[b]	Rate of Ca^{2+}-Induced Ca^{2+} Release nmol Ca^{2+}/min x mg[c]
#1 = 4- 8,000 x g	33.9	23.4
#2 = 8-10,000 x g	34.8	22.2
#3 = 10-12,000 x g	44.5	19.2
#4 = 12-20,000 x g	50.6	10.2
#5 = 20-40,000 x g	42.3	1.2

[a]The fractions 1-5 were pellets obtained at the centrifugal forces listed.

[b]Rate of Ca^{2+} uptake was measured under conditions described in legend to Table 3.

[c]See Figure 5 for details of measuring Ca^{2+}-induced Ca^{2+} release.

A synopsis of our data is presented in Table 5. The isolation procedures have varied as we have attempted to produce a SR fraction from human muscle that has consistent Ca^{2+}-induced Ca^{2+} release activity. Conditions that have varied include homogenization times and forces. In each preparation the purpose has been to produce heavy and light SR fractions and to measure Ca^{2+} transport and release activity in each. The broad range of

TABLE 5. Calcium Transport Rates Among Sarcoplasmic Reticulum Fractions from Malignant Hyperthermia Diagnostic Human Skeletal Muscle.

	PHENOTYPE K		PHENOTYPE H		PHENOTYPE N	
	HSR	LSR	HSR	LSR	HSR	LSR
	11.6[a]	16.5	2.8	7.1	2.7	25.8
	10.7	15.4	13.3	15.7	---	14.0
	17.7	20.7	2.7	4.8	2.5	3.6
	3.5	10.3	6.1	15.7	1.8	4.1
	3.6	4.5	2.7	5.0	23.1	15.2
	2.4	5.0	15.5	16.2	3.0	5.0
	3.5	4.2	1.7	12.9		
	2.2	2.4	3.0	9.0		
	2.8	3.3	0.9	3.2		
	8.2	6.1	8.8	18.6		
	5.3	11.2	0.5	1.0		
	4.0	8.6	5.8	9.2		
			9.2	18.2		
			11.1	17.8		
			---	21.9		
			15.8	19.0		
$\overline{X}$	6.3	9.0	6.7	12.2	6.6	11.3
S.D.	4.8	5.9	5.3	6.6	9.2	8.8
N	12	12	15	16	5	6
Range	(2.2-17.7)	(2.4-20.7)	(0.5-15.8)	(1.0-21.9)	(1.7-23.1)	(3.6-25.8)

[a]See Table 3 for experimental details.

values within fractions and diagnostic groups and the large variance associated with each mean value indicate the problems associated with testing hypothesis of differences in SR transport function among MHS diagnostic outcome groups. Attempts are underway in our laboratory to produce a uniform method of SR isolation from human skeletal muscle that will eliminate this large variation. Blanck et al. (26) have stated that "the absolute Ca^{2+} activity varies considerably from patient to patient depending on age, physical status, and purity of SR preparation". To overcome this problem, these workers have compared Ca^{2+} uptake activity in muscle homogenates. In spite of the variation of SR results observed from patient to patient, the Ca^{2+} transport activity is usually higher in LSR compared with HSR (Table 4). This is consistent with previous observations for rabbit skeletal SR (40) and with the distribution of Ca^{2+} transport protein in human HSR versus LSR fractions (Figure 3).

TABLE 6. Oxalate-Facilitated Ca^{2+} Transport in Heavy and Light Fractions of Sarcoplasmic Reticulum Isolated from MHS and Control Pig Skeletal Muscle.

	PHENOTYPE K		PHENOTYPE H		PHENOTYPE N	
	HSR	LSR	HSR	LSR	HSR	LSR
	16.0[a]	33.0	19.0	49.1	28.5	28.0
	18.0	26.0	37.3	41.6	29.5	40.9
	28.9	45.8	37.5	37.4	44.2	46.8
	23.9	33.7	---	38.8	30.9	48.5
	30.7	38.7	18.8	---	13.6	15.7
	24.0	32.0	20.7	32.0	7.2	38.6
	24.0	35.9	16.7	34.4	43.3	41.8
	26.9	40.3	18.3	29.7	23.9	24.4
	24.4	35.1	19.4	26.7	---	---
$\overline{X}$	24.1	35.6	23.5	34.9	27.5	35.6
S.E.	4.7	5.6	8.6	7.5	12.1	11.6
N	9	9	8	8	8	8

[a]Each value represents nmoles Ca^{2+}/sec x mg transported by an SR fraction isolated from an individual animal. Biopsies of gracilis muscle were obtained under thiopental anaesthesia. HSR is a heavy SR fraction collected at 8-12,000 x g and LSR is a light SR fraction collected between 12-48,000 x g. Phenotypes are diagnostic contracture classifications. N = negative for MHS, K and H are positive for MH, details provided in ref. #8.

Similar to studies on human SR, those reported on SR from MHS pig muscle vary considerably from experiment to experiment. Again, only one of these studies measured the rate of Ca^{2+} uptake in the presence of oxalate. Table 6 summarizes some of our recent studies on oxalate-facilitated Ca^{2+} transport in SR from MH and control pig skeletal muscle. The experimental conditions for isolation of heavy and light SR fractions did not vary as it did for our human SR data.

Based on the data in Table 6 there is no significant difference in oxalate-facilitated Ca^{2+} transport between SR fractions from normal versus MHS pig skeletal muscle. The Ca^{2+} transport activity of LSR fractions is greater than it is in HSR and this is true for each MH diagnostic group of pigs. On the basis of this comparative study, we conclude that the Ca^{2+} transport rate in SR from MHS Poland China pig muscle is normal and does not represent a primary aetiological mechanism in the MH syndrome in pigs.

EVIDENCE FOR DEFECT IN Ca^{2+} RELEASE FROM SR IN MHS MUSCLE

a) Origin of Ca^{2+} in MHS Muscle

The commonly stated hypothesis for MH aetiology is that free Ca^{2+} concentration is increased within the muscle cell. The origin of this Ca^{2+} could be from either inside the cell, from outside the cell or from both sources. Previous studies have demonstrated that the abnormal in vitro contracture response of MHS skeletal muscle is dependent upon extracellular Ca^{2+} (7,41) and also that Ca^{2+} content of MHS and control muscle varies with the extracellular Ca^{2+} concentration (15). It is possible, therefore, that the source of Ca^{2+} for halothane-induced contracture of MHS muscle is intracellular, but that this intracellular pool of Ca^{2+} is readily exchangeable with and dependent upon the presence of extracellular Ca^{2+}.

If we assume that MH is triggered by release of intracellular Ca^{2+} stores, then we must consider where this Ca^{2+} originates within the cell. Another chapter in this book discusses the mitochondria in MH and its possible role will not be considered here. Since skeletal SR serves as the primary pool for Ca^{2+} storage and release, then it becomes a logical source of Ca^{2+} for the MH Ca^{2+} theory. The anaesthetic agents that trigger MH supposedly do so by causing a release of Ca^{2+} from the SR that occurs normally, and then inhibiting the reuptake of Ca^{2+} by an abnormal SR Ca^{2+} transport system in the MHS muscle. On the other hand, these triggering anaesthetic agents may only produce Ca^{2+} release from abnormal SR in MHS muscle and the Ca^{2+} transport system functions normally to remove this Ca^{2+} from the myoplasm. This would represent a "leak" in the SR and a continual expenditure of energy to transport the "leaked" Ca^{2+} back into the SR membrane. Let us consider then the evidence for abnormal Ca^{2+} release from the SR in MHS skeletal muscle.

b) **Ca^{2+} Channels in Skeletal SR**

In spite of detailed biochemical and biophysical studies that have broadened our knowledge and understanding of the Ca^{2+} transport protein of skeletal SR, little is known or understood about details surrounding Ca^{2+} release from the SR. The exact mechanisms for excitation-contraction coupling in skeletal muscle have challenged scientists for many years. Recent experiments have demonstrated that excitable cells not only have specialized channels for Na^{2+} and K^{+}, but that Ca^{2+} channels also exist (42) and it is thought that the SR of skeletal muscle has Ca^{2+} channels that are opened to release Ca^{2+} for initiation of contraction. Exactly how these channels are opened is not entirely clear but at least two general theories are presented. Electrophysiologic techniques have demonstrated that Ca^{2+} channels have putative gates that are opened and closed by voltage changes, i.e., voltage-sensitive Ca^{2+} channels (42). The other type of Ca^{2+} channel is thought to be opened and closed as a consequence of agonist binding to a receptor that opens or closes the Ca^{2+} channel gate. Since electrophysiological methodology is not currently applicable to isolated SR membranes, most known voltage-sensitive Ca^{2+} channels are confined to the plasma membrane of cells. Chemical, rather than electrical, depolarization of SR membranes can be performed, as long as one does not neglect the changing osmotic forces imposed by these types of experiments. One study that illustrates these two putative Ca^{2+} channels in skeletal SR is that reported by Miyamoto and Racker (43). These investigators have proposed that not only did their data support two different Ca^{2+} channels (i.e., voltage-gated and receptor-mediated) but that Ca^{2+} release from the receptor (Ca^{2+}-induced Ca^{2+} release) mediated

Ca^{2+} channel created a change in membrane potential that opened the voltage-dependent Ca^{2+} channel as well. The concepts of Miyamoto and Racker's studies are summarized in Table 7 and will be used to formulate theories on the role of SR in MH.

c) **Ca^{2+}-Induced Ca^{2+} Release in MH**

Ca^{2+}-induced Ca^{2+} release from skeletal SR is a phenomenon first described by Endo et al. (44). Using single skinned muscle fibers, these workers have shown that Ca^{2+} itself

TABLE 7. Theoretical Ca^{2+} Interactions in Excitation-Contraction Coupling

Ca^{2+} Source	Trigger for Ca^{2+} Release	Effect of Ca^{2+} Released
Transverse Tubule	Sarcolemma depolarization	Ca^{2+} binds to receptor on Ca^{2+} channel and opens channel for Ca^{2+} release
Terminal Cisternae	Ca^{2+}	Ca^{2+} efflux alters the electrical potential across membrane of the terminal cisternae
Terminal Cisternae	Change in electrical potential of SR membrane	Ca^{2+} flux from SR for activation of contractile elements

can produce Ca^{2+} release from the SR in a regenerative manner. Endo has discussed as unlikely the possibility that Ca^{2+}-induced Ca^{2+} release plays a physiological role (45), but Miyamoto and Racker (43) have reconsidered this possibility in skeletal muscle while others consider it physiological in cardiac muscle (46). The study of Takagi et al. (47) first suggested that Ca^{2+}-induced Ca^{2+} release may play a role in the pathogenesis of MH. Using single, skinned skeletal muscle fibers from MH and control humans, these investigators have observed that muscle from MHS patients produces contracture to halothane concentrations of 0.06% (v/v) or less while normal muscle fiber tension is never developed by halothane below 0.06% (47). Based on previous studies suggesting that halothane releases Ca^{2+} from the SR by enhancing Ca^{2+}-induced Ca^{2+} release, Takagi et al. (47) have speculated that the Ca^{2+}-induced Ca^{2+} release mechanism may be more than normally easily activated in SR of MH skeletal muscle. Abnormal caffeine-induced contracture of skinned single fibers from human MHS muscle reported by Wood et al. (16) and Britt et al. (17) may also relate to the Ca^{2+}-induced Ca^{2+} release mechanism since caffeine also activates the Ca^{2+}-induced Ca^{2+} release mechanism (48).

Subsequent to the skinned muscle fiber studies, methods have been established for investigating Ca^{2+}-induced Ca^{2+} release from fragmented SR and these have been applied to SR from MH pig muscle (49,50). Ohnishi et al. (50) have shown at equal, but lower Ca^{2+} preloads, i.e., 26 nmol/mg SR, that halothane, 200 M, produced Ca^{2+} release from MH - SR but not from control SR. At higher Ca^{2+} preloads, i.e. 60 nmol/mg SR, halothane produces Ca^{2+} release from both MH and control SR. It has also been shown in this study that the amount of Ca^{2+}-induced Ca^{2+} release is greater in MH-SR at the same Ca^{2+} preload of 40 nmol/mg SR. The authors have concluded that these results support a role for the SR in the pathogenesis of MH. In a similar study, light and heavy SR fractions have been compared for caffeine and Ca^{2+}-induced Ca^{2+} release activity between MH and control pig muscle (49). Caffeine-induced Ca^{2+} release does not differ between LSR fractions from MH and control pigs (49). In HSR fractions it has been discovered that during the process of attempting to load the SR with Ca^{2+}, via the Ca^{2+} transport mechanism, the HSR fractions release some of the Ca^{2+} taken up before the bulk of added Ca^{2+} load has been removed.

In attempts to avoid this problem we decided to add Ca^{2+} in small consecutive Ca^{2+} pulses, only after the previous load was completely removed by the SR. In the process it was discovered that after accumulating a critical Ca^{2+} load, the HSR would release Ca^{2+} upon addition of the next Ca^{2+} pulse. This experiment is illustrated in Figure 5. Three variables are routinely measured in

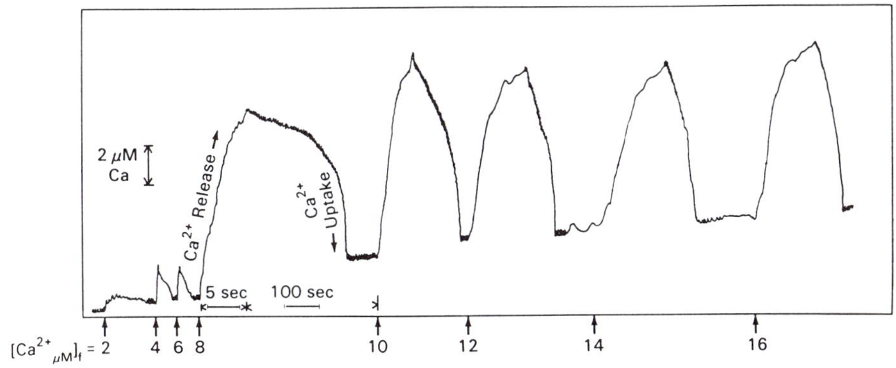

FIGURE 5. Ca^{2+} pulse-induced Ca^{2+} release from a heavy SR fraction from human vastus lateralis muscle. Changes in Ca^{2+} concentration were measured by dual wavelength (Absorbance 650 – Absorbance 700 nm) spectrophotometry using arsenazo III as a Ca^{2+} indicator. A 30°C cuvette with final volume of 300 l contained SR 1 mg/ml, HEPES 10 mM (pH 6.8), NaN_3 5 mM, KCl 150 mM, arsenazo III 30 M, MgATP 1 mM, and an ATP regenerating system of creatine phosphate 5 mM and creatine phosphokinase 20 g/ml. Each Ca^{2+} pulse was 2 M and the initial 3 pulses were completely taken up by the SR. The 4th Ca^{2+} pulse (8 M total added Ca^{2+}) resulted in release of Ca^{2+} from the SR through a putative Ca^{2+} channel. The released Ca^{2+} is removed after the Ca^{2+} channel closes and subsequent Ca^{2+} pulses can produce additional Ca^{2+} release and reuptake cycles.

these Ca^{2+} pulse-induced Ca^{2+} release experiments. The total amount of Ca^{2+} added until the first Ca^{2+} release phase occurs is referred to as the threshold and is expressed as nmol Ca^{2+}/mg SR. In addition, rate and amount of Ca^{2+} released are measured. When these experiments are performed on HSR from MHS and control pig muscle, it is observed that the threshold for Ca^{2+}-induced Ca^{2+} release is lower in HSR from MHS pig muscle (49). No significant differences in rates and amounts of Ca^{2+} released are observed between SR from MHS and control muscle. Caffeine lowers the threshold for Ca^{2+}-induced Ca^{2+} release in both MH and control HSR but the ED_{50} is lower (ED_{50} = 0.5 mM) for MH-SR than for controls (ED_{50} = 1.0 mM caffeine).

Other groups of investigators have compared rates and amounts of Ca^{2+} release from a heavy (10-17,000 x g) SR fraction isolated from MH pig skeletal muscle (51). Their study utilizes SR actively loaded with Ca^{2+} (ca. 30 nmoles Ca^{2+}/mg SR) and Ca^{2+} release is initiated by four different mixtures including halothane, Ca^{2+}, halothane plus Ca^{2+}, or a depolarizing solution. Each of these agents produces a faster rate of Ca^{2+} release from SR isolated from MHS pig muscle. These authors have concluded from their study that Ca^{2+} release channels in SR are likely defective in muscle of pigs susceptible to malignant hyperthermia. Studies on Ca^{2+}-induced Ca^{2+} release from MH pig skeletal SR are summarized in Table 8.

Based on the results of Ca^{2+} pulse-induced Ca^{2+} release studies in our laboratory results of studies by Ohnishi et al. (50) and by Kim et al. (51) may have some predictable elements. For example, the rate and amount of Ca^{2+}-induced Ca^{2+} release are functions of the amount of Ca^{2+} loaded into the SR before release is initiated (45) and results of our work show that the threshold Ca^{2+} preload for Ca^{2+}-induced Ca^{2+} release is lower in the SR from MHS muscle and that this is not a function of variation in endogenous Ca^{2+} levels in the different SR sources (49). It could be argued that conditions of the previous experiments (50,51) have favoured faster rates and amounts of Ca^{2+} release from MH SR because the Ca^{2+} preload of about 30 nmoles/mg is probably close to or exceeding the Ca^{2+} holding capacity of MH SR while perhaps only 50% saturating for the control SR. Under these conditions, rate and amount of Ca^{2+} release would be near maximal for MH SR and well below maximal for control SR. The physiological environment of the in situ SR membrane under resting conditions is poorly understood, and how closely these in vitro experiments reflect the in vivo status is unclear. Good estimates are available for the free Ca^{2+} of the myoplasm which is what the p-face of the SR membrane is exposed to. Reported estimates for the free Mg^{2+} concentration

TABLE 8. Comparison of Ca^{2+}-Induced Ca^{2+} Release Experiments on Heavy Sarcoplasmic Reticulum Fractions Isolated from MHS Pig Skeletal Muscle

Ref	Pig Group	Ca^{2+} Preload nmol/mg	Endogenous Ca^{2+} nmol/mg	Total Ca^{2+} nmol/mg	Trigger Ca^{2+} [a] μmol/l	Amount Released nmol Ca^{2+}/mg	Rate Ca^{2+} Release nmol/sec × mg
(49)	CONT	63	68.5	131.5	2	10.7	2.0
	MHS	20	59.4	79.4	2	12.1	4.3
(50)	CONT	60	NA	NA	20	0	NA
	MHS	60	NA	NA	20	24.0	NA
(51)	CONT	30	61.9	91.9	2	10.8	0.98
	MHS	30	63.9	93.9	2	16.2	1.25

[a] Trigger Ca^{2+} is the concentration of added Ca^{2+} that produced Ca^{2+} release from the previously Ca^{2+} loaded SR membrane.

in situ are not in close agreement, which is unfortunate because Mg^{2+} plays a significant role in regulating Ca^{2+} release from the Ca^{2+}-induced Ca^{2+} release channels (45,49).

The lumen of the SR terminal cisternae contains proteins, i.e., calsequestrin, that serve to bind Ca^{2+} and retain a pool of Ca^{2+} for release from this portion of the membrane. Is there a deficiency of these Ca^{2+} binding proteins in the SR of MH muscle? The answer to this question appears to be no since studies have shown no qualitative or quantitative differences in the protein composition of MHS and control SR membranes (52,53). This is not to say that some abnormality in the Ca^{2+} binding or releasing properties of these proteins does not exist and detailed biochemical studies in this area are needed.

SUMMARY

Studies on SR isolated from malignant hyperthermic human and pig skeletal muscle vary in experimental approach and, perhaps as a result, they vary considerably in results. The functions of isolated SR membranes, i.e., principally Ca^{2+} transport or Ca^{2+} release, are determined by many variables such as muscle homogenization and methods of centrifuging the SR. Extraction of SR membranes with 0.6 M KCl is used to solubilize and remove contaminating proteins, but it has recently been reported that this step of purification may be disrupting triads (54) and thus changing the function of the fragmented SR. Based on several studies of the Ca^{2+} transport function in MHS human and pig skeletal SR, there is no compelling data to support a defect in this function of the membrane; at least not to an extent that would produce the MH syndrome. It is tempting to conclude that Ca^{2+} transport function of SR does not play a primary aetiological role in the MH syndrome.

If the Ca^{2+} pump is not defective in SR from MH skeletal muscle, then some abnormal Ca^{2+} release mechanism may be responsible. Recent studies on heavy SR fractions from MH pig muscle suggest that a Ca^{2+} channel is defective in the SR membrane. In contrast to Ca^{2+} channels that are opened and closed by changes in membrane potential (i.e., voltage-sensitive), the abnormal Ca^{2+} release from MH SR membranes appears to occur through a Ca^{2+} channel that is opened and closed by an agonist-receptor effect. In this case, Ca^{2+} itself appears to be the agonist. Calcium channels that are active in the excitation-contraction coupling process appear to be very good candidates for exploring a defective membrane phenomenon in the etiology of MH.

Methods are now available for isolating and studying intact triads (55), and since these structures are expected to contain the elements for E-C coupling, they may provide new insights into our understanding the aetiology of malignant hyperthermia.

ACKNOWLEDGEMENTS

These studies were supported in part by National Institutes of Health Grant DHHS 5R01 GM 23875-07 and General Clinical Research Centers Program of the Division of Research Resources, National Institutes of Health Grant M01 RR00073. The breeding colony maintained at The University of Texas System Cancer Center, Science Park, Veterinary Resources Division, Bastrop, Texas, 78602, was supported in part by Grant Ca-16672 from the Division of Cancer Research Resources and Centers, National Institutes of Health.

The technical assistance of Christine Dettbarn-Palade, Sing-Ling Luh and Patricia Turk is gratefully acknowledged. Surgical and anaesthesia management of MH patients by Drs. J.C. Fish, O. Winsett, E.H. Flewellen, III, M. Belt and D. Kennamer made parts of these studies possible. I am thankful to Pamela Dawson for her hard work in typing and preparing this manuscript.

REFERENCES

1. Kalow, W., Britt, B.A., Terreau, M.E. and Haist, C. Lancet 1970;2:895.

2. Rasmussen, H. and Barrett, P.Q. Physiol. Rev. 1984;64:938.

3. Bianchi, C.P., Narayan, S. and Lakshminarayanaiah, N. In: Carafoli, E., Clementi, F., Drabikowski, W. and Margreth, A. (eds.). Calcium Transport in Contraction and Secretion. 1975. North Holland: American Elsevier, pp. 503-515.

4. Rasmussen, H. and Tenenhouse, A. Proc. Natl. Acad. Sci. 1968;59:1364.

5. Weber, A. J. Gen. Physiol. 1977;4:315.

6. Nelson, T.E. and Denborough, M.A. Clin. Exp. Pharmacol. Physiol. 1977;4:315.

7. Nelson, T.E., Bedell, D.M. and Jones, E.W. Anesthesiology 1975;42:301.

8. Nelson, T.E. Flewellen, E.H. and Gloyna, D.F. Anesth. Analg. 1983;62:545.

9. Britt, B.A. and Kalow, W. Can. Anaesth. Soc. J. 1970;17:293.

10. Auerbach, V.H., Di George, A.M., Mayer, B.W., Hayden, M., Carpenter, G.G., Krumperman, L.W. and Truter, T.R. In: Gordon, R.A., Britt, B.A. and Kalow, W. (eds.). International Symposium on Malignant Hyperthermia. 1973. Springfield: Charles C. Thomas. pp. 30–51.

11. Flewellen, E.H. and Nelson, T.E. Anesth. Analg. 1984;63:693.

12. Kolb, M.E., Horne, M.L. and Martz, R. Anesthesiology 1982;56:254.

13. Britt, B.A., Endrenyi, L., Barclay, R.L. and Cadman, D.L. Br. J. Anaesth. 1975;47:647.

14. Nelson, T.E., Jones, E.W., Henrickson, R.L., Falk, S.N. and Kerr, D.D. Am. J. Vet. Res. 1974;35:347.

15. Nelson, T.E. and Chausmer, A.B. J. Pharmacol. Exp. Ther. 1981;219:107.

16. Wood, D.S., Mozo, A. and Willner, J.H. Neurology 1979;29:557.

17. Britt, B.A., Frodis, W., Scott, E., Clements, J.J. and Endrenyi, L. Can. Anaesth. Soc. J. 1982;29:550.

18. Ikemoto, N. Ann. Rev. Physiol. 1982;44:297.

19. De Meis, L. and Vianna, A.L. Ann. Rev. Biochem. 1979;48:275.

20. Britt, B.A., Kalow, W., Gordon, A., Humphrey, J.G. and Rewcastle, N.B. Can. Anaesth. Soc. J. 1973;20:431.

21. Isaacs, H. and Heffron, J.J.A. Br. J. Anaesth. 1975;47:475.

22. Isaacs, H., Heffron, J.J.A. and Badenhorst, M. Br. J. Anaesth. 1975;47:1075.

23. Heffron, J.J.A. and Isaacs, H. Klin. Wochenschr. 1976;54:865.

24. Donlon J.V., Newfield, P., Sreter, F. and Ryan, J.F. Anesthesiology 1978;49:298.

25. Blanck, T.J.J., Gruner, R., Stern, L.Z. and Thompson, M. Anesthesiology 1979;51:S246.

26. Blanck, T.J.J., Gruner, R., Suffecovl, S.L. and Thompson, M. Anesth. Analg. 1981;60:492.

27. Dhalla, N.S., Sulakhe, P.V., Clinch, N.F., Wode, J.G. and Naimark, A. Biochem. Med. 1972;6:333.

28. Nelson, T.E., Jones, E.W., Venable, J.H. and Kerr, D.D. Anesthesiology 1972;36:52.

29. Brucker, R.F., Williams, C.H., Popinigis, J., Galvez, T.L., Vail, W.J. and Taylor, C.A. In: Gordon, R.A., Britt, B.A. and Kalow, W. (eds.). International Symposium on Malignant Hyperthermia. 1973. Springfield: Charles C. Thomas, pp. 238-270.

30. Berman, M.C. and Kench, J.E. In: Gordon, R.A., Britt, B.A. and Kalow, W. (eds.). International Symposium on Malignant Hyperthermia. 1973. Springfield: Charles C. Thomas, pp. 287-297.

31. Steward, D.J. and Thomas, T.A. In: Gordon, R.A., Britt, B.A. and Kalow, W. (eds.). International Symposium on Malignant Hyperthermia. 1973. Springfield: Charles C. Thomas, pp. 409-414.

32. Britt, B.A., Endrenyi, L. and Cadman, D.L. Br. J. Anaesth. 1975;47:650.

33. Nelson, T.E. and Bee, D.E. J. Clin. Invest. 1979;64:895.

34. Nelson, T.E., Bedell, D.M. and Jones, E.W. Anesthesiology 1975;42:301.

35. Gronert, G.A., Heffron, J.J.A. and Taylor, S.R. Eur. J. Pharmacol. 1979;58:179.

36. Sreter, F.A. Arch. Biochem. Biophys. 1969;134:25.

37. Nelson, T.E. and Schochet, S.S. Jr. Can. Anaesth. Soc. J. 1982;29:163.

38. Scarpa, A., Brinley, F.J., Tiffert, T. and Dubyak, G.R. Ann. N.Y. Acad. Sci. 1978;307:86.

39. Hasselbach, W. Biophys. Struct. Mech. 1977;3:43.

40. Louis, C.F., Nash-Adler, P.A., Fudyma, G., Skigekawa, M. and Akowitz, A. Eur. J. Biochem. 1980;111:1.

41. Moulds, R.F.W. and Denborough, M.A. Br. Med. J. 1974;2:241.

42. Hagirwara, S. Membrane Potential-Dependent Ion Channels in Cell Membrane--Phylogenetic and Developmental Approaches. 1983. New York: Raven Press.

43. Miyamoto, H. and Racker, E. J. Membr. Biol. 1982;66:193.

44. Endo, M., Tanaka, M. and Ogawa, Y. Nature 1970;228:34.

45. Endo, M. Physiol. Rev. 1977;57:71.

46. Fabiato, A. Am. J. Physiol. 1983;245:C1.

47. Takagi, A. Sugita, H., Toyokura, Y. and Endo, M. Proc. Jpn. Acad. 1976;52:603.

48. Endo, M. Proc. Jpn. Acad. 1975;51:479.

49. Nelson, T.E. J. Clin. Invest. 1983;72:862.

50. Ohnishi, S.T., Taylor, S. and Gronert, G. FEBS Letters 1983;161:103.

51. Kim, D.H., Sreter, F.A., Ohnishi, S.T., Ryan, J.F., Roberts, J.T., Allen, P.D., Meszaros, L.G., Antoniu, B. and Ikemoto, N. Biochim. Biophys. Acta 1984;775:320.

52. Sullivan, J.S. and Denborough, M.A. Int. J. Biochem. 1982;14:741.

53. Oku, S., Liew, C.C. and Britt, B.A. J. Neurol. Sci. 1983;60:127.

54. Campbell, K.P., Franzini-Armstrong, C. and Shamoo, A.E. Biochim. Biophys. Acta 1980;602:97.

55. Mitchell, R.D., Palade, P. and Fleischer, S. J. Cell Biol. 1983;96:1008.

4

MITOCHONDRIA AND MALIGNANT HYPERTHERMIA

K.S. CHEAH

Agricultural and Food Research Council, Food Research Institute - Bristol, Langford, Bristol BS18 7DY, United Kingdom

ABSTRACT

Malignant hyperthermia (MH) is postulated to be due to a molecular defect in membrane permeability brought about by an enhanced phospholipase A_2 activity. This ultimately leads to an increase in the level of sarcoplasmic Ca^{2+}, which is responsible for the muscle hyper-rigidity, enhanced rate of glycolysis, rapid rate of lactate formation and a low pH in MH muscle. Skeletal muscle mitochondria are involved in the development of the MH syndrome by their ability to release Ca^{2+} faster than normal and to influence the sarcoplasmic reticulum to release additional Ca^{2+} into the sarcoplasm. This is attributed principally to an enhanced phospholipase A_2 activity in the MH-skeletal muscle mitochondria. The enhanced phospholipase A_2 activity is due to a significantly higher than normal amount of endogenous calmodulin, Ca^{2+} and fatty acids in these mitochondria. The enhanced Ca^{2+}-activated phospholipase A_2 activity is responsible for the 9^oC higher mitochondrial transition temperature, large amplitude mitochondrial swelling and increased sarcoplasmic Ca^{2+} in porcine MH syndrome.

INTRODUCTION

Malignant hyperthermia (MH), a genetically inherited disorder, affects primarily the skeletal muscle of humans and pigs. In apparently healthy humans (1-5) and in stress-susceptible pigs (6-12), the MH syndrome can easily be induced with halothane (1-12), suxamethonium (3,13-18) or halothane plus suxamethonium (4). In stress-susceptible pigs, the MH syndrome can also be triggered by physiological or environmental stress such as excitement and changes in temperature (7,18-20). Once initiated, the classical symptoms of the syndrome are gross muscular rigidity, a rapid rise in body temperature, hyperventilation, severe metabolic acidosis and elevated levels of serum metabolites (21-23). The body temperature can increase at a rate of 1°C per five minutes (4), and the metabolic rate can exceed seventeen times normal (20); if uncontrolled, death occurs (8,19,24). The occurrence of anaesthetic deaths in apparently healthy patients is about one in fifteen thousand anaesthetics (3). In healthy pigs, the incidence of MH is dependent on the breed and can be as high as 88% in highly stress-susceptible breeds (25). The MH syndrome shows striking similarities in humans and pigs (26) and genetically-selected pigs are thus frequently used as experimental models for investigating the human MH syndrome.

The development of muscle rigidity in human and porcine MH is believed to be due to an increased level of Ca^{2+} in the sarcoplasm (14,27-29). A significantly higher than normal sarcoplasmic Ca^{2+} has indeed been observed in MH-susceptible pigs early post-mortem (30).

The purpose of this article is to discuss the current understanding of the probable role of mitochondria in the molecular mechanism of MH. It reviews pertinent findings on respiration, Ca^{2+} transport, structure, membrane permeability and phospholipase A_2 activity of skeletal muscle mitochondria. It postulates that an enhanced phospholipase A_2 activity is responsible for the alteration in the membrane permeability resulting in an increased level of Ca^{2+} in the sarcoplasm, which is then responsible for the muscle hyper-rigidity and enhanced rate of glycolysis in the MH syndrome.

I. Mitochondria

Various investigators have suggested that skeletal muscle mitochondria are associated with the development of MH (8,28-34). Mitochondria are mainly responsible for the synthesis of ATP by oxidative phosphorylation (35); the ATP generated is essential for many metabolic processes such as muscle contraction and Ca^{2+} transport (36). Mitochondria can accumulate Ca^{2+} at the expense of

ATP formation (37), and these organelles can also act as efficient Ca^{2+} buffers under various physiological and pathological conditions (38-40).

a) <u>Respiration</u>

Reports on the respiratory functions of mitochondria isolated from post-mortem and biopsy skeletal muscles of MH-susceptible pigs and humans (Table I) are conflicting; present evidence suggests that there is no abnormality in isolated MH mitochondria when ADP is used to stimulate substrate oxidation. The discrepancies reported could be attributed mainly to differences in the quality of the isolated mitochondria; for example, the isolated organelles have been reported either to be mostly grossly swollen and damaged (41) or to show abnormal respiratory control with ADP (42). However, most, although not all, isolated MH-skeletal muscle mitochondria have been apparently normal. A decrease in both the State 3 and State 4 respiration and a significantly lower than normal respiratory control with ADP has been observed for the oxidation of glutamate plus malate in mitochondria from MH-susceptible pigs (43). A marked increase (44) or decrease (8,42) in the capacity of oxidative phosphorylation has also been reported in porcine MH-skeletal muscle mitochondria. The hypothesis of uncoupling of oxidative phosphorylation (8,32) and the proposed existence of an aberrant mitochondrial respiratory chain system (42) in MH-skeletal muscle mitochondria is, however, not generally supported. The principal objections against these hypotheses are: firstly, intact and tightly-coupled mitochondria can be isolated either immediately post-mortem (28,45-47) or from biopsied porcine (44,46-49), or human MH skeletal muscles (50); secondly, carefully prepared MH mitochondria show normal oxidative phosphorylation and respiratory control (Table I); and thirdly, skeletal muscle of MH-susceptible pigs contain normal amounts of mitochondria (45) and cytochrome components (45,48) in the mitochondrial respiratory chain system.

The effect of halothane on mitochondrial respiration of MH-susceptible pigs and humans has also been investigated. Halothane inhibits the mitochondrial State 3 respiration of

TABLE 1. RESPIRATION, OXIDATIVE PHOSPHORYLATION AND RESPIRATORY CONTROL OF SKELETAL MUSCLE MITOCHONDRIA ISOLATED FROM BIOPSY (b) AND IMMEDIATE POST-MORTEM (pm) SAMPLES OF MH-SUSCEPTIBLE PIGS (*) AND HUMANS (**)

Mitochondria	Substrate	Respiration (State 3)	Oxidative Phosphorylation	Respiratory Control	Reference
Skeletal muscle (b)*	Glutamate + malate		Uncouple	Decrease	8
Biceps femoris (b)*	Glutamate + malate	Decrease	Decrease	Decrease	42
Biceps femoris (b)*	Malate + pyruvate	Normal	Normal	Normal	48
M.longissimus dorsi (pm)*	Malate + pyruvate Succinate Ascorbate + TMPD	Normal	Normal	Normal	28,45
M.longissimus dorsi (b)*	Malate + pyruvate	Decrease	Normal	Normal	46
Skeletal muscle (b)*	Glutamate	Normal	Normal	Normal	49
Semimembranosus (b,pm)*	Glutamate + malate	Normal	Normal	Normal	47
Vastus lateralis (b)*	Glutamate Succinate	Increase	Increase	Normal	44
Semitendinosus (b)*	Glutamate + malate	Decrease	Normal	Decrease	43
	Succinate	Decrease	Normal	Normal	43
Vastus lateralis (b)**	Glutamate + malate Succinate	Normal	Normal	Normal	50

NAD^+-linked substrate oxidation by MH-skeletal muscle mitochondria of pigs (48,49,51) and humans (50,52), and this is accompanied by uncoupling of oxidative phosphorylation (21,42,44,49,52) in both types of MH mitochondria. The main effect of halothane is inhibition of electron transport of NAD^+-linked substrate oxidation (43,50), and succinate oxidation is either slightly inhibited (43) or unaffected (50). These observations on the effect of halothane on MH-skeletal muscle mitochondrial respiration are essentially similar to those observed for mitochondria from normal beef-cardiac muscle (53,54), rat-skeletal muscle (55), rat liver (51) and human skeletal muscle (52) following treatment with halothane.

b) <u>Ca^{2+} Transport</u>

Mitochondria can accumulate large amounts of Ca^{2+} by an active transport system through energy derived either by electron transport or by the hydrolysis of ATP (38,58-63). Mitochondria can also act as efficient Ca^{2+} storage systems in various pathological conditions linked with a large influx of Ca^{2+} into cells (39,40). The remarkable ability of mitochondria to sequester Ca^{2+}, the suggestion that mitochondria may play an active role in Ca^{2+} regulation (37,59) and that an abnormal increase of sarcoplasmic Ca^{2+} probably occurs in the MH syndrome (14) have influenced research on the Ca^{2+} uptake by MH-skeletal muscle mitochondria (Table 2). The main aim of all these investigations has been to determine whether in MH there is a malfunction in Ca^{2+} uptake by these organelles which could explain the increase in the level of sarcoplasmic Ca^{2+}, believed to be due to defective Ca^{2+} regulation in the skeletal muscle (4,5,27,29).

The Ca^{2+} transport system of MH-skeletal muscle mitochondria has been investigated using organelles isolated either from post-mortem or biopsied samples. Most of the studies have been concentrated on Ca^{2+} uptake rather than on Ca^{2+} release using either Ca^{2+}-binding dyes such as murexide or radioactive $^{45}Ca^{2+}$. Porcine MH-skeletal muscle mitochondria isolated from post-mortem samples contain twice the normal amount of endogenous Ca^{2+} (28). Significantly greater than normal amount of endogenous Ca^{2+} have also been observed by x-ray microprobe analysis in MH-skeletal

TABLE 2. **Calcium uptake of MH-skeletal muscle mitochondria of pigs (*) and humans (**) isolated from post-mortem (pm), biopsy (b), prior to MH (+) and during MH (++) reactions.**

Mitochondria	Substrate	Calcium Uptake	Reference
M. longissimus dorsi (pm)*	Succinate	Normal (20°C)	28
	Succinate	Decrease (40°C)	28
Vastus lateralis (b)*	Succinate	Decrease (25°C)	44
M. longissimus dorsi (pm)*	ATP	Normal (24°C)	41
+Semitendinosus (b)*	Glutamate + malate	Decrease (37°C)	43
++Semitendinosus (b)*	ATP	Decrease (25°C)	64
+Biceps femoris (b)*	Endogenous (in situ)	Normal	34
++Biceps femoris (b)*	Endogenous (in situ)	Increase	34
Skeletal muscle (b)**	Glutamate + malate	Normal (25°C)	65

muscle mitochondria in biopsied porcine muscle obtained after 10 minutes in vivo halothane plus succinylcholine-induced MH (34). At 20°C, the rate and the amount of exogenous Ca^{2+} accumulated by tightly-coupled MH-skeletal muscle mitochondria during succinate oxidation is normal (28). Ca^{2+} accumulation is also normal in MH-skeletal muscle mitochondria isolated either at 5 minutes post-mortem (41), in mitochondria from biopsied samples of skeletal muscle of MH-susceptible pigs treated in vivo with neuroleptic drugs (azaperone or spiperone) to delay the onset of halothane-induced MH (64), and in human MH-skeletal muscle mitochondria (65). Ca^{2+} uptake, however, decreases significantly at 40°C with tightly-coupled preparations (66), and also at 25°C with mitochondria isolated either following the development of MH (64) or at 30 minutes post-mortem (67); the altered properties of mitochondria from 30 minutes post-mortem samples are the result and not the cause of MH. The rate of Ca^{2+} uptake by MH-skeletal muscle mitochondria has also been reported to be less than 40% of normal with either succinate (43,44) or glutamate plus malate (43) as substrate at 25°C (44) and at 37°C (43). Halothane inhibits mitochondrial Ca^{2+} uptake with glutamate plus malate (43) and either stimulates (44) or has no effect (43) with succinate at 37°C.

Porcine MH-skeletal muscle mitochondria, under aerobic or anaerobic conditions, release their endogenous Ca^{2+} significantly ($P<0.001$) faster than normal in the presence of an uncoupling agent (26,68), and faster than normal exogenous Ca^{2+} release is also observed at the onset of anaerobiosis (28,30). Halothane also enhances the exogenous Ca^{2+} release from porcine MH-skeletal muscle mitochondria in both aerobic and anaerobic conditions (69). Human MH-skeletal muscle mitochondria also show a faster than normal aerobic Ca^{2+} efflux in the presence of ruthenium red (65). The significantly faster than normal Ca^{2+} release from porcine MH-skeletal muscle mitochondria is attributed to an enhanced endogenous Ca^{2+}-activated phospholipase A_2 activity (28-31,68).

c) <u>ATPase activity</u>

Very limited studies have been conducted on the mitochondrial ATPase activity of MH-skeletal muscle mitochondria. The reports of normal mitochondrial oxidative phosphorylation and respiratory control, as discussed previously, imply that there is no significant difference in the ATPase activity between normal and MH-skeletal muscle mitochondria in pigs and in humans. However, the first investigation on the (Ca^{2+} + Mg^{2+})-ATPase activity at 25°C shows significant difference between normal and MH-skeletal muscle mitochondria when these organelles are isolated at 5 minutes post-mortem (67). With the MH-skeletal muscle mitochondria, however, the ATPase activity shows a 37% reduction when isolated at 30 minutes post-mortem, while in normal mitochondria the ATPase activity remains stable even when isolated at one hour post-mortem. Comparative studies of the skeletal muscle mitochondrial

Mg^{2+}-ATPase and the (Ca^{2+} + Mg^{2+})-ATPase activities, and their sensitivities towards oligomycin have been also conducted using tightly-coupled organelles at 35°C (31). In both types of mitochondria, no significant differences have been observed in the Mg^{2+}-ATPase, the (Ca^{2+} + Mg^{2+})-ATPase and the oligomycin-sensitive Mg^{2+}- and (Ca^{2+} + Mg^{2+})-ATPase activities (31). However, mitochondria from porcine MH-skeletal muscle contains significantly (P 0.001) higher oligomycin-insensitive (Ca^{2+} + Mg^{2+})-ATPase activity than normal. This has been suggested to be due to a higher mitochondrial endogenous Ca^{2+}-stimulated phospholipase A_2 activity (28,29), which could destroy the oligomycin sensitivity of the mitochondrial ATPase activity (70).

d) Transition Temperature and Energy of Activation

A change in mitochondrial membrane permeability in MH-susceptible pigs can be inferred from changes in the transition temperature obtained from the Arrhenius plots of either the ADP or Ca^{2+}-stimulated respiration of substrate oxidation (71). The Arrhenius plots of the Ca^{2+}-stimulated respiration of mitochondria isolated from longissimus dorsi muscle of MH-susceptible and normal pigs show a distinct difference in the transition temperature for succinate oxidation (28). Porcine MH-skeletal muscle mitochondria show a transition temperature of 26.3 ± 0.8°C, and this is 9°C higher than the transition temperature of 17.3 ± 0.8°C for normal pigs. The values for the energy of activation above and below the transition temperature are 40.4 ± 15.7 kJ/mole and 102.0 ± 24.8 kJ/mole respectively as compared with the normal values of 50.7 ± 2.9 kJ/mole and 127.0 ± 26.5 kJ/mole, respectively. The Arrhenius plots of the Ca^{2+}-stimulated respiration of succinate oxidation show that there is a significant difference ($P < 0.001$) in the transition temperature between the two types of pigs, but with only small changes in the energy of activation above and below the transition temperature. The transition temperature of MH-skeletal muscle mitochondria can be restored to normal by spermine and Mg^{2+}, and the energy of activation above and below the transition temperature under these conditions is practically identical to that of normal pigs (28). No difference in the transition temperature has been observed between the two types of pigs when ADP has been used instead of Ca^{2+} to stimulate the oxidation of succinate (28).

e) Phospholipase A_2

At 40°C but not at 25°C, intact and tightly-coupled porcine MH-skeletal muscle mitochondria accumulate only half the normal amount of exogenous Ca^{2+} during succinate oxidation. A further increase in the amount of exogenous Ca^{2+} in the medium results in uncoupling and large amplitude swelling (28,66). In the presence of phospholipase A_2 inhibitors at 40°C, however, MH-skeletal muscle mitochondria accumulate the same amount of exogenous Ca^{2+} as normal mitochondria and do not show Ca^{2+}-induced uncoupling and large amplitude swelling (28,72). The 9°C higher transition temperature for porcine MH-skeletal muscle mitochondria observed in the Ca^{2+}-stimulated respiration is restored to normal with either spermine, a phospholipase A_2 inhibitor (73), or Mg^{2+}, a cation specific for inhibiting Ca^{2+} binding (74-76). Spermine can also affect the kinetics of the Ca^{2+} transport by interacting with the Ca^{2+}-binding sites of the mitochondrial membranes (77), and Mg^{2+} can protect and stabilize the structural integrity of the mitochondrial membranes (78,79). On stimulation by Ca^{2+} (80-83), mitochondrial phospholipase A_2 causes the hydrolysis of the phospholipids in the mitochondrial membranes, and this results in the liberation of unsaturated long chain fatty acids and the formation of lyso-derivatives of the mitochondrial phospholipids (82,84,85). The fatty acids released induce uncoupling of the mitochondria (86,87), and the free fatty acids and the lyso-derivatives of the mitochondrial membrane phospholipids cause the destabilization of the mitochondrial membranes, resulting in swelling of the mitochondria (88-91). Mitochondrial phospholipase A_2 cannot be activated by ADP (81), and this is the reason for the lack of a difference in the mitochondrial transition temperature between MH-susceptible and normal pigs when ADP instead of Ca^{2+} is used to stimulate succinate oxidation (28), and also the failure of ADP to induce large amplitude swelling in MH-skeletal muscle mitochondria (28).

Porcine MH-skeletal muscle mitochondria of longissimus dorsi contain a normal amount of phospholipase A_2 protein but its activity is at least twice that of normal mitochondria (31). These mitochondria also contain a significantly ($P < 0.001$) higher than normal amount of endogenous Ca^{2+} (28) and fatty acids (31), both of which are known activators of mitochondrial phospholipase A_2 activity (92). The endogenous phospholipase A_2 activity, estimated by the amount of long chain fatty acids formed following incubation of isolated mitochondria of MH-susceptible pigs at 40^oC for 20 minutes (31), is inhibited by spermine and by low concentrations of trifluoperazine (Cheah and Cheah, unpublished). Seventy-nine percent inhibition of phospholipase A_2 activity is observed with 21 nmole spermine per mg protein, and 45% inhibition with 22 nmole trifluoperazine per mg protein (93). Under the latter conditions, the amount of fatty acids formed is very similar to that produced by normal skeletal muscle mitochondria in the absence of the inhibitor, indicating that the enzyme activity of porcine MH-skeletal muscle mitochondria is reduced to a normal level. This concentration of trifluoperazine is also sufficient to prevent the Ca^{2+}-induced uncoupling and large amplitude swelling of porcine MH-skeletal muscle mitochondria. Thus, both the abnormal changes in these mitochondria are due to an enhanced Ca^{2+}-activated phospholipase A_2 activity.

f) <u>Calmodulin</u>

Mitochondria isolated from longissimus dorsi muscle of MH-susceptible pigs contain a significantly ($P < 0.01$) higher amount of endogenous calmodulin [44.0 ± 1.0 ng/mg protein] than that [25.0 ± 6.0 ng/mg protein] of normal mitochondria (29). The mitochondrial calmodulin has been determined on the basis of its ability to stimulate calmodulin-deficient phosphodiesterase activity by estimating the amount of phosphate released by 5'-nucleotidase (94,95) in boiled mitochondrial extracts. The heat-stable activator of the phosphodiesterase activity is Ca^{2+}-dependent and can be inhibited by trifluoperazine (Cheah and Cheah, unpublished), an inhibitor of calmodulin-dependent enzyme (96,97). The activation profile of the Ca^{2+}-dependent phosphodiesterase activity of the

FIGURE 1. Effect of extracted mitochondrial calmodulin on the calmodulin-deficient cyclic AMP phosphodiesterase activity.

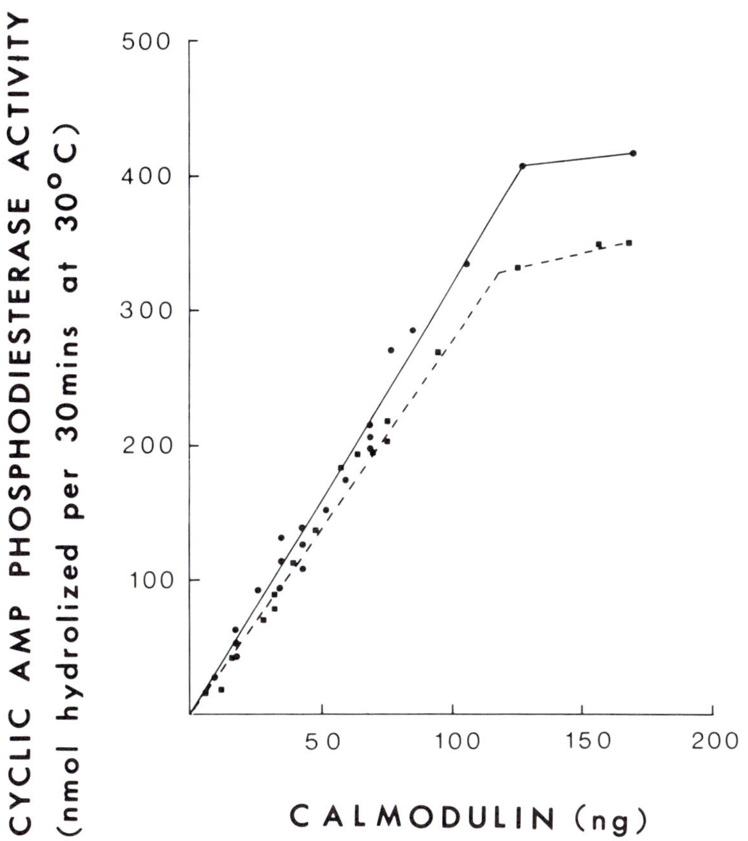

LEGEND: The activation profiles of the calmodulin-dependent cyclic AMP phosphodiesterase activity by the extracted skeletal muscle mitochondrial calmodulin from MH-susceptible (●) and normal (■) pigs were determined in the presence of 1.0 mM Ca^{2+} at 30°C.

isolated calmodulin of MH-skeletal muscle mitochondria is similar to that of normal pigs (Figure 1), and both are identical to those of purified rat liver mitochondrial matrix calmodulin (98) and purified bovine cardiac mitochondrial matrix calmodulin (99). The concept (29) that a significantly higher than normal amount of endogenous calmodulin could be responsible for the abnormal features of porcine MH-skeletal muscle mitochondria through activating the Ca^{2+}-dependent phospholipase A_2 activity is supported by the results obtained with normal mitochondria. When these isolated organelles are treated with exogenous calmodulin to bring it to a level equivalent to the endogenous level of MH-skeletal muscle mitochondria, Ca^{2+}-induced uncoupling at 40°C is observed as for MH-skeletal muscle mitochondria, and this is prevented by a low concentration (7 nmole/mg mitochondrial protein) of trifluoperazine (Cheah and Cheah, unpublished). Inhibition of the Ca^{2+}-induced uncoupling by trifluoperazine under these conditions is due to the inhibition of the calmodulin-dependent phospholipase A_2 activity.

g) Structure

A marked difference in membrane permeability has been observed in porcine MH mitochondria when these organelles are treated with exogenous Ca^{2+} but not with ADP during succinate oxidation at high temperature. At 40°C but not at 25°C, large amplitude swelling of mitochondria is observed in the presence of 1220 ± 300 nmole Ca^{2+} per mg protein, and this is accompanied by a loss of respiratory control (28). No swelling has been detected with normal mitochondria under the same condition, even when treated with at least twice the amount of exogenous Ca^{2+} (28). Large amplitude swelling of mitochondria in longissimus dorsi muscle of MH-susceptible pigs is also observed at 40°C but not with mitochondria of similar muscle from normal pigs (Cheah and Cheah, unpublished). Gross mitochondrial swelling is detected in biopsied skeletal muscle samples of MH-susceptible pigs during muscle rigidity but prior to temperature increase (22), and at the onset of the MH syndrome induced by halothane plus succinylcholine (34), and in biopsied samples of human MH-skeletal muscle (100-108). No mitochondrial swelling has been observed in biopsied skeletal muscle of normal pigs subjected to halothane plus succinylcholine treatment (34). Thus, in porcine and human MH, the skeletal muscle mitochondria tend to exhibit large amplitude swelling with disrupted cristae (Table 3) but when isolated intact these mitochondria show no apparent impairment in their capacity to carry out oxidative phosphorylation (28,43,45-50). The Ca^{2+}-induced large amplitude swelling of porcine MH-skeletal muscle mitochondria can probably be attributed to an enhanced calmodulin-dependent Ca^{2+}-activated phospholipase A_2 activity since swelling is prevented by phospholipase A_2 inhibitors (28) and trifluoperazine (Cheah and Cheah, unpublished).

TABLE 3. **Configuration of MH-skeletal muscle mitochondria from post-mortem (pm) and biopsied (b) samples of pigs (*) and humans (**)**

Muscle	Preparation	Configuration	Reference
M.longissimus dorsi (pm)*	In vitro (isolated)	Swollen	67
M.longissimus dorsi (pm)*	In situ	Swollen	105,106
Skeletal (b)*	In situ	Swollen	22
Biceps femoris (b)*	In situ	Swollen	34
Skeletal (b)**	In situ	Rupture	99-101
Skeletal (b)**	In situ	Swollen	102,103
Vastus lateralis (b)**	In situ	Swollen	104
Palmaris longus (b)**	In situ	Swollen	107

II. ROLE OF MITOCHONDRIA IN MALIGNANT HYPERTHERMIA (HYPOTHESES)

It is generally agreed that in porcine and in human MH there is an apparent loss of Ca^{2+} control in the skeletal muscles (14,27–29), resulting in a rise in sarcoplasmic Ca^{2+}. This could occur because of either a malfunction in the Ca^{2+}- regulating organelles or a defect in the cell membrane. The increase in the level of sarcoplasmic Ca^{2+} (30,34) could be due either to Ca^{2+} efflux from the mitochondria and/or sarcoplasmic reticulum or to an increased influx of Ca^{2+} from the extracellular fluid into the muscle cells via the sarcolemma. It is thus obvious that a molecular explanation for the increase in the level of sarcoplasmic Ca^{2+} is essential to comprehend the development of the MH syndrome.

In porcine MH, the data suggest that the increase in sarcoplasmic Ca^{2+} is due to Ca^{2+}-regulating organelles rather than to a Ca^{2+} influx from the extracellular fluid (34) since the total muscle Ca^{2+} starts to decrease within 5 minutes of the induction of porcine MH with halothane, and is accompanied by an increase in plasma Ca^{2+} (21). An increase in total serum Ca^{2+} (9,23) and whole blood ionized Ca^{2+} (4) also occurs during halothane-induced porcine MH.

Mitochondria have been suggested to play a role in the development of MH mainly by their ability to sequester Ca^{2+} (see section I[b]). Uncoupling of mitochondria was initially proposed as a cause of MH since symptoms similar to MH could be experimentally induced by the uncoupling agent, 2,4-dinitrophenol, in halothane-anaesthetized dogs (32). This hypothesis is, however, not widely accepted. Halothane, at low concentration, uncouples oxidative phosphorylation, but inhibits respiration at high concentration (see section I[a]). Uncoupling of oxidative phosphorylation alone can only partially account for all the observed heat produced during the MH syndrome (4,27,109). Inhibition of respiration would result in the reduction of oxygen consumption and CO_2 production, all of which are known to be elevated in MH (4,9,24).

The hypothesis of an aberrant respiratory chain system (42) cannot be substantiated since MH-skeletal muscle mitochondria contain a normal amount of cytochrome components (45,48) and a normal electron transport chain system (see section I[a]). Furthermore, the majority of investigators agree that isolated mitochondria from MH-skeletal muscles show normal oxidative phosphorylation and respiratory control (see section I[a]).

An enhanced phospholipase A_2 activity has been postulated (28,29,31,72) to be the cause of the increase in the sarcoplasmic Ca^{2+} during a MH reaction. This enzyme (28,66,72) is responsible for the Ca^{2+}-induced uncoupling, large amplitude swelling and enhanced Ca^{2+} release in MH-skeletal muscle mitochondria (see section I[e]). Phospholipase A_2 and its by-products of long chain unsaturated fatty acids from the mitochondrial phospholipids are capable of inducing Ca^{2+} release from sarcoplasmic reticulum (31), and MH-skeletal muscle mitochondria; or the end-products of phospholipase A_2 activity could inhibit the Ca^{2+} uptake of both the normal and MH sarcoplasmic reticulum (31). No difference in the level of endogenous Ca^{2+} has been observed in the isolated sarcoplasmic reticulum of normal and MH-susceptible pigs (110,111), but significantly higher than normal amounts of endogenous Ca^{2+} have been detected in the isolated MH-skeletal muscle mitochondria (28,34). These observations suggest that skeletal muscle mitochondria are implicated in the MH syndrome. MH-skeletal muscle mitochondria have a lower than normal capacity to retain Ca^{2+} (28) because of their significantly higher than normal endogenous Ca^{2+}-activated phospholipase A_2 activity (see I[e]). This results in large amplitude mitochondrial swelling accompanied simultaneously by an enhanced Ca^{2+} release (28). The excess Ca^{2+} and the long chain unsaturated fatty acids released would then induce additional Ca^{2+} release from the sarcoplasmic reticulum (31) probably by the mechanism of calcium-induced calcium release (112,113) and fatty acid-induced Ca^{2+} release (31), ultimately leading to an increase in the level of sarcoplasmic Ca^{2+} (30,34). The increase in sarcoplasmic Ca^{2+} is responsible for the increase in muscle rigidity and the enhanced rate of glycolysis during a MH reaction since muscle contraction is dependent on the concentration of free Ca^{2+} in the sarcoplasm (114,115) and the rate of glycolysis is enhanced by Ca^{2+} stimulating myofibrillar ATPase (116) and phosphorylase kinase (117-119). The enormous rise in temperature to

$42^{o}C$ or higher (4,9) is due to a combination of several processes. Heat production in the initial 10 minutes in porcine MH syndrome can be explained by an increase in aerobic metabolism (17) associated with shortening and contraction of muscles (120), and in subsequent periods mainly by an increase in lactate formation (9,17,21).

Skeletal muscle mitochondria may play an important role in halothane-induced porcine MH reactions without the involvement of Ca^{2+} (33,121). Halothane adversely affects mitochondrial function by completely inhibiting the transport of NADH and the dehydrogenation of pyruvate. The inhibition of mitochondrial respiration by halothane would result in an enhanced production of lactate, a reduction in ATP synthesis and subsequently a decrease in muscle creatine phosphate and pH. The decrease in muscle pH due to increased lactate may stimulate the release of catecholamines which would then further accelerate glycogenolysis (121). However, this hypothesis is difficult to reconcile with the evidence on three counts. Firstly, oxygen consumption increases rather than decreases during a MH crisis (4,9,24). Secondly, drug-induced porcine MH results in an increase in sarcoplasmic Ca^{2+} in vivo (34). Thirdly, halothane inhibits the release of catecholamines (122). It has been suggested that halothane causes gross changes in the defective mitochondrial membranes and thereby inhibits the dehydrogenation of pyruvate (33,121) but no explanation has been given for the cause of the defective mitochondrial membranes present only in MH-susceptible pigs (121). The lesion responsible for the occurrence of defective mitochondrial membranes could be explained by an enhanced Ca^{2+}- activated phospholipase A_2 (see section I[e]). It is conceivable that in drug-induced porcine MH the inhibition of mitochondrial dehydrogenation of pyruvate can result in a further increase in the rate of lactate formation (121). However, this cannot be the primary lesion since porcine MH (7,18–20) and to a certain extent human MH (4,123) can be initiated for example by high environmental temperature without the use of any volatile anaesthetic or succinylcholine. Under these conditions, an increase in membrane permeability through activation of phospholipase A_2 would occur, ultimately resulting in an increase in sarcoplasmic Ca^{2+} as previously discussed.

Human and porcine MH can be effectively prevented by dantrolene (4), a specific skeletal muscle relaxant (124) which acts by reducing the amount of Ca^{2+} released into the sarcoplasm (125-127) Dantrolene (20 μM) has no effect on the Ca^{2+} transport system in purified sarcoplasmic reticulum from both MH-susceptible and normal pigs (128) but inhibits mitochondrial Ca^{2+} release in MH-sysceptible pigs by 28% and 44% with 20 μM and 40 μM dantrolene, respectively (Cheah and Cheah, unpublished). Furthermore, skeletal muscle mitochondrial phospholipase A_2 activity is inhibited by 17% and 29% by 20 μM and 40 μM dantrolene, respectively (Cheah, unpublished). These findings further support the concept (28,29,66) that skeletal muscle mitochondria are most likely to play an important role in the development of the MH syndrome.

REFERENCES

1. Denborough, M.A., Forster, J.F.A., Lovell, R.R.H., Maplestone, P.A. and Villiers, J.D. Br. J. Anaesth. 1962;34:395.

2. Britt, B.A., and Kalow, W. Ann. N.Y. Acad. Sci. 1968;151:947.

3. Britt, B.A. and Kalow, W. Can. Anaesth. Soc. J. 1970;17:293.

4. Gronert, G.A. Anesthesiology 1980;53:395.

5. Denborough, M.A. Pharmacol. Ther. 1980;9:357.

6. Harrison, G.G., Biebuyck, J.F., Terblanche, J., Dent, D.M., Hickman, R. and Saunders, S.J. Br. Med. J. 1968;3:594.

7. Topel, D.G., Bicknell, E.J., Preston, K.S., Christian, L.L. and Matsushima, C.Y. Mod. Vet. Pract. 1968;49:40.

8. Sybesma, W. and Eikelenboom, G. Neth. J. Vet. Sci. 1969;2:155.

9. Berman, M.C., Harrison, G.G., Bull, A.B. and Kench, J.E. Nature 1970;225:653.

10. Hall, L.W., Trim, C.M. and Woolf, N. Br. Med. J. 1972;2:145.

11. Nelson, T.E., Jones, E.W., Venable, J.H. and Kerr, D.D. Anesthesiology 1972;36:52.

12. Clark, M.G., Williams, C.H., Pfeifer, W.F., Bloxham, D.P., Holland, P.C., Taylor, C.A. and Lardy, H.A. Nature 1973;245:99.

13. Ryan, J.F. and Papper, E.M. Anesthesiology 1970;32:196.

14. Moulds, R.F.W. and Denborough, M.A. Br. Med. J. 1974;2:241.

15. Hall, L.W., Woolf, N., Bradley, J.W.P. and Jolly, D.W. Br. Med. J. 1966;4:1305.

16. Gronert, G.A. and Theye, R.A. Br. J. Anaesth. 1976;48:513.

17. Hall, G.M., Bendall, J.R., Lucke, J.N. and Lister, D. Br. J. Anaesth. 1976;48:305.

18. Allen, W.M., Berrett, S., Harding, J.D.J. and Patterson, D.S.P. Vet. Rec. 1970;87:64.

19. Allen, W.M., Hebert, C.N. and Smith, L.P. Vet. Rec. 1974;94:212.

20. Williams, C.H., Houchins, C. and Shanklin, M.D. Br. Med. J. 1975;3:411.

21. Berman, M.C. and Kench, J.E. In: Gordon, R.A., Britt, B.A. and Kalow, W. (eds.). International Symposium on Malignant Hyperthermia. 1973. Springfield: Charles C. Thomas. pp. 287–297.

22. Brucker, R.F., Williams, C.H., Popinigis, J., Galvez, T.L., Vail, W.J. and Taylor, C.A. In: Gordon, R.A., Britt, B.A. and Kalow, W. (eds.). International Symposium on Malignant Hyperthermia. 1973. Springfield: Charles C. Thomas. pp. 238–270.

23. Van den Hende, C., Lister, D., Muylle, E., Ooms, L. and Oyaert, W. Br. J. Anaesth. 1976;48:821.

24. Williams, C.H., Shanklin, M.D. and Houchins, C. In: Henschel, E.O. (ed.). Malignant Hyperthermia: Current Concepts. 1977. New York: Appleton-Century-Crofts. pp. 149–156.

25. Webb, A.J. Anim. Prod. 1980;31:101.

26. Wingard, D.W. and Gatz, E.E. In: Aldrete, J.A. and Britt, B.A. (eds.). Second International Symposium on Malignant Hyperthermia. 1978. New York: Grune and Stratton, Inc. pp. 363–372.

27. Britt, B.A. Fed. Proc. 1979;38:44.

28. Cheah, K.S. and Cheah, A.M. Biochim. Biophys. Acta 1981;634:70.

29. Cheah, K.S. Biochem. Soc. Trans. 1984;12:358.

30. Cheah, K.S., Cheah, A.M., Crosland, A.R., Casey, J.C. and Webb, A.J. Meat Sci. 1984;10:117.

31. Cheah, K.S. and Cheah, A.M. Biochim. Biophys. Acta 1981;638:40.

32. Wilson, R.D., Nichols, Jr., R.J., Dent, T.E. and Allen, C.R. Anesthesiology 1966;27:231.

33. Mitchell, G. and Heffron, J.J.A. S. Afr. J. Sci. 1980;76:546.

34. Stadhouders, A.M., Viering, W.A.L., Verburg, M.P., Ruitenbeek, W. and Sengers, R.C.A. Acta Anaesthesiol. Scand. 1984;28:14.

35. Chance, B. and Williams, G.R. Advances in Enzymology 1956;17:65.

36. Mela, L. In: Bronner, F. and Kleinzeller, A. (eds.). Current Topics in Membrane Transport. 1977. New York: Academic Press. pp. 321-366.

37. Lehninger, A.L. Biochem. J. 1970;119:129.

38. Saris, N.E. and Akerman, K.E.O. Current Topics in Bioenergetics 1981;10:103.

39. Martonosi, A.N. Physiol. Rev. 1984;64:1240.

40. Duncan, C.J. Experientia 1978;34:1531.

41. Greaser, M.L., Cassens, R.G., Briskey, E.J. and Hoekstra, W.G. J. Food Sci. 1969;34:125.

42. Eikelenboom, G. and Van den Bergh, S.G. J. Anim. Sci. 1973;37:692.

43. Gronert, G.A. and Heffron, J.J.A. Anesth. Analg. 1979;58:76.

44. Britt, B.A., Endrenyi, L., Cadman, D.L., Ho, M-F. and Fung, H.Y.K. Anesthesiology 1975;42:292.

45. Cheah, K.S. J. Sci. Fd. Agric. 1973;24:51.

46. Brooks, G.A. and Cassens, R.G. J. Anim. Sci. 1973;37:688.

47. Campion, D.R., Olson, J.C., Topel, D.G., Christian, L.L. and Kuhlers, D.L. J. Anim. Sci. 1975;41:1314.

48. Ruitenbeek, W., Verburg, M.P., Janssen, A.J.M., Stadhouders, A.M. and Sengers, R.C.A. Acta Anaesthesiol. Scand. 1984;28:9.

49. Denborough, M.A., Hird, F.J.R., King, J.O., Marginson, M.A., Mitchelson, K.R., Nayler, W.G., Rex, M.A., Zapf, P. and Condron, R.J. In: Gordon, R.A., Britt, B.A. and Kalow, W. (eds.). International Symposium on Malignant Hyperthermia. 1973. Springfield: Charles C. Thomas. pp. 229-237.

50. Britt, B.A., Kalow, W. and Endrenyi, L. In: Gordon, R.A., Britt, B.A. and Kalow, W. (eds.). International Symposium on Malignant Hyperthermia. 1973. Springfield: Charles C. Thomas. pp. 387-398.

51. Mitchelson, K.R. and Hird, F.J.R. Amer. J. Physiol. 1973;225:1393.

52. Britt, B.A., Kalow, W., Gordon, A., Humphrey, J.G. and Rewcastle, N.B. Canad. Anaesth. Soc. J. 1973;20:431.

53. Harris, R.A., Munroe, J., Farmer, B., Kim, K.C. and Jenkins, P. Arch Biochem. Biophys. 1971;142:435.

54. Grist, E.M. and Baum, H. FEBS Letters 1974;48:41.

55. Britt, B.A., Kalow, W. and Endrenyi, L. Biochem. Pharmacol. 1972;21:1159.

56. Miller, R.N. and Hunter, Jr., F.E. Mol. Pharmacol. 1970;6:67.

57. Hall, G.M., Kirtland, S.J. and Baum, H. Brit. J. Anaesth. 1973;45:1005.

58. Bygrave, F.L. Biol. Rev. 1978;53:43.

59. Carafoli, E. and Crompton, M. Curr. Top. Membr. Transp. 1978;10:151.

60. Fiskum, G. and Lehninger, A. Fed. Proc. 1980;39:2432.

61. Carafoli, E. Membranes and Transport 1982;1:611.

62. Carafoli, E. In: Carafoli, E. (ed.) Membrane Transport of Calcium. 1982. New York: Academic Press. pp. 109-117.

63. Akerman, K.E.O. and Nicholls, D.G. Rev. Physiol. Biochem. Pharmacol. 1983;95:149.

64. Somers, C.J. and McLoughlin, J.V. J. Comp. Path. 1982;92:191.

65. Heffron, J.J.A. Biochem. Soc. Trans. 1984;12:360.

66. Cheah, K.S. and Cheah, A.M. FEBS Letters. 1978;95:307.

67. Greaser M.L., Cassens, R.G., Briskey, E.J. and Hoekstra, . J. Food Sci. 1969;34:120.

68. Cheah, K.S. and Cheah, A.M. FEBS Letters 1979;106:265.

69. Cheah, K.S. and Cheah, A.M. J. Sci. Food Agric. 1976;27:1137.

70. Kagawa, J. and Racker, E. J. Biol. Chem. 1966;241:2461.

71. Raison, J.K. Bioenerg. 1973;4:285.

72. Cheah, K.S. and Cheah, A.M. In: First European Bioenergetics Conference (Urbino). 1980. Patron Editore, Bologna, Italy. pp. 397-398.

73. Sechi, A.M., Cabrini, L., Landi, L., Pasquali, P. and Lenaz, G. Arch. Biochem. Biophys. 1978;186:248.

74. Akerman, K.E.O. Arch. Biochem. Biophys. 1978;189:256.

75. Jacobus, W.E. and Brierley, G.P. J. Biol. Chem. 1969;244:4995.

76. Sordahl, L.A. Arch. Biochem. Biophys. 1975;167:104.

77. Akerman, K.E.O. J. Bioenerg. Biomembr. 1977;9:65.

78. Packer, L., Utsumi, K. and Mustafa, G. Arch. Biochem. Biophys. 1966;117:381.

79. Leblanc, P. and Clauser, H. Biochim. Biophys. Acta 1974;347:87.

80. Bjornstad, P. J. Lipid Res. 1966;7:612.

81. Waite, M., Scherphof, G.L., Boshouwers, F.M.G. and Van Deenen, L.L.M. J. Lipid Res. 1969;10:411.

82. Nachbaur, J., Colbeau, A. and Vignais, P.M. Biochim. Biophys. Acta 1972;274:426.

83. Zurini, M., Hugentobler, G. and Gazzotti, P. Eur. J. Biochem. 1981;119:517.

84. Scherphof, G.L., Waite, M. and Van Deenen, L.L.M. Biochim. Biophys. Acta 1966;125:406.

85. Honjo, I. and Ozawa, K. Biochim. Biophys. Acta 1968;162:624.

86. Vazquez-Colon, L., Ziegler, F.D. and Elliot, W.B. Biochemistry 1966;5:1134.

87. Weinbach, E.C., Garbus, J. and Glaggett, C.E. J. Biol. Chem. 166;241:3708.

88. Lehninger, A.L. and Remmert, L.F. J. Biol. Chem. 1959;234:2459.

89. Avi-Dor, Y. Biochim. Biophys. Acta 1960;39:53.

90. Zborowski, J. and Wojtczak, L. Biochim. Biophys. Acta 1963;70:596.

91. Aleksandrowicz, Z., Swierczynski, J. and Wrzolkowa, T. Biochim. Biophys. Acta 1973;305:59.

92. Waite, M. and Sisson, P. Biochemistry 1971;10:2377.

93. Cheah, K.S., Cheah, A.M. and Waring, J.C. Meat Sciences 1986;17:37.

94. Butcher, R.W. and Sutherland, E.W. J. Biol. Chem. 1962;237:1244.

95. Teo, T.S., Wang, T.H. and Wang, J.H. J. Biol. Chem. 1973;248:588.

96. Levin, R.M. and Weiss, B. Molec. Pharmacol. 1977;13:690.

97. Weiss, B., Prozialeck, W., Cimino, M., Barnette, M.S. and Wallace, T.L. In: Watterson, D.M. and Vincenzi, F.F. (eds.). Calmodulin and Cell Functions. Ann. N.Y. Acad. Sci. 1980;356:319.

98. Hatase, O., Tokuda, M., Itano, T., Matsui, H. and Doi, A. Biochem. Biophys. Res. Commun. 1982;104:673.

99. Hatase, O., Tokuda, M., Sharma, R.K., Wang, J.H. and Green, D.E. Biochem. Biophys. Res. Commun. 1983;113:633.

100. Carpenter, G.C. Soc. Paediat. Res. 1966;175:29.

101. Isaacs, H., Frere, G. and Mitchell, J. Brit. J. Anaesth. 1973;45:860.

102. Schiller, H.H. and Mair, W.G.P. J. Neurol. Sci. 1974;21:93.

103. Reske-Nielsen, E., Haase, J. and Kelstrup, J. Acta Path. Microbiol. Scand. Sect. A. 1975;83:651.

104. Isaacs, M. In: Aldrete, J.A. and Britt, B.A. (eds.). Second International Symposium on Malignant Hyperthermia. 1978. New York: Grune and Stratton, Inc. pp. 89-102.

105. Hull, M.T., Muller, J. and Albrecht, W.H. Anesthesiology 1978;48:223.

106. Bergmann, V. Exp. Path. 1979;17:243.

107. Cloke, J.D., Davis, E.A., Gordon, J., Hsieh, S-I., Grider, J., Addis, P.B. and McGrath, C.J. Scanning Electron Microscopy, III. 1981;435-446, SEM Inc., AMF O'Hare, Chicago, U.S.A.

108. Reske-Nielsen, E. In: Aldrete, J.A. and Britt, B.A. (eds.). Second International Symposium on Malignant Hyperthermia. 1978. New York: Grune and Stratton, Inc. pp. 287-327.

109. Wang, J.K., Moffitt, E.A. and Rosevear, J.W. Anesthesiology 1969;30:439.

110. McIntosch, D.B., Berman, M.C. and Kench, J.E. Biochem. J. 1977;166:387.

111. Ohnishi, S.T., Taylor, S. and Gronert, G.A. FEBS Letters 1983;161:103.

112. Endo, M. Physiol. Rev. 1977;57:71.

113. Miyamoto, H. and Racker, E. FEBS Letters 1981;133:235.

114. Ebashi, S., Endo, M. and Ohtsuki, I. Quart. Rev. Biophys. 1969;2:351.

115. Weber, A. and Murray, J.M. Physiol. Rev. 1973;53:612.

116. Scopes, R.K. Biochem. J. 1974;142:79.

117. Ozawa, E., Hosoi, K. and Ebashi, S. J. Biochem. (Tokyo) 1967;61:531.

118. Heilmeyer, L.M.C., Jr., Meyer, F., Haschke, R.H. and Fischer, E.H. J. Biol. Chem. 1970;245:6649.

119. Brostrom, C.O., Hunkeler, F.L. and Krebs, E.C. J. Biol. Chem. 1971;246:1961.

120. McMahon, T.A. In: McMahon, T.A. (ed.). Muscles, Reflexes and Locomotion. 1984. New Jersey: Princeton University Press. pp. 27-52.

121. Mitchell, G., Heffron, J.J.A. and van Rensburg, A.J.J. Anesth. Analg. 1980;59:250.

122. Rorie, D.K., Tyce, G.M. and Mackenzie, R.A. Anesth. Analg. 1984;63:1059.

123. Britt, B.A. New Engl. J. Med. 1974; 290:1140.

124. Snyder, H.R., Davis, C.S., Bickerton, R.K. and Halliday, R.P. J. Med. Chem. 1967;10:807.

125. Hainaut, K. and Desmedt, J.E. Nature 1974;252:728.

126. Morgan, K.G. and Bryant, S.H. J. Pharmacol. Exp. Ther. 1977;201:138.

127. Desmedt, J.E. and Haunaut, K. J. Physiol. 1977;265:565.

128. White, M.D., Collins, J.G., and Denborough, M.A. Biochem. J. 1983;212:399.

5

PORCINE MALIGNANT HYPERTHERMIA—THE SAGA OF THE "HOT" PIG

G.G. HARRISON

Department of Anaesthetics, University of Cape Town and Groote Schuur Hospital Medical School, Observatory 7925, South Africa

ABSTRACT

Much of what we understand today of the pathogenesis and therapeutic management of the anaesthetic induced malignant hyperthermic syndrome in man followed from the happy chance discovery, approximately twenty years ago, that the well-known stress syndrome of swine was a very close facsimile of this puzzling and fatal syndrome in man. In both, the syndrome was genetically based, a fact that allowed for the breeding of a valid animal experimental model of human malignant hyperthermia. So extensive has been the use of the MHS pig model that today an account of porcine hyperthermia per se would indeed be the whole story of the evolution of our present understanding of the human MH syndrome, the subject of this book.

This chapter confines itself to the more purely porcine aspects (if that is possible) of the syndrome. It includes descriptions and discussions of:

1. The syndrome in the pig including clinical signs, biochemical accompaniments and variations in the syndrome.
2. The identity of the separately described stress myopathies of swine--porcine malignant hyperthermia, porcine stress syndrome and pale, soft, exudative pork.
3. Modes of inheritance.
4. The MHS pig and other experimental models of human MH.
5. Pharmacological triggering, attenuation, blocking and reversal of porcine malignant hyperthermia.
6. Pathogenesis of the vicious cycle of malignant hyperthermia in relation to the stress syndromes of the pig and man.

INTRODUCTION

"All animals are equal; only some animals are more equal than others". The pig Napoleon, Animal Farm (George Orwell).

Of all the complications of general anaesthesia perhaps the most dramatic and indeed frightening is that veritable metabolic storm which characterizes the unexpected, in the past frequently fatal, syndrome of anaesthetic-induced malignant hyperthermia. First described in 1960 in Denborough's et al report (1) (later elaborated in 1962 (2)) of a family which displayed an apparent hereditary predisposition to death attributable to general anaesthesia, this syndrome became definitively characterized over the next six years through the reports of Saidman et al (3) and those in the Canadian Anaesthetists' Society Journal of September 1966 (4) which presented the collected experience of six groups of authors of eleven cases of this fatal condition. Variously referred to up to this date as "hyperpyrexia", "hyperthermia" or even "extreme hyperthermia during anaesthesia", in this publication we find this syndrome referred to for the first time as "Malignant Hyperpyrexia" (5,6), later changed by common usage to "Malignant Hyperthermia" (MH). Since this time little has been added to the clinical manifestations we recognize of the syndrome.

As on so many other occasions in the long history of medicine, serendipity now provided the deus ex machina that would ultimately help solve this dangerous clinical conundrum--the "hot" pig. The possibility that strains of certain breeds of pig might serve as a valid animal model of MH was suggested, also in 1966, by the observation reported by Hall and co-workers (7) of muscle rigor and hyperthermia in littermate pigs, anaesthetized with halothane in response to the injection of succinylcholine, followed shortly by Harrison and co-workers' (8) report of the same reaction but now in response to halothane alone. As with the human cases this response appeared to be pharmacogenetic.

These discoveries were seminal. Today, just less than twenty years later, though the minutiae of the pathogenesis of malignant hyperthermia have yet to be elucidated, knowledge that has come from studies worldwide in malignant hyperthermia susceptible (MHS) swine has led not only to effective pharmacological control of the syndrome, but also to valuable spin-offs in many biomedical areas, in particular those of muscle and membrane physiology, calcium transport and kinetics and the meat industry.

THE CLINICAL SYNDROME IN PIGS

The malignant hyperthermia syndrome in swine has two prerequisites:

1. Possession of the operative gene or genes.
2. Exposure to a specific stressor for an adequate time.

The specific stressors originally described--halothane anaesthesia alone or succinylcholine administration during halothane anaesthesia--identified genetically based differing grades of severity of lesion and response.

Once established the syndrome rapidly becomes self-sustaining and irreversible, continuing independently of withdrawal of the stressor which can be then seen to have acted as a trigger (8,9).

Clinical Signs

The syndrome is manifested by a group of inexorably progressive clinical signs together with a myriad of biochemical changes which immediately precede them (10).

The clinical features include:

1. Cyanosis of the snout and blotchy cyanosis of the skin, most marked on the anterior abdominal wall. This blotching disappears as the syndrome becomes established, often "pinking up" before it does so.
2. Twitching and sporadic fasciculation of muscles, most obvious in the thigh, proceeding usually within minutes to progressive extreme rigor of muscle. This rigor causes the limbs, most markedly the hind limbs, to become rigidly extended. Rarely, rigor may hold a limb in flexion. Triggering of the syndrome by succinylcholine is characterized by extremes of coarse muscle fasciculation and myotonia.

3. Sinus tachycardia proceeding often to extreme rates of the order of 240 beats per minute. This is accompanied in the early stages of the syndrome by a rise in blood pressure. Cardiac arrhythmias, usually of the nature of ventricular tachyarrhythmias, appear variably, sometimes very early, at others terminally.

4. An explosive sustained rise in core temperature. Change in temperature may be earliest observed by means of a thermistor probe sited deep in the thigh muscle mass. The rate of temperature rise varies between animals. In highly susceptible strains rates of rise of $1^{o}C$ every five to seven minutes are common, reaching $43-45^{o}C$ ante mortem.

5. Open-mouthed breathing and hyperventilation. This manifests only in pigs allowed to breathe spontaneously. Apnoea supervenes within ten to fifteen minutes.

Biochemical Accompaniments (11)

The underlying hypercatabolism of the syndrome and its effects are reflected in the conventional indices of acid/base, blood gas, electrolyte and substrate status. Of these, changes observed in the first two can be considered diagnostic. These changes in biochemical parameters are characterized by a rapid progression, preceding and reaching extreme values ahead of the rise in core temperature. Profound, rapidly worsening metabolic acidosis--lactacidosis--with concomitant rise in PCO_2 from its buffering, is the hallmark of the syndrome. Blood and tissue pH may fall from 7.4 to 6.8 and lower in as little as ten minutes (10). The hypercapnia, readily monitored by capnography, is the earliest and most sensitive marker of initiation of the syndrome (12). Capnography, therefore, should be considered an essential monitoring device in the clinical situation in humans. Reduction in PaO_2 accompanies the acidosis. While this need not be great, the reduction in PO_2 of the mixed venous blood is, leading to a greatly increased $P(a-V)O_2$ gradient. The A-a gradient may also widen (13), all this in the face of a marked increase in O_2 uptake (9,14).

Rises recorded in all serum electrolytes, total protein, glucose and lactate indicate a shift of water into cells, a leak of Na^{+}, K^{+}, Ca^{++}, Mg^{++} and Pi out, and glycogenolysis (11,14). Of particular import to the clinical management and prognosis is the rise in serum K^+. As the syndrome progresses, increasing permeability of the sarcolemma is manifest by rapidly rising serum levels of the muscle enzymes creatine kinase (CK) and lactic dehydrogenase (LDH) (12,15). Within skeletal muscle the hypercatabolic state is marked by a rapid fall off in the high energy intermediates creatine phosphate (CP) and adenosine triphosphate (ATP) (11,12,15). Untreated the syndrome is inevitably fatal.

Animals permitted to breathe spontaneously may die in fifteen minutes. Artificially ventilated animals can survive up to forty, even sixty minutes.

Variations in the Syndrome

Variations in (1) sensitivity of the triggering mechanism; and (2) rate of progression of the established syndrome, occur in response to environmental, pharmacological and genetic factors.

Excitement from the stress of manhandling and struggling has a profound effect on both the sensitivity of the triggering and the speed of progression of the hyperthermia crisis and may of itself spontaneously initiate the syndrome (16). Such conditions are the inevitable accompaniment of the conventional barnyard screening procedure for MH susceptibility viz brief mask anaesthesia with halothane. In these circumstances the syndrome may be fired by an exposure to halothane of as short a duration as 30-40 seconds. Skin blotching becomes very marked and muscle rigor occurs almost before the onset of anaesthesia. It is often difficult to distinguish between the stiff leg of conscious induction struggling and the extended leg of muscle rigor--one merges so imperceptibly with the other. The syndrome rapidly becomes irreversible and dantrolene (17) may be needed for salvage. Highly reactive animals should be carefully observed during recovery from a halothane challenge for fatal recrudescence of the syndrome can follow return of consciousness and muscle activity. The same may indeed occur after initially successful treatment with dantrolene.

Sedation with hypnotics, tranquillizing or anxiolytic drugs (18-20) on the other hand, reduces triggering sensitivity and delays initiation of the syndrome, though the speed of progression once the syndrome is established, may be unaffected. Variations in these same aspects of the syndrome may also be induced non-specifically by other groups of drugs such as non-depolarizing relaxants (21) or those which lower blood pressure and perhaps reduce cardiac output and limb blood flow (G.G. Harrison, unpublished). In these circumstances triggering of the syndrome by halothane may be delayed by as much as 45 minutes or longer or even prevented altogether, requiring ultimately the injection of succinylcholine to fire the syndrome (vide infra: Triggering agents).

Genetic factors are the major determinants of variations in the manifested syndrome (22). Though malignant hyperthermia susceptibility was originally thought of as something of an all-or-none phenomenon (23), controlled breeding experiments have since produced phenotypes with graded responses in both the sensitivity of the triggering mechanism and thereafter in the rate and extent of progression of the established syndrome from what we may call the hair-triggered "rocket" to the "slow burner" (22,24,25). In the case of the former, the syndrome is fired by exposure to halothane alone while in the latter the addition of succinylcholine (as was the case with Hall's historic pigs (7)) is necessary. These genetically based variations in response occur in all the MH affected breeds--Landrace, Poland China and Pietrain.

PORCINE STRESS MYOPATHY--PMH, PSS AND PSEP

Following its identification it was soon established that this Porcine Malignant Hyperthermia (PMH) response to halothane and/or succinylcholine, was heritable and could be evoked only in selected strains of relatively few breeds of pigs (26). Further, clinical and biochemical features of the syndrome were identical to those of the Porcine (sudden death) Stress Syndrome (PSS) (27) long recognized in the same breeds of pig in response to stress, excitement, severe exercise and/or environmental extremes alone (28-30). The breeds, worldwide, in which this genetic trait is manifest include the Landrace, Pietrain, Poland China and crossbreeds of these strains, all breeds genetically selected for rapid growth, high feed efficiency, heavy muscling and high muscle/fat ratios. Ironically, the improved profitability of pig husbandry sought through such high grade genetic selection has often been reduced by losses incurred from the concomitant increased incidence of PSS. For example, annual losses of the order of 300 million dollars have been reported in the USA from this condition and PSEP (vide infra) (31).

Also of long time concern to the meat industry and subject of much research is the occurrence of pork with unfavourable meat quality in that it is pale, soft and watery. Studies of pork meat quality in different countries have shown its presence in as much as 20-35% of carcasses (31-33). Variously called "wasseriges fleisch", "muscle degeneration" or "pale, soft, exudative pork" (PSEP) (34) it occurs in carcasses which manifest accelerated glycolysis with concomitant myolactosis and rapid fall in pH in the immediate one to two hours post slaughter (35,36). The pale appearance, altered texture and increased waterholding quality of the muscle fibre is thought to be due to denaturation of the sarcoplasmic protein consequent on the low muscle pH while the carcass is still warm (37).

The identification of these same conditions in association with both PSS (27) and the MH syndrome (38) immediately suggests that these three separately described syndromes were simply manifestations of the same covert myopathy. This hypothesis has been amply supported by the demonstration of halothane-provoked MH syndrome in stress susceptible swine (28), the PSEP meat condition and its biochemical accompaniments post mortem in pigs dying of malignant hyperthermia (39) and finally the correlation of enzyme markers of malignant hyperthermia susceptibility with the incidence of PSEP (33,40).

The inference from this concept that susceptibility to PSS and PSEP could now be simply identified, considerably facilitated research in the field of pork quality control. By identification of susceptible animals at weanling stage, progressive pig farmers have been enabled to breed out from their herds this undesirable trait.

Screening for Malignant Hyperthermia Susceptibility

The simplest, most practical and reliable screening test for MHS remains the crude "barnyard" halothane challenge, clinical observation of the test animal's response to brief mask halothane inhalation--bearing in mind the caveat expressed earlier. Animals having a degree of susceptibility worth noting for meat quality "breeding out" or laboratory "breeding in" purposes will react with skin blotching and muscle rigor within five minutes.

Of simple blood sampling tests none have proved to be specifically diagnostic. Sadly, the variability observed in the muscle enzyme markers of this myopathy, even the most specific, CK, is such that the level recorded in an individual pig is of little diagnostic significance (33,41). For laboratory purposes collaborative diagnostic data may be sought "in vitro" from muscle biopsy. These tests (42) are based on identifying and quantifying two abnormalities manifest by MHS muscle: (1) sensitivity to caffeine induced contracture; and (2) an increased rate of ischaemically induced ATP depletion. They are time consuming and not practical for large scale screening of animals. However, Verburg's et al (12) observation of consistently low levels of creatine phosphate in MHS muscle may provide the basis for a quicker and more widely applicable screen for animals, especially if the method could be miniaturized to assay creatine phosphate in needle biopsy specimens.

Inheritance Patterns

Inheritance of this readily "in vivo" identifiable myopathic trait was thought initially to involve a simple autosomal dominant allelic gene, inconsistencies in transmission being explained as varying expressivity and incomplete penetrance (7,15,43). Subsequently, planned breeding experiments have produced evidence for the existence of two sharply differing modes of inheritance--a circumstance indeed analogous to that in humans (44). While autosomal dominant inheritance, with some modifications, is supported by most of the published evidence (including this author's Landrace breeding experience) strong evidence for a recessive mode of transmission has been reported by others.

Modifications to the concept of a single gene autosomal transmission have followed from the nicely graded clinical manifestations of the MH syndrome correlated with "in vitro" muscle caffeine and caffeine/halothane responses demonstrated by the five MHS phenotypes bred by Britt and co-workers (24) from Poland China (MHS) and Yorkshire (MH resistant) crosses. These gradations in phenotype response followed from the offspring of crosses manifesting response characteristics that were the mean of those of the individual parents and not segregating out into groups which followed the response characteristics of one or other individual parent. The identification of five phenotypes in two generations indicated that two gene alleles (at least) must be involved and the gradations observed meant that such genes must be co-dominant.

Support for this hypothesis of co-dominance in porcine MH alleles comes from Nelson's et al (25) report of similar graded phenotype reactions in first generation MHS (Poland China) and MH negative mini swine (Pittman-Moore/Hartford) crosses, similarly identifiable clinically, with corroborative graded "in vitro" halothane and caffeine/halothane contractures.

Observing in addition to graded clinical severity of the syndrome, graded elevations in basal metabolic rate in MHS phenotypes, Williams and co-workers (22) also proposed the existence of more than one co-dominant gene, but accepted the possibility of a strongly modified single dominant gene. Like Nelson et al they warn against the validity of accepting the barnyard halothane challenge alone in identifying MHS phenotypes in breeding studies and themselves only consider an animal completely MH negative when its response to succinylcholine administration as well as to halothane is negative and basal metabolic rate is not raised.

The likelihood of a multifactorial inheritance pattern is further suggested by evidence of cross-linkage of this myopathic trait with other characteristics. Rasmussen and Christian (45) and Jorgensen et al (46) report the association of MH susceptibility with only one porcine blood group, the H system. Susceptibility to MH occurs in only two of the phenotypes of this complex seven allele system while resistance or negative response is associated with three others. Andersen (47) goes further and infers from his data a linkage in Landrace swine of the MHS gene with that for phosphohexose isomerase.

While all the above is compatible with the general thesis of autosomal dominant or codominant inheritance, good evidence of the antithesis--a recessive pattern of transmission--has also been published.

Eikelenboom and co-workers (48) especially, as well as Christian (49) and Ollivier et al (50) have reported from their breeding programs with Dutch Landrace and Pietrain swine, strong evidence of recessive transmission of MH susceptibility, proposing the involvement of a single major recessive gene with complete penetrance.

An explanation for the co-existence of two distinct patterns of inheritance is not as yet forthcoming. That these differences may be associated with breed is obviously not the case--certainly in broad terms--for although studies involving the Poland China breed have all evidenced the dominant pattern of inheritance, both dominant and recessive modes have been demonstrated in Landrace, Landrace crossbreeds and Pietrains.

Though intellectually dissatisfying, it seems that for the moment we must accept that in swine, as in the human, there are two subclinical or covert stress-associated myopathic states which may manifest the malignant hyperthermic syndrome as the functional final common pathway. Considering the subsets of musculoskeletal disorder as well as the various overt myopathies and muscular dystrophies associated with the susceptibility of humans to malignant hyperthermia (51), it is perhaps not surprising that the inheritance pattern of MH susceptibility in swine has turned out to be considerably more complex than appeared at first sight.

Anaesthetic-induced syndromes clinically similar to the malignant hyperthermia syndrome of swine and humans have been described rarely in other species such as the horse (52,53), dog (54) and cat (55). In contradistinction to the porcine and human syndrome, genetic factors do not appear to play an obvious role in these biological curiosities and the repeatability of the reaction in the individual animal has not been established (invariably the affected animal has died).

Though the topic of MH in other species is discussed by Klein and Rosenberg elsewhere (see Chapter 6), I wish to allude briefly here to a common regularly appearing stress-related myopathy in other species which has some superficial resemblance to that of swine though the initiating factors are entirely environmental and genetic factors appear to play no part. This is the so-called "capture myopathy" of wild animals, in particular ungulates (described in nineteen species) and birds (56). Indeed, a similar condition has been observed also in sharks (Thurman, South African Association of Marine Biologists, personal communication). Precipitated by the alarm, stress and extreme muscular exertion of flight that precedes the darting, sedation, immobilization and capture of wild game, the syndrome presents as shock, prostration, muscle stiffness ("spastic paresis" is another name) with moderate hyperthermia proceeding to coma and death. Sharks, taken live but exhausted from entanglement in beach protection nets and released in aquaria, swim with ever increasing muscle stiffness until completely incapacitated.

The principal biochemical derangement in all these species is a profound lactacidosis which together with raised serum levels of muscle enzymes, CK, LDH, alanine amino transferase (ALT) and aspartate amino transferase (AST), point to the initiation of anaerobic metabolism in muscle and membrane damage (57). Skeletal muscle lesions very similar to those of PSEP appear post mortem. Once induced, the mortality from this syndrome, as is that from PMH, is exceptionally high--rates of up to 100% of affected animals having been reported. For this reason an understanding of the syndrome and its prevention is now of enormous importance for game conservation as attempts at capture of remaining nuclei of rare species for relocation or restocking can well be, and indeed have been, rendered abortive by this unacceptably high mortality rate. Fortunately, the steps in the reaction chain lack the extreme run-away characteristics of the porcine syndrome. Adequate sedation by a variety of tranquillizers or anxiolytics, if achieved with a minimum of alarm, stress and pursuit, prevents the condition. Harthoorn's demonstration (58) that control of the acidosis alone with sodium bicarbonate will abort the established condition, as well as being of great practical importance indicates that in all probability the pathogenesis of this condition differs fundamentally from the PSS it superficially resembles (vide infra Pathogenesis). Further understanding of this condition and its relationship, if any, to PSS must surely follow from observation of the effects of halothane and succinylcholine on stressed and unstressed animals, as well as from the effects of dantrolene on the established condition (doubtless a very expensive exercise). But all this is still for the future.

MODELS OF MALIGNANT HYPERTHERMIA

Proposed in 1968 as an animal model for the study of the MH syndrome in man (8), use of the MHS pig has stood the test of time. With few exceptions, all the features of the syndrome observed in man--clinical, biochemical and ultracellular--have been documented in susceptible swine and vice versa. Some differences there may be but, being largely quantitative rather than qualitative, they do not appear to be fundamental.

The clear association of the porcine syndrome with stress, both psychological and environmental, was once held to distinguish it from human MH. This distinction has since been rendered considerably less clear by Wingard's (59,60) identification of instances of unexpected sudden deaths in families of MH probands as exemplifying a "Human Stress Syndrome".

Another difference that has been postulated twixt MH in pig and man stems from differences recorded in levels of serum Ca^{++} observed at various stages of the syndrome (11,61). However, as levels recorded by various workers have been very variable, a real systematic difference is difficult to sustain (51).

While in the most reactive animals the trigger mechanism--being responsive to environmental as well as pharmacologic activation--may be regarded as more sensitive than in man, not all the drugs associated in man with initiation of the syndrome have been demonstrated to fire the syndrome in the MHS pig. This may well be because the association of particular drugs with the triggering of MH in man is largely anecdotal usually involving greater or lesser degrees of the polypharmacy of clinical anaesthesia, whereas in the pig model drugs can be deliberately tested singly. However, as MHS humans have been shown to develop the syndrome in response to all the drugs demonstrated to trigger MH in susceptible swine, this model does provide a useful positive screen for drugs that would be potentially harmful to the MHS human.

In using the MHS pig model, especially for drug screening, it is important that the caveats previously expressed (vide supra Variations in the syndrome) be observed if misleading, even spurious, results are to be avoided. The conclusion that certain phenothiazine and neuroleptic drugs displayed MH blocking properties (19,20) is an example of such an error. These deductions were based on animal halothane exposure times of minutes only. Experimental protocols should be so designed that drug induced attenuation or delay in onset of MH can be clearly distinguished from MH blocking or prevention. Uneventful exposure to halothane of adequate duration, sixty to ninety minutes, terminated by an MH free succinylcholine challenge is necessary before true MH blocking properties be attributed to any drug. (vide infra Triggering, Attenuation, Blocking and Reversal).

Other animal models of MH both "in vivo" and "in vitro", based primarily on two concepts of its pathogenesis, have been proposed. The first of these sought to give experimental expression to Wilson and co-workers' (62) proposal that "...this explosive thermal idiosyncracy" was due to "some disturbance of biological oxidation, possibly uncoupling of oxidative phosphorylation". Dinitrophenol pretreated dogs (63) and rats (64) were utilized to study halothane induced or exacerbated uncoupling of oxidative phosphorylation. This mechanism, and thus model, was soon discounted by the demonstration of several workers (10) that mitochondria isolated from MHS muscle, both porcine and human, display normal respiratory control and that halothane induced depression of NAD linked substrate metabolism is no different in MHS or normal muscle.

The second more relevant group of animal models is based on Kalow and co-workers' (65) early fundamental observations that muscle from survivors from MH was more sensitive "in vitro" to caffeine induced contracture than normal muscle, a sensitivity enhanced by exposure to halothane; observations that carried with them the inference that the functional lesion of MH was a caffeine-like disturbance of SR Ca^{++} flux. This was given further expression by the demonstration 'in vitro' that muscle strips from laboratory animals including frogs (66) and rats (67) pretreated with subcontracture producing concentrations of caffeine responded to halothane and indeed other MH triggering drugs in a manner similar to MHS muscle. These conditions were produced also "in vivo" in the rabbit (68). These models were proposed inter alia as suitable screens for the MH trigering or blocking potential of drugs. The caffeine-like effects of ryanodine were also used to generate models of MH both "in vivo and vitro" (69) but these were less successful.

Lastly, the myotonic/hyperpyrexic response of domestic poultry to succinylcholine which has been shown to be blocked by the same drugs that attenuate MH (70) has also been proposed as constituting a suitable MH drug screen.

However, none of these models can be said to emulate the reactions of human MH as accurately as does the MHS pig. At best they may each be said to mirror some fragment, and some fragment only, of the MH responses. As such one must question their relevance.

Anaesthesia for the MHS Pig

The conventional method of inducing anaesthesia in the pig for laboratory procedures, i.e. physical restraint followed by the injection of thiopentone into an ear vein, carries with it the risk that, in the most sensitive MHS strains, the physical manhandling and struggle necessary to achieve this may initiate MH. This may be avoided by the preliminary intramuscular injection--into the dorsal muscles of the neck--of ketamine 10 mg/kg, a procedure that requires a minimum of manhandling or skill. Within five minutes the animal will be recumbent and unresponsive, allowing easy access to ear veins. It must be allowed that we have not found ketamine to be a particularly suitable anaesthetic in the pig for general surgical procedures but for this purpose it serves well. General anaesthesia is then induced with thiopentone (up to 10 mg/kg), endotracheal intubation is performed and IPPV with nitrous oxide/oxygen (FiO_2 0.3) instituted. Thereafter anaesthesia is maintained with nitrous oxide and oxygen supplemented by intermittent thiopentone as required. ECG and intramuscular temperature monitoring are established. Arterial and venous access for arterial and venous pressure and biochemical monitoring as well as drug and fluid administration are achieved by means of jugular vein and carotid artery canulation. If the time taken for establishment of the planned experimental protocol or for achieving steady state conditions is prolonged, in order to avoid the MH attenuating effects of large doses of thiopentone (vide infra) its use is discontinued to be replaced by continuous infusion of etomidate (0.5-1.0 mg/kg/minute). At the conclusion of the experimental procedure the endotracheal tube is left in situ with the pig breathing spontaneously until it is conscious once more.

THE PHARMACOLOGICAL TRIGGERING, ATTENUATION, BLOCKING AND REVERSAL OF PMH

Anaesthetics

Triggering of anaesthetic induced MH in susceptible swine is a property displayed by potent volatile anaesthetics only. Intravenous anaesthetics not only fail to initiate the syndrome but in many instances modify, even block, its onset in response to the active volatile agents.

Although the volatile agents most active in the MHS pig include the halogenated alkanes and ethers, no correlation of their activity as triggers of MH is evident with any physical characteristic--solubilities, various partition coefficients or MAC values. Prompt initiation of the syndrome in susceptible swine follows their exposure to halothane, chloroform, enflurane (10), isoflurane (G.G. Harrison, unpublished), and sevoflurane (71). Fluoroxene and cyclopropane have been documented as weak and inconsistent triggers (10). However, not all of the conventional inhalational anaesthetics have been shown to act as triggers of MH in susceptible swine. Strangely, methoxyflurane (MOF) as well as trichloroethylene, diethyl ether and nitrous oxide have failed to initiate the syndrome in exposed animals. Each of these, particularly MOF, has been documented as triggering MH in man (72).

Triggering of MH in swine has not been ascribed to any of the conventional intravenous anaesthetic agents, including ketamine. On the contrary, one, the steroid, althesin (73), administered as a continuous infusion, blocks initiation of the syndrome in response to halothane alone, while another, thiopentone, slows and attenuates its onset (18,21). In neither circumstance is the trigger action of succinylcholine affected. Surprisingly, the water soluble analogue of althesin, minaxolone (74) has no similar attenuating effect.

Somewhat out of line with other intravenous agents is etomidate (75) which, a recent report alleges, enhances speed of onset and progression of the halothane induced syndrome though itself is not a trigger.

Muscle Relaxants

Of the relaxant drugs tested in MHS swine, it is the depolarizing drugs only--succinylcholine and decamethonium--that have been shown to trigger MH. In the most sensitive MHS strains, those that respond with MH to halothane alone, succinylcholine alone is also a potent trigger though two pulse doses may be necessary (23). In the less sensitive strains that fail to respond with MH to halothane alone, succinylcholine acts as the specific additive stressor to trigger MH though alone it, too, fails to fire the syndrome (10). In these circumstances triggering of the syndrome appears to depend on the widespread endplate depolarization with resultant induced muscle contraction, because if this is blocked by preliminary administration of a non-depolarizing relaxant the syndrome fails to develop (76).

A corollary of this action of succinylcholine would appear to be the triggering of the syndrome in sensitive strains by excitement and exercise alone, as is also the exercise-mediated reduction in response time and increase in the severity of the syndrome in response to halothane (16,28).

Of the non-depolarizing muscle relaxants, none have been documented as triggers of MH in swine with the possible exception of pancuronium. There is one report of the apparent precipitation of MH by this drug which is invalidated however by the concomitant use of halothane (77). Further it has not been confirmed by other workers (10).

Atracurium, the non-depolarizing relaxant that differs markedly from others in its mode of biodegradation, has also now been screened in MHS swine (78,79) and shown to lack any MH triggering potential.

In contrast to their blocking effects on succinylcholine-induced MH, non-depolarizing relaxants do not prevent initiation of the syndrome by the potent volatile anaesthetics. Some do attenuate or delay onset of the syndrome. In particular, attenuation of halothane-induced MH by pancuronium (21,80) has led to its being recommended as the muscle relaxant of choice in both MHS animals and man (81).

Attenuation, to the extent of prevention, has been claimed for the pre-administration of metocurine (82) though examination of the documented results validates attenuation only.

Catecholamines

Induction of MH by stress, fright and severe exercise immediately calls in question the part played by catecholamine response. The answers are still far from clear cut. Early on the Bristol group demonstrated that increased levels of circulating catecholamines accompanied induction of MH in swine (83). Thereafter they reported initiation of the syndrome in response to the infusion of alpha adrenergic agonists (84) and the converse, the prevention of MH in susceptible animals by alpha adrenergic blockade (85). Williams and co-workers (29,31) have consistently and vigorously proposed circulating norepinephrine as the key triggering agent, yet the Bristol group could induce MH in susceptible Pietrains with infusion of epinephrine only in the presence of beta adrenergic blockade (84).

At the same time there is incontrovertable evidence that the biochemical markers and muscle rigor of MH can be induced in isolated perfused muscle (86) and "in vitro" in muscle strips (87) in the absence of any catecholamine intervention. This circumstance was nicely illustrated by Gronert's et al (88) demonstration "in vivo" that total sympathetic denervation failed to block initiation of the syndrome with no accompanying rise in catecholamines and the converse that dantrolene blocked initiation of MH in response to a stressor but not the accompanying rise in the circulating catecholamines.

While it is most likely that the surge of catecholamines that accompanies initiation of MH is a secondary phenomenon which plays an amplifying role, a primary triggering role for the catecholamines is possible if their effects on skeletal muscle (89)--facilitation of prejunctional acetylcholine release, amplification of glycogenolysis energy transduction and 3/5 cyclic AMP mediated change in the rate of SR Ca^{++} uptake--are considered in the light of present concepts of pathogenesis (vide infra).

Calcium Channel Blocking Drugs

Of the many factors involved in the pathogenesis of MH, evidence of sarcolemmal dysfunction involving halothane-induced depolarization (90) and increased calcium permeability (91-93) has led to the postulate that calcium channel blocking drugs might be effective therapeutically or at least prevent or attenuate initiation of the syndrome by halothane.

Indeed, in the case of verapamil (92) and diltiazem (94) some in vitro evidence does appear to support this hypothesis. However, later studies in vivo in MHS swine failed to show therapeutic effect for any calcium channel blocking drugs. Gallant et al (95) clearly demonstrated that verapamil has no therapeutic or prophylactic effect. These observations were supported by a study in this author's laboratory (96) which showed that in addition to verapamil, other calcium channel blocking drugs--nifedipine, diltiazem, nisoldipine and flunarazine--also lacked any therapeutic effect on the established MH syndrome.

However, in the latter study, nifedipine and diltiazem were shown to manifest the property of blocking and attenuating, respectively, halothane initiation of the syndrome. While diltiazem merely delayed the time of onset of MH in response to halothane, in the case of nifedipine attenuation was of an order that prevented onset of the syndrome in response to halothane for the planned sixty minute period of exposure in 36% of animals and markedly delayed the onset in the remainder.

These effects were achieved only with doses far in excess of the human therapeutic dose.

Negative Responses of Note

Besides drugs that trigger MH in the MHS pig model, it is relevant to comment on some that unexpectedly do not. Such observations bring in question past strictures placed on the use in the MH syndrome or MHS patients of certain drugs--strictures based originally on anecdotal clinical association ("personal communication" and "unpublished" data) backed by theoretical considerations of the effects of such drugs on membrane Ca^{++} fluxes.

An important example of such negative response is that now reported to amide-linked local anaesthetic drugs. The use of this group of drugs, perhaps the most widely used local anaesthetic agents in clinical practice and in many hospitals the only local anaesthetic agents available, has long been regarded as contra-indicated in the MHS patient. Such strictures are now called in question by the demonstration by two groups of workers (97,98) that the intravenous infusion into the MHS pig of lidocaine and bupivacaine, up to monitored toxic levels failed to trigger the MH syndrome. This negative response indicates that this important group of drugs, in conventional doses, is safe to use in the MHS patient.

Digitalis, equally long claimed as deleterious in the MH syndrome and contra-indicated in the MHS patient, is another such drug. Gronert et al (99) have screened digitalis in the MHS pig model supported by cardiopulmonary bypass to obviate its cardiotoxic effects. In this model digitalis infused to the supratoxic levels of 61 ng/ml failed to stimulate any increase in whole body metabolism or by inference that of skeletal muscle. In the same preparation calcium chloride too, infused to serum levels in excess of 15 mEq/L, similarly failed to trigger the syndrome.

Blocking and Reversal

In considering specific pharmacological reversal of the established syndrome in the MHS pig--the bottom line in terms of human therapy--two drugs only, procaine and dantrolene warrant consideration. The action of each in these circumstances supports Kalow's original broad hypothesis (65) of the central role of increased myoplasmic Ca^{++} ion flux in the genesis of MH.

Procaine, the classical inhibitor "in vitro" of caffeine induced muscle contracture, was early shown "in vivo" to block initiation of MH in the MHS pig by halothane (76). For this effect large doses were necessary. Along with all the other drugs to which attention has been drawn for demonstrably attenuating onset of the syndrome in response to halothane, procaine also failed to prevent triggering of MH by the administration of succinylcholine additional to halothane though it did modify this reaction. If administered in large doses early in the progress of the established syndrome, procaine has been documented as terminating it (100), an action supported by similar "in vitro" action on halothane induced muscle contracture (87). In spite of the disappointing inconsistency of this reaction documented in the MHS pig (101,102) and the fact that large doses with grave cardiac negative inotropic effects were required, procaine did acquire a history of successful therapeutic use in MH in humans (61).

Dantrolene, a drug which reduces the rate and total amount of SR Ca^{++} release on appropriate stimulus (103) is the only drug to have demonstrated the property of blocking both halothane and halothane/succinylcholine induced MH "in vivo" in the MHS pig (17).

Moreover it is the only drug to date to reverse the established MH syndrome consistently, even if administered fairly late in its progress. Like the "turn-on" quality of the triggering mechanism of MH, dantrolene's action on the established syndrome and all its biochemical accompaniments "in vivo" has all the "turn-off" attributes of a switch (104-107).

PATHOGENESIS--THE VICIOUS CYCLE

The physical and biochemical events that characterize MH are witness to the induction of a profound hypercatabolic state in skeletal muscle which continues inexorably until total substrate depletion (10).

The fundamental lesion responsible is functional and in its simplest terms cannot be better described than it was in 1970 by Kalow et al (65), "...an inborn error in skeletal muscle which renders the muscle susceptible to disturbance in intracellular calcium distribution". In short, control of the myoplasmic flux of excitation-contraction coupling Ca^{++} is abruptly lost, resulting in the cytosolic $[Ca^{++}]$ being sustained at contracture producing levels. Drugs such as caffeine which serve to raise myoplasmic $[Ca^{++}]$ enhance this process, while those such as dantrolene and procaine which serve to lower cytosolic $[Ca^{++}]$ counteract the rigor and lead to survival.

Brief consideration of the role of calcium in muscle excitation/contraction coupling and energy transduction (108) indicates that membrane dysfunction, genetic or drug induced, at any or all three sites--the sarcolemma, the SR or the mitochondria--could cause the postulated rise in myoplasmic Ca^{++}. The extensive investigations that have been undertaken into these functional domains, together with the insight they give into the pathophysiology of malignant hyperthermia are discussed elsewhere in this volume in chapters by Britt (Chapter 2), Cheah (Chapter 3) and Nelson (Chapter 4).

However, to complete the picture here of PMH, PSS and its reultant PSEP and perhaps to distinguish it from capture myopathy, it remains necessary to summarize briefly the MH cascade to

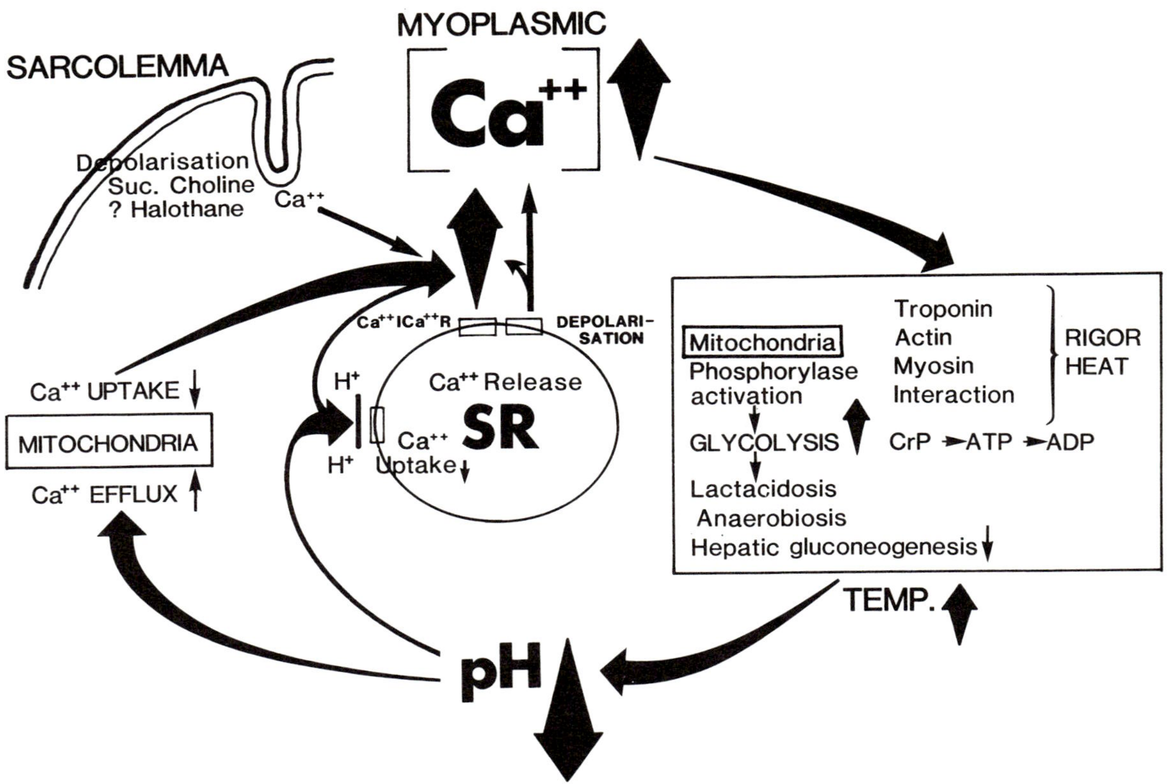

FIGURE 1. The intracellular events of the vicious cycle of MH. SR = sarcoplasmic reticulum; $Ca^{++}ICa^{++}R$ = calcium induced calcium release; H^{+} = acid inhibition of SR calcium uptake

irreversibility and, in particular, the recent work which brings us a little closer to an understanding of the trigger mechanism (Figure 1).

Though it has long been established that physiological contraction of skeletal muscle is elicited by Ca^{++} released from the SR and the molecular mechanisms of how the released Ca^{++} triggers off the contractile reaction--the troponin/myosin/actin interaction--have been fairly well worked out, the mechanism by which calcium is released from the SR is the least understood part of the entire series of events through which spread of an action potential in a muscle cell leads to contraction (109).

Following Bianchi's proposal that the coupling of T tube depolarization to SR Ca^{++} release was by means of "trigger calcium" possibly released from sarcolemmal sites, Endo (109) and Ford and Podalsky (110) independently identified a mechanism of Ca^{++} induced SR Ca^{++} release. This mechanism was demonstrated in both skeletal and cardiac muscle. As the threshold cytosolic $[Ca^{++}]$ for contraction in skeletal muscle was 6-9 x 10^{-7}M peaking to 10^{-5} M at peak twitch while the threshold of Ca^{++} induced Ca^{++} release was 10^{-4}M, they concluded that the mechanism did not play a primary role in normal physiological contraction of skeletal muscle and possibly constituted an independent calcium channel. However, the mechanism did appear to be relevant in the excitation/contraction coupling of cardiac muscle.

Consideration of the factors which affect Ca^{++} induced Ca^{++} release suggested a strong relevance for this mechanism in the genesis of MH. Of the physiological variables, Mg^{++} was paramount--Ca^{++} induced Ca^{++} release being inversely and its threshold positively correlated (111). The action of caffeine-enhancement of twitch response progressing with increasing concentration to contracture--as well as halothane-enhancement of muscle twitch--were both shown to result from an enhancement of the Ca^{++} induced Ca^{++} release phenomenon. On the other hand inhibition of Ca^{++} induced Ca^{++} release with a rise in its threshold was shown to be the basis of the classical caffeine inhibitory reactions of procaine and tetracaine.

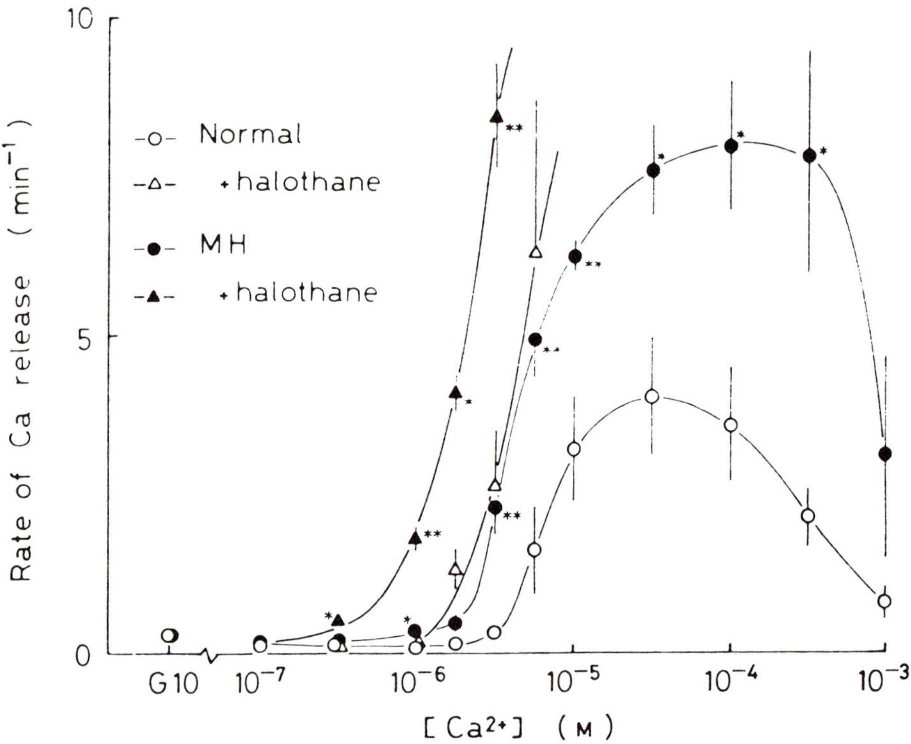

FIGURE 2. **Dependence of Ca^{++} release from SR on free [Ca^{++}] in normal (open symbols) and MH (closed symbols) fibres at 37°C.**
Without halothane: circles (n=4–11)
With 0.01% halothane (v/v) applied during Ca^{++} induced Ca^{++} release (n=22–5)
Mean and SE shown.
***P 0.05**
(From Endo et al., reference 112, with permission).

Endo's group have now reported from studies of human MH muscle (112) corroborative evidence of the involvement of Ca^{++} induced Ca^{++} release in the genesis of MH. They have demonstrated that muscle of a human MH patient displayed a lower threshold for Ca^{++} induced Ca^{++} release than normal and in addition, a greater maximum rate of Ca^{++} release at optimum $[Ca^{++}]$. Confirmatory findings from studies in MHS pig muscle are reported by Do Han Kim and co-workers (113).

The higher sensitivity of the Ca^{++} induced Ca^{++} release mechanism of MH fibres compared with normal, together with the greater maximum rate of Ca^{++} release at optimum $[Ca^{++}]$ is illustrated in the Figure from Endo's report (Figure 2). Halothane was shown to enhance this effect.

Of great interest in explaining many of the vagaries of the triggering of MH by various stressors as well as factors that provide attenuation of the response, are the results of Endo's calculation of simultaneous SR Ca^{++} uptake and release rates in relation to cytosolic $[Ca^{++}]$. These showed that even in circumstances of moderately enhanced Ca^{++} induced Ca^{++} release, such as in normal fibres under the influence of halothane or even MH fibres without halothane, Ca^{++} uptake rates were substantially greater than release rates. This ensured that even when $[Ca^{++}]$ was increased by release through channels other than those of Ca^{++} induced Ca^{++} release, eg. by depolarization, the rate of SR Ca^{++} uptake was still adequate to return myoplasmic $[Ca^{++}]$ to resting levels. Whereas in MH fibres influenced by halothane, the rate of Ca^{++} induced Ca^{++} release exceeded the rate of SR uptake throughout most of the $[Ca^{++}]$ range so that net Ca^{++} release proceeded spontaneously to levels of $[Ca^{++}]$ that produced contracture. Further, in circumstances of enhanced Ca^{++} induced Ca^{++} release it is not difficult to imagine circumstances in which the balance between rates of uptake and release are finely balanced, even unstable, so that any further increase in $[Ca^{++}]$ produced by other factors such as succinylcholine induced depolarization or even severe exercise, could provide the positive feedback loop to tip the balance to contracture (Figure 3).

The concept that the primary genetic abnormality which characterizes susceptibility to MH lies in the domain of SR Ca^{++} release mechanisms, fits well with the observed action of dantrolene viz that it depresses both the rate and total amount of SR Ca^{++} released in response to a stimulus yet has no effect on SR Ca^{++} uptake (114). The prompt termination of the established syndrome

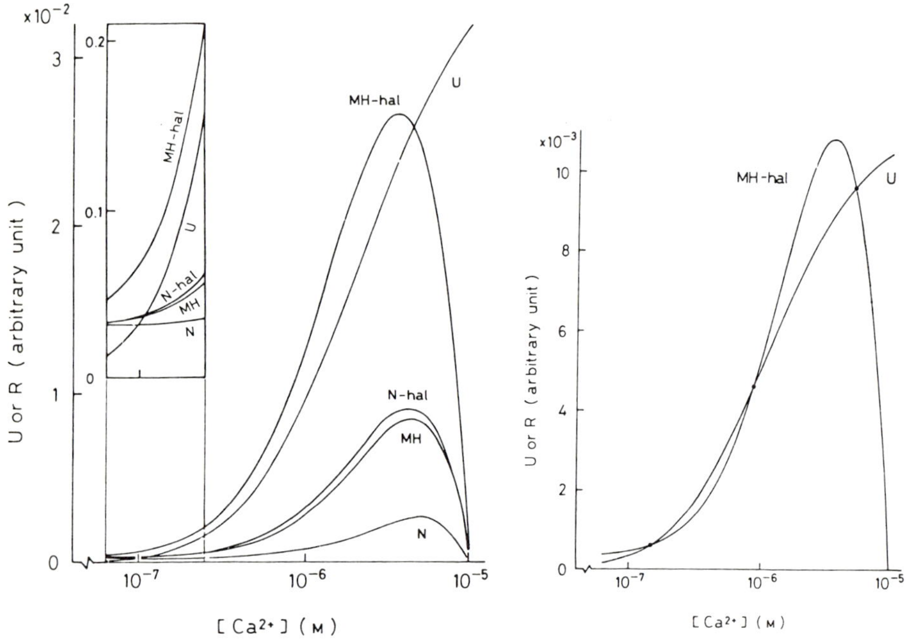

FIGURE 3. Left – dependence of rate of Ca^{++} uptake, U, and that of Ca^{++} release, R, on free $[Ca^{++}]$ in a model system. (See Ref. 109 for calculations.)
Right – demonstration of possible existence in MH muscle under certain circumstances of three crossover points between Ca^{++} uptake rate and Ca^{++} release rate which may explain succinylcholine induced aggravation of MH.
Lowest and highest points are stable. Middle point represents unstable state, from which departure of $[Ca^{++}]$ in either direction will proceed further until stable point is reached. An increased $[Ca^{++}]$ concentration will result in sustained contracture.
(From Endo et al., reference 112, with permission).

that follows administration of this drug implies continuing SR Ca^{++} release as the key factor in its genesis. Once terminated, the function of other Ca^{++} modulating mechanisms is adequate to restore myoplasmic $[Ca^{++}]$ to resting levels, permitting muscle relaxation and termination of the syndrome.

In addition to this primary mechanism other factors of note that serve to swell the cascade to irreversibility are (Figure I):

1. Acid inhibition of SR Ca^{++} transport (115), an event enhanced by halothane (116);
2. Thermal inactivation of Ca^{++}/actomyosin control mechanisms, an event enhanced by ATP depletion (117);
3. Abnormalities manifest by MHS swine:
   a) reduced mitochondrial Ca^{++} accumulating ability (118);
   b) increased anaerobically induced mitochondrial Ca^{++} efflux (119);
   c) decreased hepatic gluconeogenesis rate (120).

In these terms the variations of the MH syndrome both in the sensitivity of triggering and the rate of progression in response to genetic, environmental and pharmacological factors can be simply conceptualized as reflecting inherent variations in the threshold and magnitude of Ca^{++} induced Ca^{++} release phenomenon based in its turn possibly on an SR structural abnormality.

In slaughter animals susceptible to MH, post mortem rise in myoplasmic [Ca^{++}] due both to anaerobically depressed SR Ca^{++} uptake and increased mitochondrial efflux (greater than normal in MHS animals) exceeds the threshold of the abnormal Ca^{++} induced Ca^{++} release mechanism which then provides the milieu for the excessive and rapid glycolysis and lactacidosis essential for the production of PSEP.

Attention has been drawn to the superficial resemblance capture myopathy bears to this porcine stress myopathy. It was noted, however, that while an end stage vicious cycle was apparent, it could be interrupted by simple control of the acidosis alone. Such observation would not support the presence in this case of an abnormal Ca^{++} iduced Ca^{++} release mechanism. Acid inactivation of SR Ca^{++} transport in response to the severe lactacidosis of excessive muscular activity, together with the effects of raised circulating catecholamine levels, appears to provide an adequate explanation for the manifestations of this syndrome.

CONCLUSION

Since its original identification, elucidation of the pathogenesis of this frightening, yet fascinating, originally irreversible and fatal, hypercatabolic state that is Malignant Hyperthermia, has provided an ongoing stimulus for research by investigators from the whole field of biomedical sciences from the clinical anaesthetist to the membrane chemist, from the veterinary scientist to the geneticist.

The story recounted in this volume of the development and progression of knowledge and ideas in these various areas, from identification of muscle as the site of the lesion and abnormal SR Ca^{++} release mechanisms as a basis of the vicious cycle, to calcium flux depressant drugs as therapy, is an exciting history. Much indeed has been a gift of "pearls from swine".

REFERENCES

1. Denborough, M.A. and Lovell, R.R.H. Lancet 1960;2:45.

2. Denborough, M.A., Forster, J.F.A., Lovell, R.R.H. et al. Brit. J. Anaesth. 1962;34:395.

3. Saidman, L.J., Harvard, E.S. and Eger, E.I. J.A.M.A. 1964;190:1029.

4. Canadian Anaesthetists' Society Journal 1966;13(5).

5. Gordon, R.A. Canad. Anaesth. Soc. J. 1966;13:415.

6. Cullen, W.G. Canad. Anaesth. Soc. J. 1966;13:437.

7. Hall, L.W., Woolf, M., Bradley, J.W.P. et al. Brit. Med. J. 1966;2:1305.

8. Harrison, G.G., Biebuyck, J.F., Terblanche, J. et al. Brit. Med. J. 1968;3:594.

9. Berman, M.C., Harrison, G.G., Bull, A.B. et al. Nature 1970;225:653.

10. Harrison, G.G. In: Britt, B.A. (ed.). Malignant Hyperthermia. International Anesthesiology Clinics 1979;17:25.

11. Berman, M.C. and Kench, J.E. In: Gordon, R.A., Britt, B.A. and Kalow, W. International Symposium on Malignant Hyperthermia. 1973. Springfield: Thomas, p. 287.

12. Verburg, M.P., Oerlemans, F.T., Van Bennekom, C.A. et al. Acta Anesthesiol. Scand. 1984;28:1.

13. Britt, B.A. In: Orkin, F. and Cooperman, L.H. (eds.). Complications in Anesthesiology. 1983. Philadelphia: Lippincott, p. 291.

14. Lucke, J.N., Hall, G.M. and Lister, D. Brit. J. Anaesth. 1976;48:297.

15. Jones, E.W., Nelson, T.E., Anderson, I.L. et al. Anesthesiology 1972;36:42.

16. Van den Hende, C., Lister, D., Muylle, E. et al. Brit. J. Anaesth. 1976;48:821.

17. Harrison, G.G. Brit. J. Anaesth. 1975;47:62.

18. Hall, L.W., Trimm, M.N. and Wolfe, N. Brit. Med. J. 1972;2:145.

19. McLoughlin, J.V., Somerset, J., Ahern, C.P. et al. Acta Agric. Scand. Suppl. 1979;21:343.

20. McGrath, C.J., Rempel, W.E., Jessen, C.K. et al. Lab. Animal Sci. 1980;30:992.

21. Gronert, G.A. and Milde, J.H. Anesth. Analg. 1981;60:499.

22. Williams, C.H., Shanklin, M.D., Hedrick, H.B. et al. In: Aldrete, J.A. and Britt, B.A. (eds.). Second International Symposium on Malignant Hyperthermia. 1978. New York: Grune and Stratton, p. 113.

23. Harrison, G.G., Saunders, S.J., Biebuyck, J.F. et al. Brit. J. Anaesth. 1969;41:844.

24. Britt, B.A., Kalow, W. and Endrenyi, L. In: Aldrete, J.A. and Britt, B.A. (eds.). Second International Symposium on Malignant Hyperthermia. 1978. New York: Grune and Stratton, p. 195.

25. Nelson, T.E., Flewellen, E.H. and Gloyna, D.F. Anesth. Analg. 1983;62:545.

26. Nelson, T.E. In: Gordon, R.A., Britt, B.A. and Kalow, W. International Symposium on Malignant Hyperthermia. 1973. Springfield: Thomas, p.191.

27. Topel, D.G., Bicknell, E.J., Preston, K.S. et al. Mod. Vet. Pract. 1968;49:40.

28. Sybesma, W. and Eikelenboom, G. Neth. J. Vet. Sci. 1969;2:155.

29. Williams, C.H., Houchins, C. and Shanklin, M.D. Brit. Med. J. 1975;3:411.

30. Forest, J.C., Will, J.A., Schmidt, G.R. et al. J. Appl. Physiol. 1968;24:33.

31. Williams, C.H. In: Henschel, E.O. (ed.) Malignant Hyperthermia--Current Concepts. 1979. New York: Appleton-Century-Crofts, p. 117.

32. Naude, R.T. J. S. Afr. Vet. Ass. 1972;43:47.

33. Sybesma, W. and Eikelenboom, G. Meat Science 1978;2:79.

34. Lawrie, R.A. J. Comp. Path. 1960;70:273.

35. Lawrie, R.A., Catherun, D.P. and Hale, H.P. Nature 1958;182:807.

36. Wismer-Pedersen, J. Food Res. 1958;24:711.

37. Wismer-Pedersen, J. and Briskey, E.J. Nature 1961;189:318.

38. Nelson, T.E., Jones, E.W., Henrickson, R.L. et al. Am. J. Vet. Res. 1974;35:347.

39. Harrison, G.G. J. S. Afr. Vet. Ass. 1972;43:57.

40. Berman, M.C., Du Toit, P. and Kench, J.E. S. Afr. Med. J. 1971;45:1208.

41. Mitchell, G. and Heffron, J.J. J. S. Afr. Vet. Ass. 1975;46:145.

42. Britt, B.A. In: Britt, B.A. (ed.). Malignant Hyperthermia. International Anesthesiology Clinics 1979;17:63.

43. Allen, W.M., Brett, S., Harding, J.D.J. et al. Vet. Rec. 1970;87:410.

44. Denborough, M.A. In: Britt, B.A. (ed.). Malignant Hyperthermia. International Anesthesiology Clinics 1979;17:11.

45. Rasmussen, B.A. and Christian, L.L. Science 1976;191:947.

46. Jorgensen, P.F., Hyldgaard-Jensen, J., Eikelenboom, G. et al. As quoted by Sybesma and Eikelenboom (Ref. 33). Proc. 3rd Int. Conf. Production Disease in Farm Animals. Wageningen, The Netherlands, Pudoc, Wageningen.

47. Andreson, E. and Jensen, P. Nord. Vet. Med. 1977;29:502.

48. Eikelenboom, G., Minkema, D., Van Eldik, P. et al. In: Aldrete, J.A. and Britt, B.A. (eds.). Second International Symposium on Malignant Hyperthermia. 1978. New York: Grune and Stratton, p. 141.

49. Christian, L.L. In: Cassens, R. Geisler, F. and Kolb, . (eds.). Proc. Pork Quality Symposium. 1972. Wisconsin, Q. Madison Univ. Press, p. 91.

50. Ollivier, L., Sellier, P. and Monin, G. Ann. Genet. Sel. Anim. 1975;7:159.

51. Gronert, G.A. Anesthesiology 1980;53:395.

52. Klein, L.V. Vet. Anaesth. 1975;2:41.

53. Waldron-Mease, E. and Rosenberg, H. Vet. Sci. Comm. 1979;3:45.

54. Short, C.E. and Paddleford, R.R. Anesthesiology 1973;39:462.

55. De Jong, R.H, Heavner, J.E. and Amory, D.W. Anesthesiology 1974;41:608.

56. Haarthoorn, A.M. The chemical capture of wild animals. Baillier Tyndall. London. 1976. p. 77.

57. Haarthoorn, A.M., Louw, G.M. and Du Preez, J.J. Madoqua Series 1973;1:45.

58. Harthoorn, A.M., Van der Walt, K. and Young, E. Nature 1974;247:577.

59. Wingard, D.W. Lancet 1974;2:1450.

60. Wingard, D.W. In: Henschel, E.O. (ed.). Malignant Hyperthermia - Current Concepts. 1977. New York: Appleton-Century-Crofts, p. 79.

61. Britt, B.A. In: Cohen, P.J. (ed.). Metabolic Aspects of Anesthesia. 1975. Philadelphia: Davis, p. 61.

62. Wilson, R.D., Dent, T.E., Traber, D.L. et al. J.A.M.A. 1967;202:183.

63. Wilson, R.D., Nichols, R.J., Dent, T.E. et al. Anesthesiology 1966;27:231.

64. Gatz, E.E. In: Gordon, R.A., Britt, B.A. and Kalow, W. International Symposium on Malignant Hyperthermia. 1973. Springfield: Thomas, p. 399.

65. Kalow, W., Britt, B.A., Terreau, M.E. et al. Lancet 1970;2:895.

66. Strobel, G.E. and Bianchi, C.P. Anesthesiology 1971;35:465.

67. Harrison, G.G. S. Afr. Med. J. 1973;47:774.

68. Durban, C.G. and Rosenberg, H. J. Pharmacol. Exp. Ther. 1979;210:70.

69. Casson, H. and Downes, H. In: Aldrete, J.A. and Britt, B.A. (eds.). Second International Symposium on Malignant Hyperthermia. 1978. New York: Grune and Stratton, p. 3.

70. Korcyn, A.D. Shavit, S. and Shlosberg, I. Eur. J. Pharm. 1980;61:187.

71. Shulman, M., Braverman, B., Invankovich, A.D. et al. Anesthesiology 1981;54:259.

72. Britt, B.A. and Kalow, W. Canad. Anaesth. Soc. J. 1970;17:293.

73. Harrison, G.G. Brit. J. Anaesth. 1973;45:1019.

74. Harrison, G.G. and Morrell, D.F. Brit. J. Anaesth. 1980;52:706.

75. Suresh, M.S. and Nelson, T.E. Anesthesiology 1984;61:A253.

76. Harrison, G.G. In: Gordon, R.A., Britt, B.A. and Kalow, W. International Symposium on Malignant Hyperthermia. 1973. Springfield: Thomas, p. 271.

77. Chalstrey, L.J.L. and Edwards, G.B. Brit. J. Anaesth. 1972;44:91.

78. Lucke, J.N. Ass. Vet. Anaest. Great Britain and Ireland Proc. 1983;11:190.

79. Morrell, D.F. and Harrison, G.G. Brit. J. Anaesth. 1986; 58:444.

80. Hall, G.M., Lucke, J.N. and Lister, D. Brit. J. Anaesth. 1976;48:1135.

81. Short, C.E., Paddleford, R.R. and McGrath, C.J. Anesth. Analg. 1976;55:643.

82. Hoech, G.P., Roberts, J.T., Williams, C.H. et al. In: Cox, B. (ed.). Thermoregulatory Mechanisms and Their Therapeutic Implication. 1980. Basel: Karger, p. 137.

83. Lister, D., Hall, G.M. and Lucke, J.N. Brit. J. Anaesth. 1974;46:803.

84. Hall, G.M., Lucke, J.N. and Lister, D. Brit. J. Anaesth. 1977;49:855.

85. Lister, D., Hall, G.M. and Lucke, J.N. Brit. J. Anaesth. 1976;48:831.

86. Harrison, G.G., Berman, M.C., Hickman, R. et al. In: Shea, L. and Dwyer, B. (eds.). Proc. 3rd Asian and Australasian Congress of Anaesthesiology. 1970. Australia: Butterworth, p. 158.

87. Moulds, R.F. and Denborough, M.A. Brit. Med. J. 1972;4:526.

88. Gronert, G.A., Milde, J.H. and Theye, R.A. Anesthesiology 1977;47:411.

89. Bowman, W.C. and Nott, M.W. Pharm. Rev. 1969;21:27.

90. Gallant, E.M., Godt, R.E. and Gronert, G.A. Muscle and Nerve 1979;2:491.

91. Okumura, F., Crocker, B.D. and Denborough, M.A. Brit. J. Anaesth. 1980;52:377.

92. Greuner, R. and Blanck, T. In: Fink, B.F. (ed.). Molecular Mechanisms of Anaesthesia. 1980. New York: Raven Press, p. 423.

93. Stadhouders, A.M., Veiring, W.L., Verburg, M.P. et al. Acta Anesthesiol. Scand. 1984;28:14.

94. Iwatsuki, M., Koga, Y. and Amaha, K. Anesth. Analg. 1983;62:861.

95. Gallant, E.M., Foldes, F.F., Rempel, W.E. et al. Anesth. Analg. 1985;64:601.

96. Harrison, G.G., Wright, I.G. and Morrell, D.F. Abstract. Association of University Anaesthetists (SA Society of Anaesthetists) Congress 1986. (ed.) Potgieter, P.D., p. 35.

97. Harrison, G.G and Morrell, D.F. Brit. J. Anaesth. 1980;52:385.

98. Wingard, D.W. and Bobco, S. Anesth. Analg. 1979;58:99.

99. Gronert, G.A., Ahern, C.P., Milde, J.H. et al. Anesthesiology 1986;64:24.

100. Harrison, G.G. Brit. Med. J. 1971;3:454.

101. Gronert, G.A., Milde, J.H. and Theye R. Anesthesiology 1976;44:124.

102. Hall, G.M., Lucke, J.N. and Lister, D. Anaesthesia 1975;30:308.

103. Van Winkel, W.B. Science 1976;193:1130.

104. Gronert, G.A., Milde, J.H. and Theye, R.A. Anesthesiology 1976;44:488.

105. Kolb, M.E., Horne, M.L. and Martz, R. Anesthesiology 1982;45:254.

106. Harrison, G.G. S. Afr. Med. J. 1981;60:909.

107. Harrison, G.G. and Chapman, D.C. S. Afr. Med. J. 1982;62:503.

108. Sandow, A. Pharm. Rev. 1965;17:265.

109. Endo, M. Physiol. Rev. 1977;57:71.

110. Ford, L.E. and Podolsky, R.S. Science 1970;167:58.

111. Flewellen, E.H. and Nelson, T.E. Canad. Anaesth. Soc. J. 1980;27:363.

112. Endo, M., Yagi, S., Ishizuka, T. et al. Biomed. Res. 1983;4:83.

113. Kim, D.H., Sreter, F.A., Ohnishi, S.T. et al. Biochim. Biophys. Acta 1984;775:320.

114. Morgan, K.G. and Bryant, S.H. J. Pharmacol. Exp. Ther. 1977;201:138.

115. Berman, M.C., McIntosh, D.B. and Kench, J.E. J. Biol. Chem. 1977;252:994.

116. Diamond, E.M. and Berman, M.C Biochem. Pharm. 1980;29:375.

117. Fuchs, F. Anesthesiology 1975;42:584.

118. Britt, B.A., Endrenyi, L., Cadman, D.L. et al. Anesthesiology 1975;42:242.

119. Cheah, K.S. and Cheah, A.M. J. Sci. Food Agric. 1976;27:1137.

120. Demarco, N.W., Beitz, D.C., Young, J.W. et al. J. Nutr. 1976;106:710.

6

MALIGNANT HYPERTHERMIA IN ANIMALS OTHER THAN SWINE

L. KLEIN[1] AND H. ROSENBERG[2]

[1]University of Pennsylvania, School of Veterinary Medicine, Kennett Square, Pennsylvania 19348; [2]Department of Anesthesiology, Hahnemann University Hospital, Philadelphia, Pennsylvania, 19102

ABSTRACT

Malignant hyperthermia (MH) has been reported in several species other than man and swine. In particular, the dog and horse may provide useful information about the relation of MH to muscle abnormalities that occur with exertion. Practitioners of veterinary anaesthesia should be aware of the manifestations and treatment of MH.

INTRODUCTION

Swine have been a valuable model for the study of malignant hyperthermia. Less well-appreciated is that MH may also occur in other species. Several factors explain why more cases of non-swine animal MH have not been reported.

First, succinylcholine, a potent trigger, is rarely used. Most routine surgical procedures in companion animals are of short duration so that even when other triggering agents such as halothane are used, the syndrome may not have time to develop. Injectable agents such as thiobarbiturates and ketamine are often used for anaesthesia without supplementary inhalation agents. Some of the commonly used pre-anaesthetic and induction agents such as acetylpromazine, fentanyl-droperidol, and ultrashort-acting barbiturates may provide protection from MH even when halothane is used, especially in those individuals not highly susceptible.

Second, the lack of reports may reflect ignorance of the syndrome on the part of many general practice veterinarians who administer most anaesthetics. It is interesting to note that of nine single case reports in the dog and cat, three were written by physician anaesthesiologists and three by veterinary anaesthesiologists who were also engaged in MH research in swine.

Third, while a fulminant intra-operative episode of MH involving rigidity would be hard to ignore, body temperature is not routinely monitored during anaesthesia except possibly in horses. Slowly developing cases of MH may be recognized as post-operative hyperthermia, hyperventilation, tachycardia, mental confusion, and death. These signs may be attributed to other causes such as pyrogens in fluids, septicemia, septic or fat embolization, hypoxic brain damage, etc., by a veterinarian with a low index of suspicion of MH.

As knowledge of MH becomes more widely disseminated, it is expected that just as in the human literature, reports of MH in animals will increase.

MH IN THE CAT

In the single report of MH in the feline species (1), an episode is described which was apparently triggered by decamethonium given to a research cat after one and one-half hours of halothane-N_2O-O_2 anaesthesia. In the ten minutes following decamethonium administration, end tidal CO_2 increased from 4.5% to 9.6% and body temperature rose from 37.8°C to 41°C. Tachycardia and ST segment depression were present. Minute volume was increased but hypercarbia persisted and the cat developed a metabolic acidosis, tachycardia of 220 beats per minute and peaked T waves. One and one-half hours after decamethonium, the diagnosis of MH was considered when the cat's rectal temperature reached 44°C. Vigorous cooling measures were initiated, but halothane was not discontinued. Despite hyperventilation and bicarbonate administration, acidosis persisted and hyperkalaemia and ventricular arrhythmias developed. Muscle rigidity was not present. Finally, halothane was discontinued and procaine (10 mg/kg) was given but severe hypotension and cortical depression were present. The cat died, and within two minutes, rigor mortis developed.

MH IN THE DOG

There are eight individual reports suggestive of MH in the dog. The first report was in 1973 (2). A Pointer, anaesthetized with thiamylal and halothane for elective proctoscopy, biopsy and x-ray, developed tachycardia and tachypnoea twenty minutes after induction. The rectal temperature was found to be 42.2°C. Halothane was discontinued, the dog was placed in ice water and was treated with bicarbonate and corticosteroids. Although rectal temperature returned to normal, the dog's pupils were dilated and spontaneous respiration did not return. The dog died of cardiovascular collapse and rigor mortis developed in less than two minutes. Ante mortem rigidity was not mentioned in the report. Blood samples taken ten minutes before death showed increases in serum K^+, alanine aminotransferase (ALT,GPT), aspartate aminotransferase (AST,GOT), and lactate dehydrogenase (LDH). Creatine kinase (CK) was not measured in the dog which died, but was found to be mildly elevated in one of five littermates.

In another report, a Spaniel-type dog, which had had an uneventful halothane anaesthetic twenty months previously, developed trismus, hypotension and apnoea after sixty-five minutes of halothane-O_2 anaesthesia. The signs progressed to asystole with limb and neck muscle rigidity. Resuscitation attempts failed and the dog developed pronounced rigor mortis within five minutes of death. The rectal temperature was 43.4°C (3).

A third report describes the case of a an Irish Wolfhound anaesthetized with halothane for correction of a gastric torsion. After one and one-half hours of anaesthesia, muscle spasms of increasing intensity developed in the dog's limbs and neck, accompanied by hyperventilation. The skin was noticeably warm and rectal temperature was 40.8°C. During the next one and one-half hours, the dog was treated with surface cooling, administration of cooled IV fluids and bicarbonate, 100% O_2, lidocaine, dexamethasone and furosemide, but body temperature remained above 40°C, arrhythmias persisted, and the dog died. Serum calcium before death was 14.2 mg/dl (4).

A fourth case was that of a 52 kg young male St. Bernard dog anaesthetized with thiamylal-halothane-N_2O-O_2 for orthopaedic surgery (5). The operating room ambient temperature was 24°C and the dog's body was triple draped, including a barrier drape, for the procedure. One hour and twenty minutes after induction, the dog was treated with bicarbonate for a base excess of -11 mEq/L and was ventilated to correct a $paCO_2$ of 54 mm Hg. Three and one-half hours later, tachycardia and tachypnoea developed, and the esophageal temperature was found to be greater than 42°C. Halothane was continued for the remaining fifteen minutes of the procedure and the dog responded to surface cooling. Serum CK was only mildly elevated and the dog recovered uneventfully. A diagnosis of MH was less convincing in this case than in the previous two (indeed the authors point out its equivocal nature) in that heavy-coated members of the giant dog breeds are frequently uncomfortable with an ambient temperature of 24°C. During anaesthesia the major canine mechanism of heat elimination, panting, is lost and body heat is retained in the breathing circuit. Draping the body further limits dissipation of heat.

The other four reports suggestive of MH relate to Greyhounds. When one considers that the Greyhound represents a very small proportion of the dog population in North America, this appears to be a high incidence rate.

Two of the dogs were research animals used in cardiovascular studies. Greyhounds had been used without incident in previous studies conducted by the investigators, however, halothane and succinylcholine had not been used for anaesthesia. In the first hyperthermic case (6), tachycardia and a decrease in blood pressure occurred in a ventilated, halothane-anaesthetized dog during placement of an aortic flow probe approximately thirty minutes after a succinylcholine infusion had been started. Rectal temperature had risen to 42°C from a preoperative value of 38°C, and end tidal CO_2, which had been 4.8% prior to the succinylcholine infusion, was > 10%. Despite discontinuing halothane and succinylcholine, initiation of surface cooling, administration of bicarbonate and chilled saline by the intravenous and peritoneal routes, the temperature rose to 45°C before apparent control was achieved. Adequate oxygenation and CO_2 elimination were maintained and blood pH was normal. The animal died ninety minutes after the hyperthermia was initially discovered. Ante mortem rigidity was not mentioned in this case.

During a subsequent study (7), the same authors gave succinylcholine to another Greyhound prior to intubation and noticed intense and protracted fasciculations. Relaxation followed and anaesthesia with halothane was continued. After sixty minutes, tachycardia, increased rectal temperature, and hypercarbia developed. Halothane was discontinued and 20 mg sodium dantrolene was given, resulting in a slow decrease of temperature and pCO_2. Rigidity was not observed. The authors did not mention if the two dogs were related.

The other two reports of MH in the Greyhound breed describe the syndrome occurring postoperatively in dogs anaesthetized with non-triggering agents. In these cases, stress probably played an important role. One of the animals was a ten year old pet requiring minor surgery. (The dog was a littermate of the first Greyhound described above.) The relationship to the dog which died of MH was known to the owner and to the veterinary anaesthesiologist, and the dog was given 200 mg of dantrolene P.O. the day preceding and the day of surgery. Preoperative medication consisted of atropine and meperidine and the dog was anaesthetized with hydromorphone-diazepam-N_2O-O_2. The rectal temperature fluctuated around or below normal during the procedure and the post-anaesthetic night. Two hundred mg of dantrolene was again given orally the morning after surgery and the dog was to be discharged later that day. Soon after discharge, the dog was brought back to the hospital unable to stand, with all four legs rigidly extended, panting, and with a rectal temperature of 42.2°C. Cooling was begun and he was treated with intravenous dantrolene, bicarbonate, O_2 by mask, corticosteroids and given oxymorphone for sedation. CK was >1,500 IU (this was the upper limit of the test; dilution of the serum was not done). The dog recovered from the episode, was discharged, and remained in good health (8).

A fourth case report described a Greyhound which developed hyperthermia (41.7°C), tachycardia, and muscle tremors twenty hours after anaesthesia with fentanyl-droperidol and sodium pentobarbital. This episode was successfully treated with tranquilization, administration of acetominophen, and surface cooling with ice. Nevertheless, hyperthermia recurred twenty-four hours later. Additional cooling again resulted in a decrease in temperature. Dantrolene, 100 mg, was given orally twice a day and no further episodes occurred. CK was 196,000 IU the day after surgery and remained elevated for several days (9).

Although documentation of MH by muscle biopsy is lacking in the four Greyhound cases, the first two almost certainly represent MH since the dogs developed classical signs of the syndrome with known triggering agents. The third and fourth Greyhounds developed hyperthermia and muscle rigidity or tremors on the day following successful anaesthesia with non-triggering agents. The fact that Greyhound #3 was a littermate of Greyhound #1 suggests strongly that MH was the reason for the rigidity and hyperthermia. In Greyhound #4, it might be argued that the fever could be due to any number of causes other than MH, were it not for the very grossly elevated CK on the day after surgery. Droperidol occasionally produces delayed behavioural abnormalities in dogs, but not hyperthermia and muscle damage.

Interestingly, exertional and stress-related muscle rigidity and rhabdomyolysis have been described in racing and coursing Greyhounds (10-15). In the older literature, the condition is called "cramp", "pinched back", or "drag". Various forms of stress including high ambient temperature, racing or training at distance or speed in excess of the individual's level of fitness, and extreme nervousness before a race seem to predispose to the condition. Muscles in the hindquarters and lumbar area become rigid during or after the run with a progression in some, to paraplegia with atrophy developing later on. The animals show evidence of muscle pain and have darkly pigmented urine. Individual case reports indicate that muscle-derived enzymes are grossly elevated. It is not known whether exertional and stress-induced myopathy in the Greyhound are related to anaesthetic-induced MH as it is in swine.

Although the numbers of case reports of MH in dogs are small, the representation by the Greyhound is far in excess of the breed's prevalence (less than 3%) of the registered purebred dog population, suggesting a susceptibility of the breed. The occurrence in littermates suggests a hereditary defect.

In addition to single canine case reports of MH, a family of MH-susceptible Doberman Pinscher-German Shepherd cross dogs has been identified and documented by halothane caffeine contracture test (16). The discovery of MH susceptibility in this family occurred when one of the dogs, used for a veterinary student surgical exercise laboratory, developed hyperthermia, tachycardia, and hyperpnoea after 3.5 hours of halothane-0_2 anaesthesia. Cardiac and respiratory arrest followed and resuscitative efforts were unsuccessful. Generalized muscle rigidity, including trismus, developed immediately after death.

A litter of seven puppies from the same sire and dam were acquired and allowed to interbreed when mature. The offspring were studied for MH susceptibility. The dogs were described as having nervous temperaments and hypertrophied muscles with abnormally high tone and strength. Body temperature, hemoglobin, and AST levels were high normal, or elevated in many of the dogs. Two dogs died at maturity under circumstances suggesting a stress-triggered syndrome although the cause of death remains speculative. Necropsy was not done and they had not been halothane-challenged to determine MH susceptibility.

In the susceptible dogs, methoxyflurane did not trigger MH but halothane did. Of various non-triggering anaesthetic regimes used for muscle biopsy in the dogs, the most satisfactory drug combination was oxymorphone, acetylpromazine, thiamylal, and N_20. Results of semitendinosis biopsy indicated some dogs had a high proportion of type II fibers and fibers with greater than normal diameters and a high percentage of internal nuclei. In vitro caffeine halothane test results were compared to those from a group of unrelated control dogs. Muscle from the susceptible dogs had significantly lower caffeine specific concentrations (caffeine concentration needed to cause a one gram contracture) than controls for caffeine alone (7.6 $\pm$ 1.38 [x $\pm$ SEM] vs 15.5 $\pm$ 2.52 mM,

$p < 0.025$) and in the presence of 1% halothane (3.6 ± 1.44 vs 10.6 ± 2.19 mM, $p < 0.05$). Overlap between normals and MH-susceptible dogs was "minimal" without halothane; more marked in the presence of 1% halothane. Variation in clinical susceptibility, biopsy results, and genetic studies suggested a multifactorial inheritance pattern. In further studies with this family, attempts at mating two susceptible siblings resulted in two litters of nine and seven puppies, all of which died within five days of birth of undetermined causes. A third mating resulted in fetal death. Outcross breeding of the male produced seven puppies of which one died at birth and four of six survivors proved to be MH-susceptible. Erythrocytes from all of the puppies had abnormal responses to osmotic fragility testing at six weeks, but were normal at one year of age. Two dogs had abnormal erythrocyte hexose monophosphate enzyme activity. The four MH reactors and one of the two non-reacting puppies had abnormally high serum CK and GOT values at one year of age (16-19).

MH IN THE HORSE

Much interest in MH has been generated in the equine veterinary community. There have been at least seven case reports highly suggestive of anaesthetic-induced MH in horses. Also, a common cause of post-anaesthetic distress and even death in horses is a syndrome of isolated or generalized myopathy characterized by muscle pain and swelling with myoglobin and enzyme release. Equine athletes are at risk, as are Greyhounds, of developing exertional rhabdomyolysis, referred to by veterinarians and horsemen as "tying up". Like racing and coursing Greyhounds, equine athletes are bred for speed, strength, and lean muscle mass. Of interest too is that MH-susceptible swine are those bred for lean muscle mass.

In 1975, two reports of fulminant, fatal, intra-operative hyperthermia in horses appeared. A three year old male Thoroughbred race horse was given guaifenesin/thiopental followed by halothane-O_2 for repair of a severely comminuted first phalanx fracture. After five and one-half hours of uneventful anaesthesia, a widening of the pulse pressure was noticed and $paCO_2$ was found to have increased. The soda lime was changed and respiratory minute volume was increased but $paCO_2$ remained above normal. Tachycardia developed and palpebral and corneal reflexes became brisk. Spontaneous ventilatory efforts made mechanical control of ventilation difficult, and further increases in minute volume and halothane concentration did not correct the problem. Arterial blood gas and pH analysis showed paO_2 = 488 mm Hg, $paCO_2$ = 63 mm Hg, and a base excess of -7 mEq/L. The urine was dark brown and rectal temperature was 106°F. Fresh O_2 flow was increased, a cold water bath, and promazine and dipyrone were given, but the rectal temperature rose to 108°F. At that time, halothane was discontinued and oxygen and cooling measures were continued. The blood pressure fell over the next thirty minutes and became unobtainable. Plasma K^+ before death was 7.9 mEq/L (20).

In another report, a twelve year old Quarter Horse was described which had undergone ten hours of uneventful halothane O_2 anaesthesia for repair of a fractured radius. At the end of the procedure, the horse exhibited muscle rigidity and profuse sweating and felt warm to the touch. The rectal temperature was found to be 105°F. Halothane was discontinued, but the temperature continued to rise to greater than 110°F over the next twenty minutes. The breathing circuit was hot and five quarts of CO_2 absorbent was consumed every thirty minutes. The horse was disconnected from the breathing circuit, ventilated with 100% O_2, and was given bicarbonate. He was packed in ice and snow and given an ice water enema. Despite therapy, the horse never showed signs of recovery from anaesthesia and died one and one-half hours after the hyperthermia had been discovered (21).

A report of successful treatment of MH appeared in 1981. A four year old male Thoroughbred was anaesthetized with halothane-O_2 for repair of a metacarpal condylar fracture. The horse was difficult to manage under anaesthesia and was given 40 mg of succinylcholine on two occasions after sudden, unexpected arousal with movement occurred, not related to surgical stimulation. Good relaxation was obtained both times but after the second dose, the heart rate increased to 48 BPM from a previous normal range of 36-42 BPM. Two and three-quarter hours after induction (forty-five minutes after the second dose of succinylcholine) the colour change in the soda lime cannister indicated more rapid than expected CO_2 elimination. In another twenty minutes, the rectal temperature was 38.9°C and the heart rate was 58 BPM. Despite topical alcohol, cold water hosing, packing with ice, cold IV solutions, and bicarbonate, heart rate increased to 95 BPM, rectal temperature rose to 39.4°C and the horse was "fighting" the ventilator with deep spontaneous breaths. When the rectal temperature increased to 41°C, the rectum was packed with ice and the horse was moved to a recovery raft in a pool and was surrounded with bags of ice. Dantrolene (2.5 mg/kg) was given by nasogastric tube when the swallowing reflex returned. Blood samples taken during the episode indicated serum K^+ peaked at 9.3 mEq/L and calcium decreased to 5.8 mg/dl. CK was 17,783 IU/L at thirty minutes and rose to 424,750 IU/L at seventeen hours. Myoglobinuria persisted for twenty four hours. Muscle biopsy performed one month later indicated abnormally low threshold for contracture for caffeine, halothane, and caffeine-halothane combined (22,23).

An additional report in 1983 described MH-like reactions in three horses during halothane anaesthesia (24). In the first case a thirteen year old male Quarter Horse anaesthetized with quaifenesin-thiamylal followed by halothane developed hind limb rigidity and persistent fasciculations ten minutes after a succinylcholine infusion was begun. Thirty minutes later the horse was breathing spontaneously against the ventilator, the anaesthesia machine was hot, and rectal temperature was 38°C. Succinylcholine and, later, halothane were discontinued. The breathing circuit was changed and anaesthesia was maintained with guaifenesin-sodium thiamylal, but rectal temperature increased to 39.5°C. The horse was sweating profusely and lumbar and semi-membranous muscles were rigid. Arterial blood gas analysis indicated mild hypercarbia, good oxygenation, and no metabolic acidosis. Anaesthesia was terminated at three hours and cold water and ice packs were applied during recovery. The horse could stand with assistance for only five to ten minutes and triceps, lumbar and gluteal muscles were hard and swollen. Despite supportive treatment and oral dantrolene, the horse's condition did not improve and he was euthanized the following day.

In the second case reported by the authors, an eight month old Thoroughbred filly was anaesthetized with guaifenesin-thiopental followed by halothane for repair of a radius fracture. After three and one-half hours of anaesthesia, heart rate increased from 50-60 BPM to 70 BPM. The respiratory rate also increased and the filly appeared to be too lightly anaesthetized. Delivered halothane concentration was increased from 1.5% to 1.75% for fifteen minutes until the end of the procedure. When the filly was noticed to be sweating profusely and to have a rectal temperature of 41.2^{o}C, halothane was discontinued and 0_2 was given. Cooling measures included alcohol baths, ice around the body and per rectum and cold IV solutions. During recovery, attempts to stand resulted in a disruption of the internal fracture fixation and the filly was euthanized.

The third case was a seven month old Arabian colt anaesthetized for exploration and repair of a lacerated stifle joint. Anaesthetic induction with halothane-N_20-0_2 by face mask was associated with some struggling. After intubation, the blood pressure was high (systolic = 180 mm Hg) and breathing was irregular. Initial attempts to use positive pressure ventilation (PPV) resulted in severe hypotension but a stable anaesthetic plane with PPV was achieved after one hour. Three and one-half hours after induction, tachycardia was present, and spontaneous respiratory efforts made mechanical ventilation difficult. This was interpreted as a lightening of anaesthesia and halothane delivery was increased. One and one-half hours later, rectal temperature was 41.7^{o}C and premature ventricular contractions were present. Despite surface application of cold water and ice, IV administration of cold fluids and discontinuing halothane, spontaneous respiration rate increased to 100 breaths per minute and systolic arterial pressure rose to 240 mm Hg. Subsequently, the colt had seizures, disrupted the surgical repair, and was euthanized.

Although documentation of MH by in vitro testing was lacking in these three cases because the animals were euthanized, the clinical course is highly suggestive of the syndrome. Vigorous spontaneous ventilatory efforts suggest a hypermetabolic state, which may not be explained by a passive temperature rise. In normothermic horses and those with preoperative fever, body temperature typically decreases during anaesthesia unless the operating room is very warm, the animal is heavily draped, and a closed circuit is used. Even under these conditions, if the breathing circuit seems very warm, increasing the fresh gas flow usually resolves the problem. Control of respirations is not difficult during equine anaesthesia with potent inhalation agents unless severe abdominal distention is present.

None of the three horses in the report became hypoxaemic, two were mildly hypercarbic ($paCO_2$ = 59 mm Hg) and the third had a $paCO_2$ of 29 mm Hg with a mild metabolic acidosis (base excess = -4 mEq/L). The absence of hypoxaemia and severe metabolic acidosis may reflect the fact that the delivery gas was 100% O_2 and hypoxaemia did not occur before the syndrome was recognized. In the absence of generalized rigidity, blood flow to muscle may have been adequate to maintain O_2 delivery. Serum K^+ was elevated at 5.98-6.79 mEq/L.

The same authors reported unusual reactions in a group of ponies used for succinylcholine infusion studies (25). The ponies were anaesthetized with thiamylal followed by halothane-O_2 with PPV. Succinylcholine 0.1 mg/kg was given IV as a bolus, followed by an infusion to maintain 90-99% twitch depression. Four ponies became hyperthermic (oesophageal temperature > 40°C). Two ponies became rigid after one to two hours of succinylcholine infusion. Other abnormalities included persistent tachycardia, hypercarbia, metabolic acidosis, and hyperkalaemia. These reactions suggest a slowly developing MH syndrome, except that the reaction did not progress to death even though succinylcholine and halothane were continued for up to three hours and the hyperthermia was successfully treated with surface cooling, bicarbonate and cessation of anaesthesia. The ponies recovered with no physical signs of muscle damage and although serum CK was somewhat elevated

(1000-4800 IU/L increase over pre-anaesthetic values in four of twelve studies) this is difficult to interpret because similar elevations have been shown to occur after uneventful halothane anaesthesia in horses (26). In previous reports of succinylcholine infusion studies in horses, K^+ increase was noted but no other untoward reactions were reported (27,28).

Another case report of an unusual reaction to succinylcholine appeared in 1985 (29). A three year old female Quarter Horse weighing 545 kg was anaesthetized with guaifenesin/thiamylal, followed by halothane-O_2 for cataract extraction and was given succinylcholine to prevent nystagmus. A bolus of 0.36 mg/kg was followed with an infusion at a rate of 2.2 mg/kg/hr. Fasciculations were severe and persisted for the duration of the infusion which was ten minutes. The horse never relaxed. Tachycardia and hypertension developed and the horse breathed spontaneously despite mechanical ventilation with a respiratory minute volume of 96 L (about twice that usually required for a normothermic 550 kg horse). Succinylcholine was discontinued, but halothane was continued, and rectal temperature rose 2^oC to 39°. Blood samples taken two hours after induction (ten minutes before termination of anaesthesia) indicated $paCO_2$ was 90.6 mm Hg, paO_2 = 119.6 mm Hg and a mild metabolic acidosis was present. Immediate post-anaesthetic recovery was normal but later a painful swelling of triceps, lateral thoracic and gluteal muscles developed. At 72 hours, serum CK was elevated at 9063 IU/L and GOT was 734 IU/L. The animal's condition improved over three days.

Post-operative myopathy with rhabdomyolysis, in the absence of overt intraoperative signs of MH, also occurs in horses. The complication can occur in 1.0-3.4% of the anaesthetized population of horses over one year of age (30,31) and in one retrospective study, horses dying from post-operative myopathy constituted 24.2% of the total anaesthetic-surgical mortality (32). The syndrome was once thought to be caused by pressure-induced ischaemic damage to inadequately padded down-side muscles, but the occurrence of bilateral and generalized muscle involvement in some cases suggest other causes. In one study of case material and experimental animals, biochemical alterations during uneventful anaesthesia and recovery led the authors to suggest that a significant derangement in energy metabolism occurs during anaesthesia in horses (30). Another report indicates that dantrolene is of benefit in treatment of some cases of equine post-operative myopathy (33).

In an attempt to investigate a link between this syndrome and MH, three horses with post-operative muscle rigidity, sweating, pain, and elevated CK were tested for MH susceptibility by halothane and caffeine contracture tests. They had received an uneventful halothane anaesthetic except that in one case, a 1°C rise in temperature was noted intra-operatively and in two cases, spontaneous movement unrelated to surgical stimulation had occurred. Muscle from a horse showing classical intra-operative signs of MH (22) was also tested. Significant contractures to halothane developed in muscle from all four horses, indicating MH susceptibility (23,34).

Exertional myopathy (exertional rhabdomyolysis, "tying up") occurs in horses used for racing and other strenuous activities and the condition shares some features with that seen in Greyhounds. Various stress factors such as sudden changes in weather or training schedule, changes in the immediate environment resulting in obvious nervousness, and a high energy, high carbohydrate diet may predispose to attacks. On a given day susceptible individuals may perform successfully without incident or become sluggish and stiff on initiation of exercise. Some horses "tie-up" mildly and only occasionally while in others, repeated attacks may preclude heavy training and end the performance career.

Affected muscles, usually those of the hindquarters, become hard, swollen, and painful to palpation. Myoglobinuria may be present and serum LDH, CK and GOT are elevated (35,36). In severe cases the horse may not be able to stand. Results of muscle biopsy and halothane caffeine contracture tests in a small group of chronically affected horses indicate that this condition may be related to MH in some, but not all cases (34). In some individuals, but not others, dantrolene appears to provide some protection.

CAPTURE MYOPATHY

Another condition which has generated a lot of interest because of its possible relationship to MH is the syndrome of "capture myopathy" in wild animals. Ungulates are most often affected by the problem, but the syndrome has been seen in birds, primates and aquatic mammals as well. If a strenuous chase is involved in the capture of highly susceptible species, death losses can approach 100%. Some animals die acutely, presumably from exhaustion, hyperthermia, metabolic acidosis and hyperkalaemia while others may live for several days, exhibiting depression, muscle weakness and myoglobinuria. Elevations in CK, AST and LDH attest to widespread muscle damage. Post-mortem examination reveals multiple areas of muscle degeneration and often severe myoglobin damage to kidneys (37-46).

In addition to strenuous physical exercise during capture, psychological stress seems to be very important in triggering the syndrome and in this respect, it resembles the porcine stress syndrome. One group of researchers was able to reduce losses dramatically by separating the initial corraling from the ultimate penning and shipping. The animals were "trained" for one month during which time human beings walked through the corral periodically and the animals were driven through chutes in short, gentle, "mock" capturing routines (37-39). The relationship of this syndrome to MH at this time is only speculative. Another group of researchers has apparently reversed the temperature rise associated with capture myopathy in elk, using the serotonin receptor blocker ketanserin, an agent which has been shown to protect susceptible swine from halothane-induced MH (47,48).

REFERENCES

1. DeJong, R.H., Heavner, J.E. and Amory, D.W. Anesthesiology 1974;41:608.

2. Short, L.E. and Paddleford, R.R. Anesthesiology 1973;39:462.

3. McGrath, C.J., Crimi, A.J. and Ruff, J. Vet. Med. Small Anim. Clinician 1982;77:218.

4. Steidl, von Th., Apelt, H-J. and Kube, B. Kleintier Praxis 1983;28:189.

5. Sawyer, D.C. J. Am. Vet. Med. Assoc. 1981;179:341.

6. Bagshaw, R.J., Cox, R.H., Knight, D.H. and Detweiler, D. J. Am. Vet. Med. Assoc. 1978;172:61.

7. Bagshaw, R.J., Cox, R.H. and Rosenberg, H. J. Am. Med. Assoc. 1981;178:1029.

8. Kirmayer, A.H., Klide, A.M. and Purvance, J.E. J. Am. Vet. Med. Assoc. 1984;185:978.

9. Leary, S.L., Anderson, L.C., Manning, P.J., Bache, R.J. and Zweber, B. J. Am. Vet. Med. Assoc. 1983;182:521.

10. Bjotvedt, G., Hendricks, G.M. and Weems, C.W. Vet. Med. Small Anim. Clinician 1983;78:1215.

11. Davis, P.E. and Paris, R. J. Small Anim. Pract. 1974;15:43.

12. Gotis-Graham, S. and Lomas, G. Aust. Vet. Practitioner 1976;6:258.

13. Gowing, G.M. Southwestern Vet. J. 1964;17:183.

14. Kirk, H. Vet. Rec. 1942;54:425.

15. Sanders, N. Aust. Vet. J. 1962;38:239.

16. O'Brien, P.J., Cribb, P.H., White, R.J., Olfert, E.D. and Steiss, J.E. Canadian Vet. J. 1983;24:172.

17. O'Brien, P.J., Forsyth, G.W., Olexson, D.W., Thatte, H.S. and Addis, P.B. Canad. J. Comp. Med. 1984;48:381.

18. O'Brien, P.J. Minnesota Vet. 1985;25:43.

19. Cribb, P.H., Olfert, E.D. and Reynolds, F.B. Proc. 2nd Int. Cong. Vet. Anesthesia. October, 1985, Sacramento, California, p. 170.

20. Klein, L.V. Vet. Anesth. 1975;2:41.

21. McClure, J.J. Minnesota Vet. 1975;15:11

22. Waldron-Mease, E., Klein, L.V., Rosenberg, H. and Leitch, M. J. Amer. Vet. Med. Assoc. 1981;179:896.

23. Waldron-Mease, E. and Rosenberg, H. Vet. Science Communications 1979;3:45.

24. Manley, S.V., Kelly, A.B. and Hodgson, D. J. Amer. Vet. Med. Assoc. 1983;183:85.

25. Hildebrand, S.V. and Howitt, G.A. Amer. J. Vet. Res. 1983;44:2280.

26. Waldron-Mease, E. J. Equine Med. Surg. 1977;1:106.

27. Benson, G.J., Hartsfield, S.M., Smetzer, D.L. and Thurmon, J.C. Am. J. Vet. Res. 1979;40:1411.

28. Benson, G.J., Hartsfield, S.M., Manning, J.P. and Thurmon, J.C. Am. J. Vet. Res. 1980;41:754.

29. Riedesel, D.H. and Hildebrand, S.V. J. Amer. Vet. Med. Assoc. 1985;187:507.

30. Johnson, B.D., Heath, R.B., Bowman, B., Phillips, R.W., Rich, L.D. and Voss, J.L. J. Equine Med. Surg. 1978;2:109.

31. Klein, L.V. Proc. 24th Ann. Convention Am. Assoc. Equine Pract., December, 1978, St. Louis, Missouri, pp. 89-94.

32. Tevik, A. Nord. Vet. Med. 1983;35:175.

33. Short, C.E. and White, K.K. Proc. 24th Ann. Convention Am. Assoc. Equine Pract., December, 1978, St. Louis, Missouri, pp. 101-106.

34. Waldron-Mease, E. Proc. 24th Ann. Convention Am. Assoc. Equine Pract., December, 1978, St. Louis, Missouri, pp. 95-99.

35. Fujii, Y., Watanabe, H., Yamamoto, T., Niwa, K., Mizuoka, S. and Anezaki, R. Bull. Equine Res. Inst. 1983;20:87.

36. Lindholm, A., Johansson, H.E. and Kjaersgaard, P. Acta Vet. Scand. 1974;15:325.

37. Harthoorn, A.M. In: Nielsen, L., Haigh, J.C. and Fowler, M.E. (eds.). Chemical Immobilization of North American Wildlife. 1982. Wisconsin Humane Society, pp. 63-71.

38. Harthoorn, A.M. In: Nielsen, L., Haigh, J.C. and Fowler, M.E. (eds.). Chemical Immobilization of North American Wildlife. 1982. Wisconsin Humane Society, pp. 150-164.

39. Spraker, T.R. In: Nielsen, L., Haigh, J.C. and Fowler, M.E. (eds.). Chemical Immobilization of North American Wildlife. 1982. Wisconsin Humane Society, pp. 83-118.

40. Pertz, C. and Sundberg, J.P. J. Am. Vet. Med. Assoc. 1978;173:1243.

41. Bartsch, R.C., McConnell, E.E., Imes, G.D. and Schmidt, J.M. Vet. Pathol. 1977;14:314.

42. Chalmers, G.A. and Barrett, M.W. J. Am. Vet. Med. Assoc. 1977;171:918.

43. Colgrove, G.S. J. Am. Vet. Med. Assoc. 1978;173:1121.

44. Haigh, J.C., Stewart, R.R., Wobeser, G. and MacWilliams, P.W. JAVMA 1977;171:924.

45. Harthoorn, A.M. and Young, E. Vet. Rec. 1974;95:337.

46. Wobeser, G., Bellamy, J.E.C., Boysen, B.G., MacWilliams, P.S. and Runge, W. J. Am. Vet. Med. Assoc. 1976;169:971.

47. Stanley, T.H., Port, J.D., van der Maatin, J. and Kimball, J. Anesthesiology 1984;61:A384.

48. Ooms, L.A.A., Degryse, A., Verheyen, A., Jageneau, T. and Marsboom, R. Proc. 1st Int. Congress Vet. Anaesth. September, 1982, Cambridge, England, pp. 227-246.

7

INHERITANCE OF MALIGNANT HYPERTHERMIA—A REVIEW OF PUBLISHED DATA

W. KALOW

Department of Pharmacology, University of Toronto, Toronto, Ontario M5S 1A8

ABSTRACT

There are many diverse clinical and biochemical observations that have implications for the genetic interpretation of cases of malignant hyperthermia (MH). Familial and solitary cases deserve distinctive evaluations. Dominant inheritance of the predisposition to MH characterizes half or more of all cases; but even these may not be uniform in the sense that some biochemical characteristics or anatomical peculiarities may be absent. In vitro measured pharmacological responses of excised muscle specimens strongly suggest that MH is a disease which depends in most instances on more than one genetic factor.

This review entails some recommendations. For instance, testing should include both parents of an affected subject, not only the suspected carrier of the dominant gene, as was often done in the past. All non-familial cases with gross deformities (eg. King syndrome) should be checked for chromosomal abnormalities. The occasional association between MH and Duchenne muscular dystrophy encourages linkage studies with established genetic markers.

INTRODUCTION

About a quarter of a century has passed since Denborough and his colleagues described the entity which we now call malignant hyperthermia (MH) (1,2). The pedigree which they presented indicated convincingly that the predisposition to MH was a dominantly inherited trait (2). In the meantime, it has become clear that dominance is not always there (3,4); but to replace this negative with well-founded affirmative statements is difficult.

Nevertheless, it seems worthwhile to try to put as much order as possible into the numerous published observations which have a bearing on the genetics of MH. Several benefits could result: first, the basis for counselling of relatives of subjects who have suffered an MH attack might be improved. Risk estimates differ widely for different modes of inheritance, and there is potential harm not only in the failure to recognize MH susceptibility, but also in falsely branding somebody as susceptible. Second, genetics may serve as a tool to come closer to an understanding of those strange and disturbing phenomena which are referred to as MH; most enzyme deficiencies are inherited as co-dominant or recessive traits while dominance tends to result from a failure of a control system (5). Third, a new inventory of current information may help to put future data into better perspective.

OBSERVATIONS AND ASSUMPTIONS THAT MAY AFFECT THE GENETIC INTERPRETATION OF HUMAN MH DATA

I. Porcine MH

MH in pigs has sometimes been considered to be an autosomal dominant disorder (6,7) but probably never with strong conviction (8). Britt et al (9) have provided early evidence for a genetically multifactorial predisposition to MH among pigs. That is, breeding experiments have indicated at least five grades of susceptibility among crosses and back-crosses of susceptible and non-susceptible pigs and their offspring. At the same time, Eikelenboom et al (10) have found evidence for autosomal recessive inheritance of susceptibility to MH among Dutch Landrace swine. More recently, Courreau (11) has stated in a review that autosomal recessive inheritance of MH susceptibility is widely accepted among pig breeders. Linkage data suggest that the location of the susceptibility-conveying gene is on chromosome 15. He has further stated that this does not exclude the occurrence of more complex determinants in some populations, particularly since the prevalence of MH susceptibility varies widely among breeds.

Thus, various modes of inheritance of MH in pigs have been claimed. It could be that different modes are occurring here or there, or that there are transitional modes, eg. if one of several contributing genes has an overwhelming effect in some but not all settings.

Unfortunately, however valid the porcine data on MH inheritance may be, they are not necessarily applicable to humans. A similarly looking trait may have a different genetic background in different species. A recent example from pharmacogenetics is the occurrence of the metabolic deficiency of debrisoquine hydroxylation in man (12) and in DA rats (13). The deficiency in man is an autosomal recessive trait due to a structural variant of an enzyme, while in rats, the enzyme deficiency seems to be multifactorial and to represent a low concentration correctable by injections of testosterone.

2. Clinical Evidence for Heterogeneity of MH

MH is not a uniform entity (3), in that there are differences in severity of the acute episodes, and not all features are present in every case (14,15).

Bell et al (16) have postulated a role of certain HLA antigens in the heritability of the trait which, however, could not be shown by Lutsky et al (17). In some cases, MH is associated with Duchenne muscular dystrophy (18,19) or with osteogenesis imperfecta (20). A distinct entity is the King Syndrome (21) in which MH is associated with a progressive myopathy and multiple anomalies. Then there are families in which cardinal symptoms are missing: non-rigid MH (22) (absence of rigidity even on exposure to rigidity favoring drugs (23)), normothermic MH (24), or MH in which the constellation of attack-triggering drugs is unusual (25); perhaps these cases represent a transition to the neuroleptic syndrome (26), a syndrome which in its classical form certainly differs from typical episodes of MH. Then, there are "drug-related heatstrokes" (27). In short, the term malignant hyperthermia or MH seems to circumscribe a group of similar and poorly distinguishable syndromes, rather than a definite disease entity.

3. The Genetic Complexities of Muscle Disease as Illustrated by the Muscular Dystrophies

It may be appropriate to recall in the present context the muscular dystrophies (MD) (28). There are nine to ten clinically distinguishable disorders, all representing probably distinct pathophysiological conditions. Several of these like the Duchenne MD, are inherited as sex-linked traits, several like the fascioscapulohumeral type are dependent on autosomal dominant genes. The limb/girdle MD is transferred as an autosomal recessive trait in some families. However, the latter may be a mixed, nonuniform entity, which in other families seems to be multifactorially transferred. Thus, MH and MD are each a group of separate muscle diseases but with some transitions. The precise lesion is not known for any of these. Both groups have members which are inherited as autosomal dominants and for MH, dominant inheritance is the only form of inheritance which is unequivocally established although it may not apply to all cases.

4. Conceptual Similarities Between Genetic Models of Susceptibility to MH and to Coronary Occlusion

Many human diseases have a genetic element. There seems to be a certain parallelism between the genetic problems faced by investigators studying MH and, for instance, by those exploring myocardial infarction. Both study a serious, usually unexpected, event of yes-or-no character, while most laboratory measures of predisposition consist of a continuum of variables. In the case of MH, these may be in vitro muscle responses to caffeine, halothane, or both (Figure 1) (3,4), and in the case of myocardial infarcts, these are plasma cholesterol levels (Figure 2) (5,29). The population distributions of the bulk of both sets of data is approximately normal (30,31), i.e. Gaussian or unimodal, but in each case there is a small group of outliers, obviously not belonging to the bulk of data. These clearly represent pre-disposed, abnormal subjects. However, the identification of the rare outliers constitutes only a partial solution of a broader problem. In respect to coronary occlusions, the outliers, that is

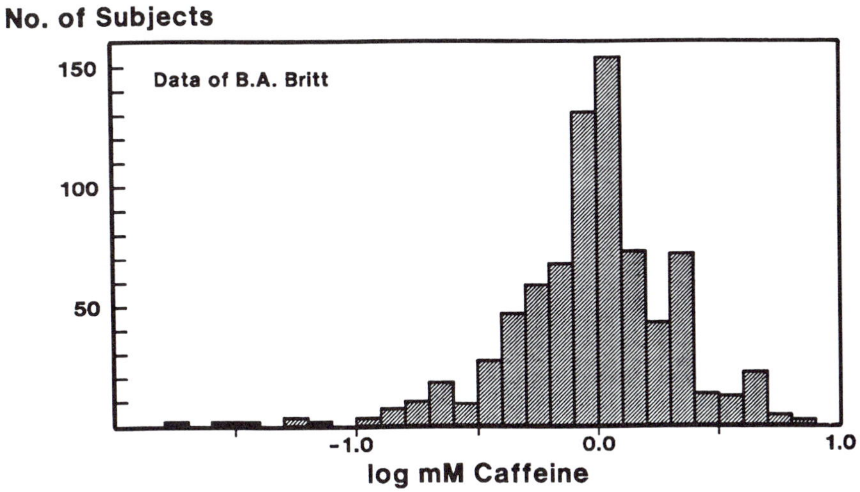

FIGURE I. Caffeine Concentration Required to Produce a Standard Contraction of Muscle Specimens In Vitro; 1% Halothane was Present in the Gas Phase.

The biopsy specimens were obtained from 780 subjects which included MH patients and relatives, and some unrelated subjects.

The histogram demonstrates that the majority of data follow a normal distribution. There is a small group of outliers to the left representing highly sensitive specimens but MH susceptibility is not confined to these.

Note that the caffeine concentration varies over a 400 fold range between the most sensitive and most resistant muscle specimens. Illustration reproduced from reference 32 with permission of the publisher. A full genetic analysis of these and additional data is in progress.

persons with extremely high cholesterol levels, tend to have a dominantly inherited receptor defect which hampers cholesterol removal and they may suffer heart attacks at an early age, often in their twenties. Besides this, the investigators of coronary occlusion have two kinds of information crucial to their further progress. First, they have established a correlation between the occurrence of heart attacks and cholesterol levels also among the normally distributed bulk of the data. Second, they have established a certain heritability of cholesterol levels also in the

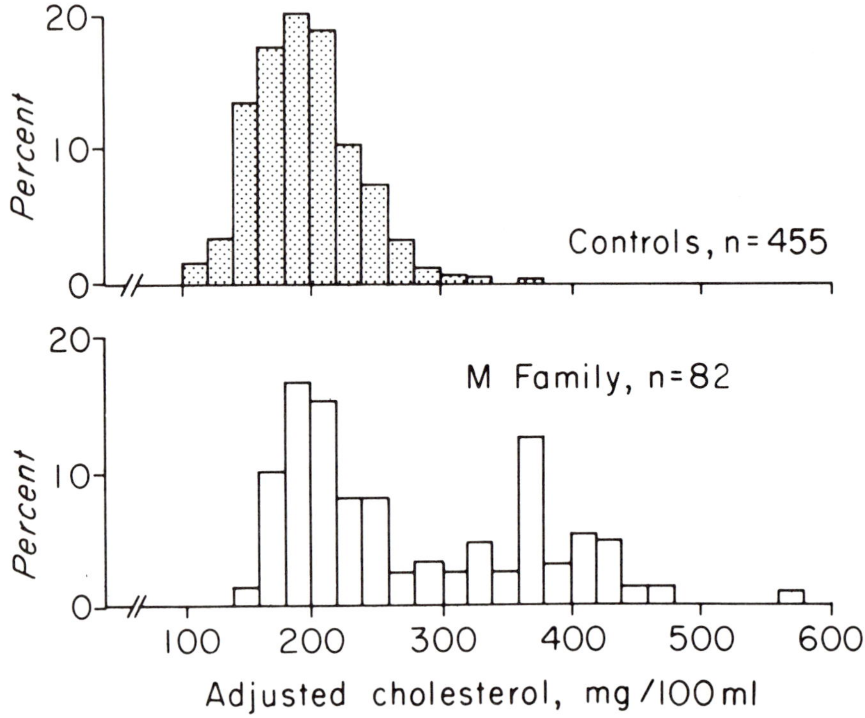

FIGURE 2. **Distribution of age- and sex-adjusted total plasma cholesterol levels**

a) Upper histogram: Control subjects.

b) Lower histogram: Members of a large family with genetic hypercholesterolemia and early heart attacks.

The controls show an essentially normal distribution with outlying data to the right, representing values as high as in the affected family.

Figures reproduced from reference 31 with permission of authors and publisher.

range of normal distribution. Persons with cholesterol levels at the upper edge of the normal distribution curve have, therefore, been subjected to special investigations (32), and many have been found to have a genetic variant of apolipoprotein E, a deviation which represents the recessively inherited basis for another substantial group of persons being predisposed to coronary occlusion.

If this is an example applicable to studies of MH, it should remind us that identical appearances of MH attacks may not have identical causes. In any case, the research strategies used by investigators of heart infarcts may very well benefit some studies of MH.

5. **The Role of Diagnostic Tests in Genetic Studies**

The purpose of this discourse is not an evaluation of the various methods proposed at one time or another to diagnose the predisposition to MH. Important in the present context are the implications of the fact that there is not a simple universal relationship between the outcome of any of the diagnostic tests and the clinical episodes of MH (3,4). This lack of an obligatory relationship may or may not reflect an inadequacy of any given test but rather might result if different tests relate to different pathophysiological causes of MH. In any case, the genetics of a test-revealed lesion and the modes of inheritance suggested by clinical observations may be divergent because of the multiplicity of factors which apparently have to convene in order to allow a MH episode to occur. These factors (33,34), always or at least very often, are: genetic predisposition; temporary predisposition based on immunological or neuro-endocrine factors of muscle mass; and an external trigger often provided by a drug. In any case, pedigrees based on the occurrence of MH episodes and those based on the determination of genetic predisposition, should not be confused.

FAMILIAL PATTERNS OF MH EPISODES

In this section, we will consider occurrences of MH crises, under-emphasizing or even disregarding the array of true and supposed indicators of MH susceptibility. The reason for dissecting the published information in this manner is the possibility (or likelihood) that MH-crises represent the common endpoint of diverse genetic and biochemical lesions. Here, we therefore ask the question how much and what kind of inheritance is recognizable if we only consider clinically established MH-crises. In doing so, we are foregoing available information on MH susceptibility and we are looking at a pattern of inheritance which must be fragmented because hereditary MH susceptibility alone is insufficient to ensure a crisis. However, we maximize freedom from any potential bias.

Considering Denborough's original pedigree (2), there were MH cases among the offspring of three of eight sibs (Generation II; the ninth sib being without offspring). These three had among them fifteen offspring (Generation III) of whom seven died of MH; two of those who died (one male, one female) had children who also died of MH. This pedigree indicates dominant inheritance because of the MH attacks in successive generations, and because of the near 50% incidence among sibs. Furthermore, two obligatory carriers of the gene had male MH offspring, indicating autosomal inheritance. For this conclusion, it does not matter that the record omits spouses and registers only one of the two parents of any affected subject.

The widely quoted pedigree of a Wisconsin family (22) also suggested an autosomal dominance of MH. This pedigree may not represent the most common variety of MH because MH in that family was characterized by absence of rigidity in spite of the use of rigidity-promoting drugs, and was characterized by unusual in vitro tests. The same pedigree was more recently explored by McPherson and Taylor (35) after the occurrence of two additional cases in the kindred. These authors extended the number of observed persons of that kin to 650. They found most MH cases in branches descended from consanguineous marriages or from a union between members of two independent sibships with apparently occurences of MH in the relatives of each. These observations speak for a recessively inherited factor causing or contributing to the occurrence of MH. The authors state "although a dominant gene for MH susceptibility may be segregating in this family, persons who receive no additional genes for MH appear to develop MH only rarely, while those receiving MH genes from both parents are at risk". They, therefore, classify this pedigree as either modified dominant or multifactorial.

There are many papers concerned with the genetics of MH (3,4,35-37) but few containing sufficient data on multi-case families which allow some genetic assessment without reliance on auxiliary test data. The large multi-case pedigree pictured by King et al (38) is the same as described originally by Denborough et al (2). Thus, the number of pertinent independent reports suitable for analysis without recourse to test data is limited. A report compiled some time ago in Toronto (33) lists multi-case families which can be subjected to a new analysis. The pertinent data (Table I) can be summarized as follows:

There were 23 families in which two or more episodes of MH occurred, comprising a total of 71 subjects. Twelve of the probands were male, 11 female. In 17 of these 23 families, a first-degree relative of the proband had had an attack. In 11 families, the first-degree relative was a parent, in 3 an offspring of the proband. The table lists altogether 19 parent-child pairs, among them 4 father-son pairs.

Hence of the 71 listed subjects, 38 were either an affected parent or child, and 10 could be recognized as having an affected sib. Other affected subjects included aunts and uncles, nephews, nieces and cousins of different degree.

The following points deserve consideration in order to attach a genetic interpretation to this collection of pedigrees. A rare event occurring on the average in three members in each of 23 affected sibships is almost certainly not only familial but has a genetic basis.

If a rare trait is determined by multifactorial inheritance, i.e. by a rare combination of a group of particular alleles (one from each of a number of independent genes) many repeats of the same constellation of alleles may be an exceptional event even within one family. There is only one outstanding Mozart or one Beethoven in their respective families. However, recurrence of a rare constellation is possible within a sibship, as there were gifted musicians in the Bach family for a century before Johann Sebastian and in the generation after him. (The modern mathematical treatment of multifactorial data with consideration of threshold effects for disease liability, of frequency and of penetrance, is discussed in reference 5, appendix 4).

TABLE 1: **Malignant Hyperthermia: Case Study**

Division of 55 Index Cases by presence or absence of MH attacks among relatives (Data from reference 34)

32 Solitary Cases

17 Familial predisposition suggested by in vitro tests

15 No evidence of familial predisposition

23 Cases with MH attacks among relatives

Cases Per Family	Relationship to Propositus of subjects with MH crisis
3	Two uncles
3	Daughter and nephew
2	Father
3	Father and his sister
3	Two first cousins
4	Son, his daughter and a grand niece
3	Father and a distant cousin
2	Father
2	Mother
2	Father
2	Third-degree cousin
4	First cousin and two distant cousins
2	Mother
4	Brother, a first cousin and a second cousin
3	Father and his sister
2	Sister
2	Great uncle
5	Son, brother, sister, father
7	Brother, father and his mother, mother and her mother and sister
3	Son and aunt
4	Three distant relatives
2	Brother
4	Mother and two brothers
71 71	**Sum Total of Familial cases**
104	**Cases**

If a trait is rare, as suggested by the infrequency of MH attacks, and if its heritable element is recessive, the likelihood of the trait occurring in parents or offspring of a proband is small--

how small is a function of the rarity of the trait. By contrast, the sibs of a proband have a good chance of being also affected, the likelihood being 25% for each sib.

If a trait is dominantly inherited, one of the parents of an affected subject must also possess the trait; whether the presence of the trait is recognizable is a different matter, but the occurrence of many parent-offspring pairs with a given trait is a strong indication of dominant inheritance. Furthermore, if the dominantly inherited trait is rare, only one parent of a proband is likely also affected.

Applying all these considerations to the cited data (33), it is clear that autosomal dominant transmission of the MH predisposition fits the data far better than anything else. If we accept this as true, the data lead to another consideration: all the 23 probands must have had one affected parent, but only 11 suffered an attack. Thus, 12 subjects must have carried the predisposing gene but--for whatever reasons--never manifested a MH episode. Some may never have had anaesthesia, some may not have been exposed to triggering agents, and some may not have suffered an attack when all conditioning elements seemed to be present. We register that a certain number of the predisposed subjects did not show an attack.

The term penetrance in genetics refers to the proportion of subjects of a particular genotype that manifest its genotypic effect in a given environment. For genes conveying MH susceptibility, true penetrance will be rarely measurable because triggering environmental conditions are variable and often absent. However, let us call "apparent penetrance" the proportion of subjects who are genetically predisposed to MH, and who actually suffer an MH attack. The apparent penetrance thus defined must be an underestimate by unknown magnitude of the true penetrance. For the data of Table 1, the apparent penetrance can be estimated as follows. The 23 familial cases can be divided into three groups. The first group consists of five cases classified as familial only because of an affected parent but without any other affected relatives. In order to reduce bias, we will omit this group from consideration. The remaining 18 cases were called familial because of an MH attack in a relative other than the parent; in 6 of these 18, a parent was also affected but not so in 12 of the 18. We can, therefore, estimate the apparent penetrance as 6/18 = 0.33.

Besides the group of pedigrees in which more than one member of any sibship had an attack, there were 32 solitary cases in different families (34). Some of these had relatives who were classified as MH susceptible on the basis of laboratory tests but this is immaterial for the present argument. If these 32 cases were of the same kind as the familial cases with apparent penetrance of 0.33, there should have been 11 affected parents with clinical manifestations of MH. There were none. This kind of difference between expectation and observation permits the assumption that many of these 32 solitary cases did not occur on the basis of the same inherited trait for MH-predisposition as did the familial cases.

In support of this assumption, Dr. Britt made available some of her data which are being prepared for publication. Her criteria for classifying a clinical event as an MH attack were particularly strict in this series. There were 63 affected families. In 15 of these, the propositus had a parent with an MH crisis but no other relative thus affected. There were 48 propositi with MH-affected relatives who were not their parents. Assuming dominant inheritance in all families, there must have been 48 MH susceptible parents but there were only 14 who had suffered a crisis. Hence, the apparent penetrance is 14/48 = 0.29, an estimate indistinguishable from that quoted above. It must be kept in mind that the apparent penetrance thus defined includes subjects who had no anaesthesia and thus had no occasion to have an MH attack. Hence the apparent penetrance of about 0.3 is an overall value, surprisingly high but perhaps useful as a rough guide to what may be expected.

The number of solitary cases in this newer collection of data was 114.

We can only assume that all solitary cases have a genetic basis. If so, many or most of them could represent multifactorial or recessive inheritance, without certain means of distinction.

Thus, the solitary cases, that is, cases without relatives who have suffered an MH crisis are of particular interest for further research.

The report by King et al (38) does not explicitly state the number of solitary cases because of the reliance upon other test data, but, according to the authors, at least 8 of 18 propositi must have been solitary cases; 5 of these showed the physical abnormalities, which in later publications were referred to as the "King syndrome", an entity with some abnormalities similar to the Turner - XX-XY phenotype (Noonan syndrome) (39). Thus some of these solitary cases with serious deformities not found in their relatives may represent fresh mutations or chromosomal abnormalities, but there is no reason to assume that all solitary cases are of such nature.

McPherson and Taylor (35) paid special attention in their literature review to sporadic cases. They reported that only about one third of MH patients tend to give a positive family history, but that 80% of families investigated after occurrence of an index case contain relatives who are classified as susceptible by other tests. They compared demographic factors, clinical manifestations and mortality between almost 200 sporadic cases and a similar number of familial cases, and found no significant differences.

FAMILIAL PATTERNS OF MH SUSCEPTIBILITY AS JUDGED BY VARIOUS CRITERIA

Many different methods have been (and are being) tried to diagnose susceptibility to MH (40). The following criteria have been among the most successful, or most widely used to assess susceptibility to MH. Plasma levels of creatine kinase have been determined in the largest number of subjects (35,41) as a non-invasive procedure and are worth consideration in the present context although the clinical use of this test for diagnosing MH susceptibility deserves to be discouraged (42). Measurements of contracture responses of biopsy specimens of skeletal muscle involve invasive procedures and have been of unquestioned diagnostic value (43-45). Indispensible has been the scrutiny of subjects for myopathies through investigations which could take the form of clinical, electrophysiological or microscopic examination of muscle (3,4,36).

I. Creatine Kinase

A full history of the use of creatine kinase ("CK", formerly "CPK") in MH studies is contained in the paper by Britt et al (41): The original discovery of the potential usefulness of CK determinations for studies of MH patients and relatives came from Denborough. This observation was quickly utilized by Isaacs and Barlow and then a number of other investigators. A final summary by Britt et al of CK data in 56 MH families (41) seems to be a good verdict:

"In only 45 percent of families did serum CK elevations prove to be of real diagnostic value. In 20 percent of families serum CK levels were normal in all members. Inconsistent levels were observed in the remaining 35 percent of families. Nevertheless, accuracy of identification of MHS individuals was considerably greater when both a high CK and muscle or musculoskeletal abnormalities were present in the same person. Such a combination was most apt to occur in males between the ages of 10 and 49". This observation of elevated CK levels in some but not all affected families has been emphasized also by King et al (38) and Kelstrup et al (46).

Another question is the mode of inheritance of CK elevations in different families. In the 45 percent of families in which CK levels were elevated in affected subjects and some relatives, the elevations were usually very high, unquestionably were different from normal, and were consistent in any individual. The inheritance pattern was clearly dominant. Any muscular or skeletal abnormalities in these families were confined to subjects with elevated CKs.

In the 35 percent of families with inconsistent CK elevations, the elevations were often modest and not always reproducible. "In some instances, both parents of the proband had elevated serum CK levels. In other cases, the CK levels in the offspring of a proband with a high serum CK and a normal individual with a normal serum

CK were intermediate between the two parental levels" (41). These statements are clearly meant to indicate that there seemed to be some families with recessive and some with multifactorial elevations of CK. Unfortunately, there are no further statements which would allow an estimation of the frequency of such inheritance patterns among the 56 families.

These impressions conveyed by Britt et al (41) are consistent with other observations. I checked published pedigrees containing pertinent information on the question of whether both parents of an MH subject, or only one parent had an elevated CK level. In cases reported in references 46-49 only one parent was affected, and only in the family reported by La Cour et al (50) were CK values elevated in both parents.

Kelstrup et al (46) stated that CK inheritance in their sibship followed a distribution which does not fit any of the known patterns of inheritance.

2. CK and Biopsy

McPherson and Taylor (35) reviewed 93 separate kindreds, taking as criteria a combination of MH crisis, high CK and biopsy data. They arrived at the following classification of these kindreds:

38% clearly autosomal dominant

14% possibly autosomal dominant

3% associated with other well-known dominant myopathies

17% no familial involvement detectable with testing. Some of these cases had multiple congenital abnormalities (The King Syndrome?)

12% siblings only affected, perhaps indicating recessive inheritance

9% kinships with two or more cases which do not share a common ancestor, suggesting either recessive or multifactorial inheritance practically excluding dominance

6% insufficient information

3. Myopathies and Halothane Response of Isolated Muscle

Ellis et al (51) used three criteria to evaluate MH survivors and their families, namely CK analysis, checking for evidence of myopathies by microscopic evaluation, and by halothane response of skeletal muscle fascicles. In agreement with other observers as quoted above, CK elevation was present in some and absent in other families. The authors found responsiveness to halothane ("H-type") and evidence of myopathies ("M-type") the most valuable criteria to assess MH susceptibility, criteria which often, but not always, coincided. Because of the results which showed divergence of these two criteria, they called the inheritance of susceptibility "multifactorial". The pedigrees published by Ellis et al (51) indicate several parent-child pairs of either "M-type" or "H-type" or both, observations suggestive of dominant inheritance. Furthermore, it seems that (except for their Family #7) only one parent of any MH subject was tested, a lack which hampers interpretations; in Family #7, the father was M and H positive, the mother negative, an observation not suggestive of multifactorial inheritance. An important conclusion from these data is that not all MH susceptible subjects can be recognized by the diagnostic halothane test in spite of the established merits of this test.

Sporn (52) summarized 30 MH cases from 24 families which had occurred in Austria up to 1980. He used essentially the classification method of Ellis et al into H-type and M-type subjects; he also liked the ATP-depletion tests while the caffeine test failed in his hands. His main conclusions were that no single test will identify all MH susceptible subjects, and that the inheritance of susceptibility is often dominant, but in some cases appears to be multifactorial. In "Family A" reported by him, the mother was "H and M positive", the father only "M positive", all four sibs "H positive", only two of them "M positive".

4. Direct Evidence for More Than One Lesion Contributing to MH

Since Kalow et al (53) in their initial work did not produce a contracture of MHS muscle by halothane alone, they used the combination of caffeine and halothane to elicit this response, and compared it with that of caffeine alone. Halothane turned out to potentiate the contracture of skeletal muscle produced by caffeine. Potentiation between drugs (a synergistic effect distinct from an additive one) is proof that their pharmacological points of attack are not identical. Furthermore, dose-effect curves for caffeine and halothane are not parallel, their maximal effects differ, and the standard deviations of population distributions of caffeine-specific-concentrations (3) in the presence and absence of halothane are significantly different (54,55). Hence, the receptors for caffeine and halothane cannot be identical; different receptors are presumably controlled by different genes.

If a given biopsy specimen of human muscle was divided to yield two preparations which could be separately tested for responsiveness to caffeine (54), the correlation between the test results was r=0.991, even if somewhat different procedures were used to produce the two independent responses of each specimen. If, on the other hand, one of the two preparations was exposed to caffeine, and the other to caffeine and halothane, the correlation coefficient was 0.65 (a value significantly different from both zero and 0.991), strongly suggesting a large measure of independence of the two types of test result. One could rationalize this independence by imagining that one effect represents the response to caffeine, the other mostly an effect of halothane. The poor correlation between the two tests indicates that not every MH muscle is uniformly sensitive to both halothane and caffeine. Hence, more than a single lesion must be able--jointly or separately--to predispose to MH. This view has received strong confirmation. Ellis (36) found a poor correlation coefficient (r=0.38) between the response to caffeine + halothane test correlated with r=0.70. Ording et al (45), reporting for Danish and Swedish investigators, cited MH cases who were responsive to halothane but not to caffeine, and vice versa; that is, most responded abnormally to both drugs, but some to halothane only, and some to caffeine only.

Kalow et al (33,55) provided evidence that both parents showed an abnormality in some families in which MH susceptibility seemed to be dominantly inherited at first sight. These authors assessed susceptibility to MH by contracture response of muscle biopsies to caffeine in the presence and in the absence of halothane. These tests were done at stabilized room temperature (21°C), and control data were established by tests on 41 muscle specimens obtained through amputations or other types of surgery (55). Particularly, the caffeine-halothane data followed very closely a normal distribution in the controls so that the standard deviation could be defined with precision. Response data two standard deviations below the mean were arbitrarily classified as abnormal.

Five nuclear families were located over a time span of a few years in which at least one child had an MH crisis, and in which both parents were available for investigation (55). In three of these families, the father's muscle had a high, grossly abnormal sensitivity to caffeine. In these same three and a fourth family, the mother's muscle showed a normal response to caffeine, but the caffeine-halothane response was abnormal, registering 3 to 4 standard deviations below the mean of the control specimens; in this fourth family, the father's response was normal to caffeine, but two standard deviations below the control value for caffeine-halothane. In short, there was in four of the five families an abnormality of both parents in at least one biopsy test (55). In a later study on a larger collection of families (33), additional observations of abnormalities in both parents of affected offspring were made, but regrettably without an attempt to define the frequency; the results of these tests represent a continuous variable, and new control for studies constructed at 37°C. Data more current, were not available. Nevertheless we felt justified to conclude that a dominantly inherited rare abnormality may predispose to MH particularly when supported by a recessive or multifactorial trait which could have a frequency of several percent in the population. Under certain circumstances, these traits could be genetically or environmentally reinforced and lead to a MH crisis in the absence of the dominantly inherited gene which is responsible for MH susceptibility in most cases. New efforts to find evidence in favour of this view seem to be justified.

5. Genetic Models

The genetic element in predisposition for more and more diseases is being studied at the molecular level; for instance, the DNA segment at the locus for Duchenne muscular dystrophy has been identified recently (56). Prior to any search at the DNA level for genes conveying MH susceptibility, investigators will have to decide what to look for. This process can be helped by conceptual models which may put order into the complex observations. In the following, some consequences of two sets of arbitrary assumptions will be presented as examples. The first four assumptions are common to both models:

Assume a normal gene h, with variant H conveying an abnormal muscle response to halothane.

Assume a normal gene c, with variant C conveying an abnormal sensitivity to the normal effects of caffeine on muscle.

Assume autosomal gene frequencies $h + H = 1$ and $c + C = 1$; that is, the two genes are independent.

Assume genotype frequencies $h^2 + 2hH + H^2 = 1$ and $c^2 + 2cC + C^2 = 1$.

The genotype frequencies n terms of the two independent genes are combined to yield 9 genotypes of frequencies: $h^2(c^2 + 2cC + C^2) + 2hH(c^2 + 2cC + C^2) + H^2(c^2 + 2cC + C^2) = 1.0$

Arbitrary choices of gene frequency then allow the construction of genetic models. These lead to predictions which can be compared with observations. The following are two such models:

Model I: Two moderately rare gene frequencies are assumed, that is, H = 0.01 and C = 0.02. The approximate genotype frequencies are

hhcc	= 94%	hHcc	= 2%	HHcc	= 1:10,000
hhcC	= 4%	hHcC	= 1:2,250	HHcC	= 1:250,00
hhCC	= 1:2,500	hHCC	= 1:100,000	HHCC	= 1:250,000,000

Assessment: In assessing this model, it has to be considered that the majority of MH patients carry a dominantly inherited predisposition. Dominant inheritance means that it is the heterozygotes who are clinically affected (with the general implication that homozygotes are not viable and may end up early as aborted foeti). This model does not fit the observations because the rare genotypes in it are all homozygotes so that the inheritance pattern would not be dominant. On the other hand, the frequency of any of the heterozygotes is too high to account for the apparent rarity of MH predisposition.

Model 2: One rare and one common gene are assumed, that is, H = 0.00005 and C = 0.2. The approximate genotype frequencies are

hhcc = 64%	hHcc = 1:15,625	HHcc = 1:625,000,000
hhcC = 32%	hHcC = 1:31,250	HHcC = 1:1,250,000,000
hhCC = 4%	hHCC = 1:250,000	HHCC = 1:10,000,000,00

Assessment: If MH susceptibility were common to all genotypes containing hH, the criteria of rarity, and dominant inheritance would be met; if the presence of the C variant would enhance the susceptibility, we would see a grading of susceptibility of the hH genotype.

There might be an MH crisis among persons with genotype hhCC under special circumstances. The inheritance of susceptibility would be recessive but the appearance might be that of a solitary case. All other genotypes containing HH homozygotes would represent a minor to negligible proportion of MH patients. Genotype HHCC may not be viable or would be too rare to occur in every generation of the world population.

Thus, a two-gene model seems to be a means to explain most but probably not all data. Additional models, or refinements of quantitation may have to be devised.

CONCLUSIONS

Perusal and evaluation of published data on heritability of MH show up certain omissions and potential faults which might be in the way of further progress. For instance, there has been no serious attempt to evaluate the heritability of the measured responsiveness of muscle biopsies to halothane and to caffeine, or to any other test for susceptibility to MH (with some exception for CK data). Heritability is a measure of the balance between genetic and environmental determinants of any entity. There are several reasons why this should be done for the test data without regard to

MH crises: first, there are few repeat biopsies of any given person which would allow an assessment of constitutional constancy of the test results. Thus, many of the normal error controls are not available but could be partly replaced by heritability measurements. Second, no one of the currently available parameters within the CHC test are able to predict all cases of MH susceptibility; this means that heritability of a feature revealed by any test and the inheritance of predisposition to MH reactions are not necessarily identical. Consequently, knowing the heritability of test data allows a new evaluation of the biological meaning of the test; this seems to be particularly important since response to halothane and to caffeine measure different properties of muscle at the molecular level. Third, knowing the heritability of MH-predictive tests may help develop sound strategies for future research, in the same way as the knowledge of heritability of plasma levels of cholesterol have helped the investigations of genetic and dietary contributions to coronary occlusion.

In many cases, when dominant inheritance of predisposition to MH was reported in the literature, testing for susceptibility had been confined to the suspected carrier of the gene, and not to both parents of an affected offspring. Careful checking of both parents is a necessity since there is reason to believe that the penetrance of the dominant gene may be enhanced by the presence of multifactorial or recessive modifying genes which could be relatively frequent in some populations or sibships.

On the positive side, the genetic classification of MH susceptibility by McPherson and Taylor (35) seems to be most useful as a guide to future listings and appears to be compatible with the view of most or all other authors.

1) The largest group is "clearly autosomal dominant" and another group "probably autosomal dominant", comprising 50% or more of the cases. A small subgroup of these cases is associated with other well known dominant myopathies, such as Duchenne Muscular Dystrophy. There may be further subdivisions among the dominant MH cases. A possible divider is CK elevation in plasma, often dominantly inherited along with susceptibility to MH. However, there also seem to be dominantly inherited MH

susceptibilities without any elevation of CK. If so, this could perhaps mean a division between dominantly inherited cases with and without integrity of the sarcolemma. Confirmatory work on this notion is advisable; this does not mean that CK determinations should be used again to diagnose susceptibility to MH.

2) In roughly 20% of cases, no familial involvement is detectable, even after testing. To this group belong patients with multiple congenital abnormalities as described by King et al. It seems worthwhile to check such patients for chromosomal anomalies. Other solitary cases could represent new mutations, but new mutants should eventually show up in their offspring. Alternatively, solitary cases could be due to multifactorial inheritance, unknown environmental sensitizers, or a confounding with other diseases like the neurolept malignant syndrome.

3) About 20% of cases suggest recessive or multifactorial inheritance. McPherson and Taylor could virtually exclude dominant inheritance of MH susceptibility in several pedigrees, and in others, in which only siblings were affected, recessiveness seemed most likely the mode of transfer.

The solitary cases or cases in which recessive inheritance seems to occur, deserve much more research attention than they have received so far. The occurrence of such cases will remain particularly frightening for both patients and anaesthetists because they are currently less predictable than those of clearly dominant inheritance. Furthermore, the susceptibility factors of such cases are bound to occur sometimes (or often ?) in conjunction with the factors which give rise to dominantly inherited MH.

4) Only a small percentage, perhaps 10 percent of cases, usually become known with too little additional information to allow any conclusions as to their genetic status.

In summary, the inheritance patterns leading to MH susceptibility may be as varied as those leading to the distinct forms of muscular dystrophy. New efforts will be worthwhile in the search for associations between clinical and genetic characteristics of MH attacks, and in defining the heritability of factors that have proven useful to ascertain susceptibility to MH.

REFERENCES

1. Denborough, M.A. Lancet 1960;II:45.

2. Denborough, M.A., Forster, J.F.A., Lovell, R.R.H., Maplestone, P.A. and Villiers, J.D. Br. J. Anaesth. 1962;34:395.

3. Britt, B.A. In: Milton, A.S. (ed.). Handbook of Experimental Pharmacology. 1982. Heidelberg: Springer, p. 547.

4. Gronert, B.A. Anesthesiology 1980;53:395.

5. Vogel, F. and Motulsky, A.G. 1982. Heidelberg: Springer-Verlag.

6. Allen, W.M., Harding, J.D.J. and Patterson, D.S.P. Vet. Rec. 1970;87:64.

7. William, C.H., Shanklin, M.D., Hedrick, H.B., Muhrer, M.E., Stubbs, D.H., Krause, G.F., Payne, C.G., Benedict, J.D., Hutcheson, D.P. and Lasley, J.F. In: Aldrete, J.A. and Britt, B.A. (eds.). Second International Symposium on Malignant Hyperthermia. 1978. New York: Grune & Stratton, p. 113.

8. Hall, G.M., Lucke, J.N. and Lister, D. Br. J. Anaesth. 1980;52:165.

9. Britt, B.A., Kalow, W. and Endrenyi, L. In: Aldrete, J.A. and Britt, B.A. (eds.). Second International Symposium on Malignant Hyperthermia. 1978. New York: Grune & Stratton, p. 195.

10. Eikelenboom, G., Minkema, D., van Eldik, P. and Sybesma, W. In: Aldrete, J.A. and Britt, B.A. (eds.). Second International Symposium on Malignant Hyperthermia. 1978. New York: Grune & Stratton, p. 141.

11. Courreau, J.F. Recueil de medecine veterinaire 1985;161:385.

12. Kalow, W. Clin. Biochem. 1986;19:76.

13. Kahn, G.C., Rubenfield, M., Davies, D.S., Murray, S. and Boobis, A.R. Drug Metab. Dispo. 1985;13:510.

14. Gronert, G.A. Semin. Anaesth. 1983;3:197.

15. Nelson, T.E., Flewellen, E.H. and Gloyna, D.F. Anesth. Analg. 1983;52:545.

16. Bell, C.L., Yocum, D., Blank, J. and Taylor, C.A. Jr. Environ. Drugs and Thermoregulation 1983;164-166.

17. Lutsky, I., Witkowski, J. and Henschel, E.O. Anesthesiology 1982;56:224A.

18. Brownell, A.K.W., Paasuke, R.T., Elash, A., Fowlow, S.B., Seagram, C.G.F., Diewold, R.J. and Friesen, C. Anesthesiology 1983;58:180.

19. Kelfer, H.M., Singer, W.D. and Reynolds, R.N. Pediatrics 1983;71:118.

20. Rampton, A.J., Kelly, D.A., Shanahan E.C. and Ingram, G.S. Br. J. Anaesth. 1984;56:1143.

21. McPherson, E.W. and Taylor, C.A. Am. J. Med. Gen. 1981;8:159.

22. Britt, B.A., Locher, W.G. and Kalow, W. Can. Anaesth. Soc. J. 1969;16:89.

23. Kalow, W. and Britt, B.A. Lancet 1973;1:391.

24. Pellicci, P.M., Paget, S. and Tsairis, P. J. Bone Joint Surg. 1983;65A:413.

25. Lee, D.S. and Zimmerman, J.E. Can. Anaesth. Soc. J. 1985;32:268.

26. Gibb, W.R.G. and Lees, A.J. Q. J. Med. 1985;56:421.

27. Clark, W.G. and Lipton, J.M. Pharmac. Ther. 1984;26:345.

28. Appel, S.H. and Roses, A.D. In: Stanbury, J.B., Wyngaarden, J.B., Fredrickson, D.S., Goldstein, J.L. and Brown, M.S. (eds.). The Metabolic Basis of Inherited Disease. 1983. New York: McGraw-Hill, p. 1370.

29. Goldstein, J.L. and Brown, M.S. In: Stanbury, J.B., Wyngaarden, J.B., Fredrickson, D.S., Goldstein, J.L. and Brown, M.S. (eds.). The Metabolic Basis of Inherited Disease. 1983. New York: McGraw-Hill, p. 672.

30. Schrott, H.G., Goldstein, J.L., Hazzard, W.R., McGoodwin, M.M. and Motulsky, A.G. Ann. Intern. Med. 1972;76:711.

31. Kalow, W. Verh Deutsch Ges Innere Med. 1983;89:1301.

32. Sing, Ch.F., Boerwinkle, E., Moll, P.P. and Davignon, J. Ann. Biol. Clin. 1985;43:411.

33. Kalow, W., Britt, B.A. and Chan, F.Y. In: Britt, B.A. (ed.). Int. Anesth. Clin. 1978;17:119.

34. Ellis, F.R. Arch. Dis. Child. 1984;59:1013.

35. McPherson, E. and Taylor, C.A. Am. J. Med. Genet. 1982;11:273.

36. Ellis, F.R. In: Ellis, F.R. (ed). Inherited Disease and Anaesthesia. 1981. Amsterdam: Excerpta Medica, p. 163.

37. Marmor, M. Surv. Ophthalmol. 1983;28:117.

38. King, J.O., Denborough, M.A. and Zapf, P.W. Lancet 1972;I:365.

39. Hunter, A. and Pinsky, L. J. Ped. 1975;86:412.

40. Aldrete, J.A. Lancet 1986;II:620.

41. Britt, B.A., Endrenyi, L., peters, P.L., Kwong, F.H-F. and Kadijevic, L. Can. Anaesth. Soc. J. 1976;23:263.

42. Amaranath, L., Lavin, T.J., Trusso, R.A. and Boutros, A.R. Br. J. Anaesth. 1983;55:531.

43. Rosenberg, H. and Reed, S. Anesth. Analg. 1983;62:415.

44. Ellis, F.R., Hallsall, P.J., Ording, H., Fletcher, R. Ranklev, E., Heffron, J.J.A., Lehan, M., Mortier, W., Steinbereitner, K., Sporn, P., Theunynck, D. and Verburg, R. Br. J. Anaesth. 1984;56:1267.

45. Ording, H., Ranklev, E. and Fletcher, F. Br. J. Anaesth. 1984;56:1183.

46. Kelstrup, J., Reske-Nielsen, E., Haase, J. and Jorni, J. Acta Anaesth. Scand. 1974;18:58.

47. Denborough, M.A., Ebeling, P., King, J.O. and Zapf, P. Lancet 1970;IV:1138.

48. Kyei-Mensah, K., Tyrrell, J.H. and Sumner, D.W. Proc. Royal Soc. Med. 1973;66:63.

49. Lewandowski, K.B., Hollmen, A. and Reunanen, M. Ann. Chir. Gynaecol. Fenn. 1975;64:112.

50. LaCour, D., Juul-Jensen, P. and Reske-Nielsen, E. Acta Anaesth. Scandinav. 1971;15:299.

51. Ellis, F.R., Cain, P.A. and Harriman, D.G.F. In: Aldrete, J.A. and Britt, B.A. (eds.). Second International Symposium on Malignant Hyperthermia. 1978. New York: Grune & Stratton, p. 329.

52. Sporn, P. Wien. klin. Wschr. 1981;93:3.

53. Kalow, W., Britt, B.A., Terreau, M.E. and Haist, C. Lancet 1970;II:895.

54. Kalow, W., Britt, B.A. and Peters, P. In: Aldrete, J.A. and Britt, B.A. (eds.). Second International Symposium on Malignant Hyperthermia. 1978. New York: Grune & Stratton, p. 339.

55. Kalow, W., Britt, B.A. and Richter, A. Can. Anaesth. Soc. J. 1977;24:678.

56. Monaco, A.P., Bertelson, C.J., Middlesworth, W., Colletti, C.A., Aldridge, J., Fischbeck, K.H., Bartlett, R., Pericak-Vance, M.A., Roses, A.D. and Kunkel, L.M. Nature 1985;316:842.

8

CLINICAL ASSESSMENT OF MALIGNANT HYPERTHERMIA PATIENTS

S.M. MULDOON AND H.M. CARLIN

Department of Anesthesiology, Uniformed Services University of the Health Sciences, Bethesda, Maryland 20814.

ABSTRACT

A continuing goal of research in malignant hyperthermia (MH) is the identification of a reliable, non-invasive method for the diagnosis of MH susceptibility (MHS). A wide variety of approaches has been proposed; unfortunately, very few of these have been appropriately evaluated. We present an overview of these proposals, with a summary of the available evidence for or against each and an assessment of their clinical utility.

HISTORY AND PHYSICAL EXAMINATION

A variety of physical findings have been reported to be associated with MH susceptibility. Most commonly observed are increased muscle bulk (particularly of extensor muscles) (Figure 1), hyperextensible joints, scoliosis or other spinal abnormalities, strabismus, and connective tissue abnormalities commonly manifested by inguinal and/or umbilical hernias (1-5). The clinical utility of such observations is limited, however. Many of the abnormalities cited occur with some frequency in the general population, and several (scoliosis, strabismus, hernia) lead to a high likelihood of surgical intervention. These factors make it difficult to base judgments of MH susceptibility on physical findings. When such findings are noted in a patient already under suspicion, they add weight to other evidence for MH susceptibility. Huckell et al. (6) reported a 28% incidence of electrocardiographic abnormalities associated with MH. These included conduction defects, Q waves or ST-T abnormalities, and increased QRS voltage. As with the

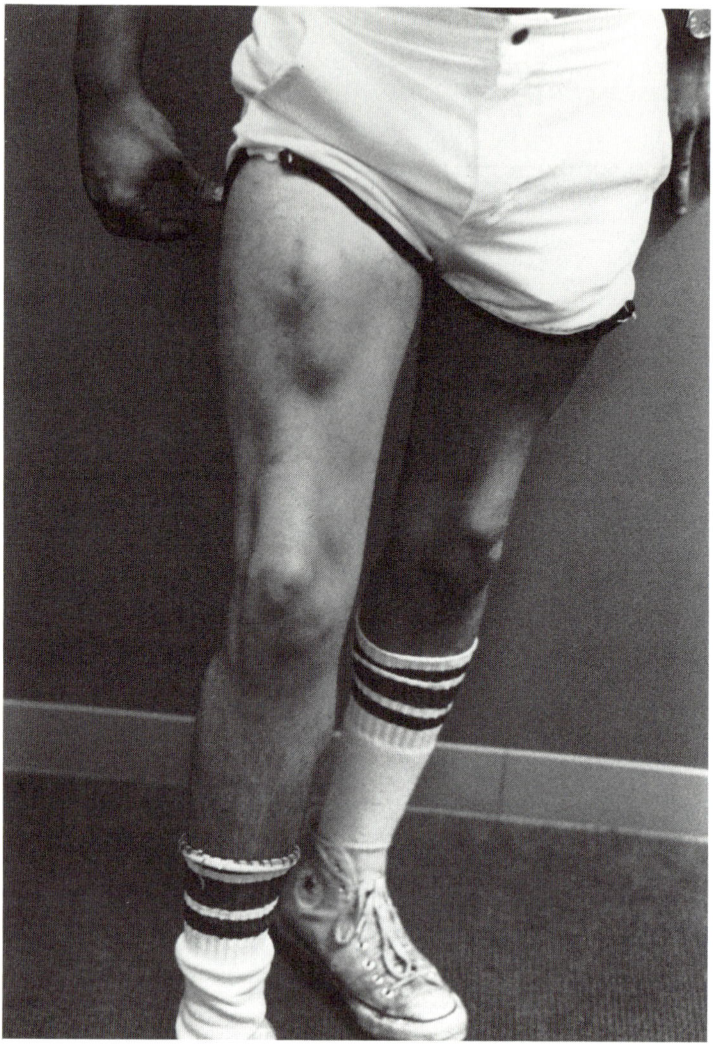

FIGURE 1. Hypertrophied extensor muscle in a MHS patient.

physical findings noted above, these are non-specific observations which occur in 5-10% of the normal population and are unlikely to be useful in identifying individuals at risk.

Patients with Duchenne's muscular dystrophy, some myotonic disorders and central core disease have a higher risk of muscle rigidity and rhabdomyolysis on exposure to volatile anaesthetics or succinylcholine (7). It is not at all clear whether this represents MH susceptibility or is simply the final common manifestation of several completely different cellular derangements.

One case of MH presenting as a heat stroke-like syndrome has been reported (8). Other anecdotal reports (9) associate heat and/or exercise intolerance with MH susceptibility. While the infrequency of such reports limits critical evaluation of the proposed link, a personal or family history of heat stroke or exhaustion, or of frequent muscle cramping with exercise may also support a previously suspected diagnosis of MH susceptibility. Campbell et al. (10) attempted to investigate the thermal response to exercise in MHS patients and in controls. They found small, inconsistent differences in the observed pattern of temperature elevation during exercise, but found no evident defect in overall thermoregulation. In a more recent study Ayling et al (11) investigated whether there were significant differences in body temperature peripheral blood flow and heat production during cooling between individuals susceptible to MH and non-affected controls. Skin temperatures decreased similarly in both groups. Core temperature increased in both groups of subjects with a significantly greater increase in the MH group.

METABOLIC STUDIES

Because acute MH episodes are characterized by gross metabolic disturbances Campbell et al. (12) performed metabolic studies on control and MH patients at rest and during mild exercise, both fasting and following a standard meal. Measurements of 0_2 consumption and of various hormones and metabolites generally associated with stress responses (insulin, glucose, serum triglyceride, lactate, pyruvate, cortisol, thyroxine and growth hormone) were performed. Resting metabolic rates and energy cost of performing a specific work load were similar in the MHS group and the control subjects. Both groups exhibited a normal thermic response to food, but the MHS group displayed a complete absence of diet-induced thermogenesis with exercise. Compared with the controls, MHS subjects had higher insulin levels for essentially the same blood glucose values. Triglycerides in the MHS group rose during the course of the experiment whereas in the controls they did not vary from the initial value. Lactate did not rise as much with exercise in the MHS group, but it did not fall to resting levels at termination of exercise. Minor differences were also found in thyroxine and cortisol levels between MHS and control groups. The authors suggest that there could be an abnormality in the activity of the sympathetic nervous system of the MH subjects but in a more recent study Ayling et al (11) reported that increases in plasma norepinephrine with increase in core temperature were greater in MH patients than in most of the controls.

SEROLOGICAL TEST

Serum Creatine Phosphokinase (CK)

In 1970 Isaacs and Barlow (13) suggested that the underlying abnormality in MH is inherited as a muscle disease and recommended the use of serum CK for its detection. The value of this simple test in identifying some MHS patients has been confirmed by many authors (3). Current opinion, however, emphasizes that elevation in resting CK is a non-specific finding, although it does raise the suspicion of MH susceptibility. False negative results occur to a variable extent--20% in one study (14), and 50% in another (15). It has been suggested that the measurement of isoenzymes of CK might more precisely define the false negative rate, but results have been inconclusive (16). False positive results also occur, and elevated CK has come to be recognized as a non-specific sign of muscle injury reported in trauma, exercise, neuromuscular disease, carrier states for muscular dystrophy, psychotics, and patients on certain drugs. CK levels also fluctuate with menstrual cycle. Recommendations that at least three CK measurements be performed before an estimate of MH susceptibility is made may help prevent incorrect diagnosis. Because of the false positive rate, it has been stated that measurement of serum CK is useful only when prior probability of MH is high. Gronert (17) has written that if a near relative has had an MH episode, then an elevated CK in itself indicates susceptibility and no further testing is required. Reasoned application of Gronert's viewpoint requires investigation of other possible causes of CK elevation before diagnosing MH susceptibility. Accordingly, it is our practice to recommend electromyography (EMG) and neurological examination before making a presumptive diagnosis of MH. If serum CKs are normal on several occasions one cannot predict susceptibility without performing skeletal muscle biopsy.

PLATELET TESTS

Platelet Nucleotide Metabolism

Because platelets and muscle cells have a similar calcium and ATP dependent contractile apparatus (18), platelet bioassays have been suggested as screening tests for MH. Solomons et al. (19) examined the synthesis of ATP, ADP and AMP in platelets following incubation of platelet rich plasma in solutions containing ^{14}C adenine. Initial results suggested that this might be a useful assay for differentiating MHS subjects from controls. However, Verburg et al. (20) have not confirmed these results. More recently, Solomons and Masson (21) used the modified energy charge ratio to examine the effects of halothane on endogenous nucleotide levels in platelets. They reported a significant difference in energy charge ratio between normal patients and those who had either documented episodes or positive muscle biopsies for MH. However, Giger and Kaplan (22) and Verburg et al. (20) were unable to duplicate these results. Solomons and Masson (23) suggest this failure is due to methodological problems in preparing the platelet rich plasma. Britt and Scott (24) in a recent comprehensive study found no significant differences in platelet-halothane nucleotide test values in MHS patients or relatives of MHS patients and normal controls, and no relationship between the nucleotide test and caffeine-halothane contracture results. These results, coupled with those of Giger and Kaplan (22) and those of Verburg et al. (20) strongly suggest that measurement of the nucleotide ratio, or the ATP ratio in platelets is not a useful test in the diagnosis or screening of MHS patients.

Aggregation Test

Platelet aggregation in response to adenosine diphosphate, epinephrine, collagen and ristocetin was examined in two MH families by Zsigmond et al. (25). The response to epinephrine and collagen in one MH survivor and two relatives was abnormal but the data presented were insufficient to permit evaluation of the test. More recently, Rosenberg et al. (26) were unable to reproduce these results.

ERYTHROCYTE TESTS

Harrison and Verburg (27) found that erythrocytes from MHS pigs showed a greater osmotic fragility than normal, but in a study of human red blood cells no difference could be found between MHS patients and their normal sibling controls (2).

On the theory that MH may be in part a generalized oxidative membrane injury Taylor et al. (28) assayed erythrocytes from MH survivors and three patients with a positive family history. The capacity of red blood cells from these patients to withstand oxidative stress was assessed by measuring levels of enzymes and substrates for the glutathione peroxidase pathway and the hexose monophosphate shunt. All patients showed deficiency in one or more of the enzymes, but the report is difficult to evaluate because of incomplete results in some patients. Further, no attempt was made to compare these results with those of standard muscle susceptibility tests.

Enzymatic properties of erythrocyte membranes in Duchenne muscular dystrophy and malignant hyperthermia were determined by Godin et al. (29). Erythrocyte membrane acetylcholinesterase (AChE) activity was markedly reduced in two individuals who were known survivors of MH crisis. No abnormalities in AChE were demonstrable in any of the Duchenne's patients. The defect in AChE was relatively selective since no abnormalities in membrane ATPase or p-nitrophenylphosphatase activity was found in these subjects. The small number of subjects tested makes it difficult to comment on the possible utility of this test.

Lymphocytes Tests

Klip et al. (30) have measured intracellular Ca^{++} in lymphocytes from malignant hyperthermia susceptible humans and pigs using the Ca^{++} chelator quin 2. They reported that halothane increased Ca^{++} in the lymphocytes from MHS pigs and humans but not from normal controls. The halothane mediated rise in Ca^{++} required external Ca^{++} and was prevented by the Ca^{++} channel inhibitor nifedipine. The effect of halothane on the releasable Ca^{++} from intracellular stores was also determined by measuring the size of the releasable pools before and after the addition of the anaesthetic. Although the number of patients studied was not large, the fraction of the releasable Ca^{++} pool available after halothane was significantly less in the MH patients than in normal patients. This study which uses newer more specific techniques that analyze the movement of intracellular Ca^{++} could represent an important new and direct approach to diagnosis of MH susceptibility.

ELECTROMYOGRAPHY (EMG)

The EMG reflects the integrity of the motor unit and can reveal pathologic states by patterns of recruitment, amplitudes, durations of motor unit action potentials and the presence or absence of spontaneous potentials--i.e. fibrillations and positive sharp waves. Britt (1) reported that electromyographic changes are seen in approximately half of all MH patients. The abnormality most frequently observed is an increased incidence of polyphasic action potentials. However, it is unclear whether these findings represent residual damage from the original MH episode or truly represent an underlying abnormality of nerve or muscle.

Electrodiagnostic Tests

Because it has been suggested that motor nerve dysfunction as well as intrinsic muscle abnormality exists in MH, Eng et al. (31) performed electrodiagnostic tests on patients known to have MH, close family members of MH patients and control subjects. Serial recordings were made of the amplitude of the evoked compound motor action potential during repetitive nerve stimulation under control conditions and after local instillations of solutions of caffeine, succinylcholine, and halothane into abductor digiti quinti muscle. The amplitudes of the evoked compound motor action potentials showed the effect of the drugs without group variation. However, EMG motor unit potentials on maximum recruitment showed a significant difference between groups upon introducing succinylcholine. Preliminary comparisons of these tests with in vitro halothane/caffeine contracture tests in nine patients showed good correlation.

Excitation-Contraction Coupling Intervals

Nelson et al. (32) have examined electromechanical coupling time intervals in muscle from MHS and control swine. They discovered a consistent delay between the muscle action potential and muscle contraction in the MH specimens. Analysis of their results shows that this delay is most likely to be caused by prolongation of the excitation-contraction coupling step. The method could be applied to humans, and offers another potential field for investigation.

Motor Unit Counting

This is another neurophysiological test which has been used to determine whether MH involves a neuropathy and myopathy. Britt et al. (33) found that most but not all MH patients demonstrate a decreased number of motor unit counts in one or more of the motor nerves supplying four muscle groups. It was also reported that half of the patients also exhibited an increase in the size of the motor unit counts. The authors point out that a deficiency of motor neuron unit counts is not specific for MH, and abnormally low counts are observed in any denervating process as well as other myopathic conditions. The results of the motor unit counting test were considered to be more accurate than concentric needle electromyography but less accurate than the caffeine halothane contracture test.

Tourniquet-Twitch Test

This test measures the ratio of electrically induced thumb twitches after ten minutes of ischaemia with those noted immediately prior to ischaemia. A "normal" tourniquet-twitch test is characterized by a gradual increase in twitch amplitude during the period of ischaemia, a decrease in amplitude to below ischaemia values after release of the cuff, and, finally, a gradual increase to the original twitch height. A positive response is characterized by a gradual increase in twitch amplitude during the period of ischaemia, a transient decrease in twitch amplitude after release of the cuff, followed by an increase in the twitch amplitude (for 3-4 minutes) to a height greater than that seen during ischaemia. Finally, a decrease in twitch amplitude to the original twitch height occurs. Britt and colleagues (34) compared this test with standard halothane and caffeine tests. They found no significant differences between normal subjects and those who have had MH reactions or relatives of such individuals. They also found no significant differences in tourniquet twitch tests results between those with normal caffeine-halothane contractures and persons with the test positive for MH. They, therefore, concluded that the tourniquet twitch test was not useful as a diagnostic or screening test for MH.

CONCLUSION

It is evident that a wide variety of physical, biochemical and electrophysiological abnormalities occur with somewhat greater incidence in MHS patients. These observations or investigations have not yet reached the stage at which a clinically useful screening test might be identified. A major deficiency to date has been the lack of well-designed, blinded studies comparing these measurements in randomly selected normal patients against those in unquestionably MH positive patients; i.e.--survivors of unequivocal MH crises. Until this type of investigation is carried out, no meaningful estimation of a test's sensitivity or specificity can be made. Without such estimates, one cannot intelligently counsel a patient concerning his or her risk of MH susceptibility. This project is one that seems ideally suited for multi-institution collaborative work.

Identification of a reasonably sensitive and specific marker for the disorder would also be of significant value in determining the natural history of the disease. With such a marker, the required longitudinal studies could be performed.

Advances have been made in identifying fertile areas for further research into non-invasive methods of detecting MH susceptibility. The time has come to move beyond descriptive reports and to begin critical evaluation of their contents.

ACKNOWLEDGEMENTS

The authors wish to acknowledge the editorial assistance of Dr. Gloria Eng and the secretarial assistance of Mary Ann McAuliffe and Patricia Hettenhouser. Based in part on work supported by Uniformed Services University of the Health Sciences grant #R08009.

REFERENCES

1. Britt, B.A. In: Britt, B.A. (ed). Malignant Hyperthermia. Boston: Little, Brown and Company, 1979, p. 63.

2. Ellis, F.R. In: Ellis, F.R. (ed). Inherited Disease and Anaesthesia. New York: Elsevier/North-Holland Inc., 1981, p. 163.

3. Gronert, G.A. Anesthesiology 1980;53:395.

4. Rosenberg, H. Hospital Practice, March 15, 1985, pp. 139-152.

5. Nelson, T.E. and Flewellen, E.H. New Engl. J. Med. 1983;309:416.

6. Huckell, V.F., Staniloff, H.M., Britt, B.A. and Morch, J.E. J. Electrocardiology 1982;15:137.

7. Takagi, A., Sunohara, N., Ishihara, T., Nonaka, I. and Sugita, H. Muscle and Nerve 1983;6:510.

8. Gronert, G.A. Anesth. Analg. 1980;59:377.

9. Wingard, D.W. In: Ellis, F.R. (ed.). Inherited Disease and Anaesthesia. New York: Elsevier/North-Holland Inc., 1981, p. 201.

10. Campbell, I.T., Ellis, F.R., Evans, R.T. and Mortimer, M.G. Acta Anaesthesiol. Scand. 1983;27:349.

11. Ayling, J.H., Ellis, F.R., Halsall, P.H. and Currie, S. Br. J. Anaesth. 1985;57:983.

12. Campbell, I.T., Ellis, F.R. and Evans, R.T. Anesthesiology 1981;55:46.

13. Isaacs, H. and Barlow, M.B. Brit. Med. J. 1970;1:275.

14. McPherson, E. and Taylor, C.A. Am. J. Med. Genetics 1982;11:273.

15. Ellis, F.R., Clark, I.M.C., Modgill, E.M., Currie, S. and Harriman, D.G.F. Br. Med. J. 1975;3:511.

16. Isaacs, H. In: Aldrete, J.A. and Britt, B.A. (eds.). Second International Symposium on Malignant Hyperthermia. New York: Grune and Stratton, 1978, p. 351.

17. Gronert, G.A. Seminars in Anesthesia, Volume 2, No. 3., September, 1983.

18. White, J.G., Clauson, C.C. and Gerrard, J.M. In: Blood, A. and Thomas, D. (eds.). New York: Churchill Livingstone, 1981, p. 33.

19. Solomons, C.C., Tan, S. and Aldrete, J.A. In: Aldrete, J.A. and Britt, B.A. (eds.). Second International Symposium on Malignant Hyperthermia. New York: Grune and Stratton, 1978, p. 221.

20. Verburg, M.P., Van Bennekom, C.A., Oerlemans, F.T. and De Bruyn, C.H.M.M. Adv. Exp. Med. Biol. 1984;165:443.

21. Solomons, C.C. and Masson, N.C. Acta Anaesthesiol. Scand. 1984;28:185.

22. Giger, U. and Kaplan, R.F. Anesthesiology 1983;58:347.

23. Solomons, C.C. and Masson, N. Anesthesiology 1984;60:265.

24. Britt, B.A. and Scott, E.A. Anesth. Analg. 1986;65:171.

25. Zsigmond, E.K., Penner, J. and Kothary, S.P. In: Aldrete, J.A. and Britt, B.A. (eds.). Second International Symposium on Malignant Hyperthermia. New York: Grune and Stratton, 1977.

26. Rosenberg, H., Fisher, C.A. and Addonizio, V.P. Anesthesiology 1980;53:S249.

27. Harrison, G.G. and Verburg, C. Brit. J. Anaesth. 1973;45:131.

28. Taylor, C.A., Druel, J.F., Schanus, E.G. and Lovrien, R. In: Lomax, P. and Schonbaum, E. Environment, Drugs and Thermoregulation. Switzerland: S. Karger, 1982, p. 160.

29. Godin, D.V., Bridges, M.A. and MacLeod, P.J.M. J. Med. 1979;10:287.

30. Klip, A., Britt, B.A., Elliott, M.E., Walker, D., Ramlal, T. and Pegg, W. Biochem. Cell Biol. 1986;64:1181.

31. Eng, G.D., Becker, M.J. and Muldoon, S.M. Muscle and Nerve 1984;7:618.

32. Nelson, T.E., Flewellen, E.H. and Arnett, D.W. Muscle and Nerve 1983;6:263.

33. Britt, B.A., McComas, A.J., Endrenyi, L. and Kalow, W. Anesthesiology 1977;47:490.

34. Britt, B.A., Scott, E.A., Kleiman, A., Jones, P. and Steward, D.J. Anesth. Analg. 1986;65:1047.

9

MUSCLE ASSESSMENT OF MALIGNANT HYPERTHERMIA PATIENTS

B.A. BRITT

Departments of Anaesthesia and Pharmacology, University of Toronto, c/o Toronto General Hospital, Toronto, Ontario M5G 2C4

ABSTRACT

For the diagnosis of malignant hyperthermia (MH) several different tests have been evaluated in a number of laboratories throughout North America. Best results are achieved with the caffeine halothane contracture (CHC) test. Addition of succinylcholine to the halothane containing chamber enhances the reliability of this test. The caffeine skinned fibre tension (CSFT) test and the ATP depletion test complement but do not replace the CHC test. The radioactive calcium uptake into thin muscle sections test does not appear to have either diagnostic or screening value.

INTRODUCTION

At present, reliable diagnosis of malignant hyperthermia (MH) requires a rather large and expensive muscle biopsy. In order to achieve as accurate results as possible, several different tests are often performed on each muscle sample. With the exception of microscopic studies, which are reported in Chapter 12 of this volume, these various tests are described on the following pages.

CAFFEINE HALOTHANE CONTRACTURE TEST (1-22)

The caffeine halothane contracture (CHC) test, which measures contractures of isolated skeletal muscle fascicles induced by caffeine and/or halothane, was first performed on a human malignant hyperthermia susceptible (MHS) patient in 1968 in Toronto, Canada. The human volunteer for the first experiment was a member of a large kindred located mainly in northern Wisconsin. Later that same year, two other individuals also consented to undergo similar biopsies. The results of these investigations were published in 1970 by Kalow and Britt in Lancet (22). In this paper, they demonstrated that caffeine induced contractures at lower doses in MHS than in normal muscle and that halothane potentiated this effect (Figures 1a-b). Additionally, it was subsequently shown that, in patients with a severe MH defect, halothane alone induced contractures, a phenomenon almost never seen in normal muscle (Figure 1c) (22).

FIGURE 1a

CAFFEINE DOSE — CONTRACTURE GRAPHS FOR A NORMAL AND A MHS HUMAN SKELETAL MUSCLE FASCICLE IN THE ABSENCE OF HALOTHANE

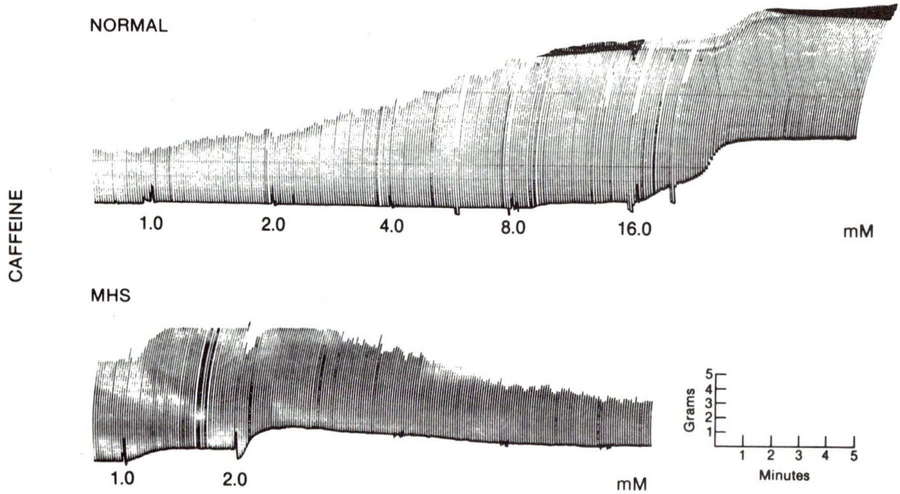

This test has now been adopted in a number of laboratories throughout North America (Appendix I) and is also performed in many European laboratories. In Europe the test is carried out somewhat differently than in North America. Therefore, methods and results obtained in European laboratories are described separately in Chapter 10.

FIGURE 1b

CAFFEINE DOSE — CONTRACTURE GRAPHS FOR A NORMAL AND A MHS HUMAN SKELETAL FASCICLE IN THE PRESENCE OF HALOTHANE

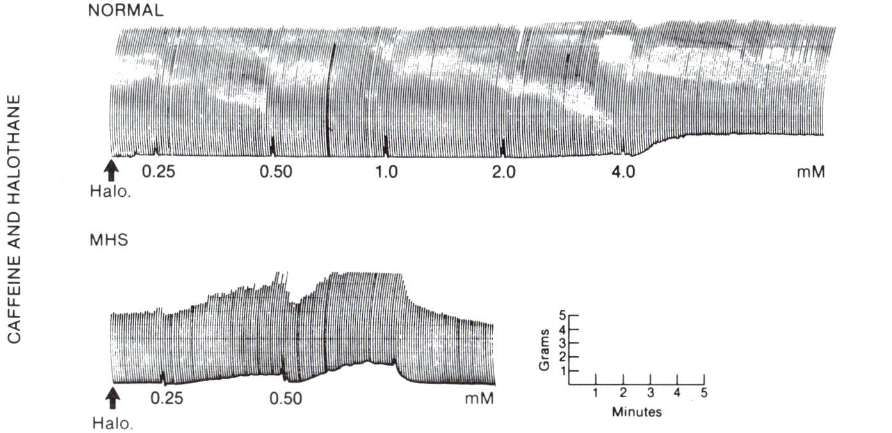

FIGURE 1c

HALOTHANE INDUCED CONTRACTURE AMPLITUDES OF A NORMAL AND A MHS HUMAN SKELETAL MUSCLE FASCICLE

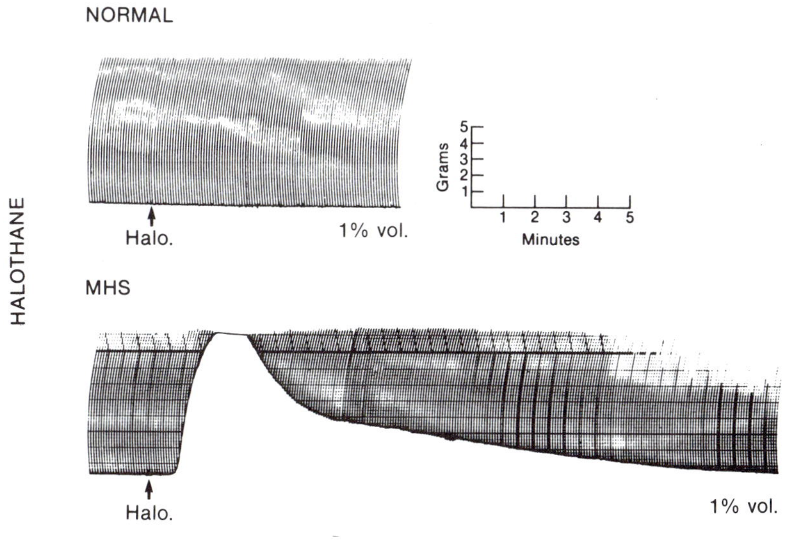

Because the equipment cannot be bought as a complete kit, parts must be separately purchased and/or custom built. The variation in the homemade components of the equipment and in the relationship of the various parts to each other, as well as in different dosages and times of application of the test drugs (caffeine and halothane) and differences in size, shape, tension, temperature and electrical stimulation of the muscle samples, has led to only moderate differences in results (Appendices 1-a - 1-p). Considering the large divergences in experimental methods among the various laboratories, the variations in the dividing points (reference values) between normal and MHS values are quite small.

METHODS

(a) Anaesthetic technique

Anaesthesia acceptable to both patient and surgeon can be achieved by a variety of techniques (Appendix I-b). If a small sample for only a diagnostic study is to be taken then a regional block of the femoral and the lateral femoral cutaneous nerve of the thigh may be adequate in adult patients. If the patient is apprehensive supplementation with narcotics and/or tranquillizers may be necessary. If larger sample sizes are contemplated for the combined performance of diagnostic and research studies then general anaesthesia may be preferable. This can safely be performed with any desired combination of thiopentone, droperidol, fentanyl, diazepam and nitrous oxide. The patient should be intubated and hyperventilated at all times to prevent hypercarbia, as hypercarbia alone has been shown in susceptible pigs to be capable of inducing MH reactions. The endotracheal tube should be one or two sizes smaller than would normally be selected because the aperture between the vocal cords is less than it would be had succinylcholine been used. Inadvertent cooling of the patient ought to be avoided because shivering (probably in combination with pain and agitation) in the immediate post operative period may trigger a stress-induced reaction. Other aspects of the management of the anaesthetic are covered in Chapter 15.

(b) Surgical technique

The most suitable skeletal muscle samples for the CHC are usually obtained from the quadriceps muscle, particularly from the vastus lateralis portion of the quadriceps (Appendix 1-a). Samples excised from the rectus abdominis and paravertebral muscles of the pectoralis major are also acceptable although these latter muscles suffer from the disadvantage that the volume of muscle that can be safely removed is less than from the quadriceps. Muscle for the CHC test ought not be removed from the gluteal or the deltoid muscles. This is because the fascicles in these latter muscles tend to be fan shaped rather than rectangular and are usually separated from each other by rather inflexible bands of fibrous material. These features reduce the responsiveness of gluteal and deltoid muscles to caffeine and halothane.

The vastus lateralis muscle is approached through an incision that varies in length between 2 and 12 cm depending on the size of the sample to be removed (Figure 2a, Appendix 1-b). During dissection through the subcutaneous tissue the antero lateral cutaneous nerve of the thigh must be identified and preserved. The muscle is more responsive to electrical and drug stimulation if electrical cautery is not used during surgery until excision of the sample has been completed and if trauma in the form of repeatedly patting the muscle with a gauze or poking the muscle with various surgical instruments is carefully avoided. Before commencement of dissection of the muscle fascicle, all areolar tissue and fat must be carefully removed. Next both ends of each fascicle are secured with a black silk suture and then meticuously dissected free from the surrounding main muscle belly while maintaining constant tension on the sutures in such a manner as to preserve the muscle cells free from contractures or excessive stretching. Care should be taken to ensure that the direction of the long cut runs parallel to the long side of the muscle cells. Except at the extreme ends of the sample cuts should never be made across muscle cells (Figure 2a).

Slight improvement in diagnostic discrimination can be gained by preventing contractures or stretching of the fascicles (Appendix 1-b). This can be done by re-tying the sutures around a balsa

FIGURE 2a

Skeletal muscle fascicle being dissected which has been tied to a grooved balsa stick with black silk sutures prior to commencement of dissection (BRITT)

FIGURE 2b

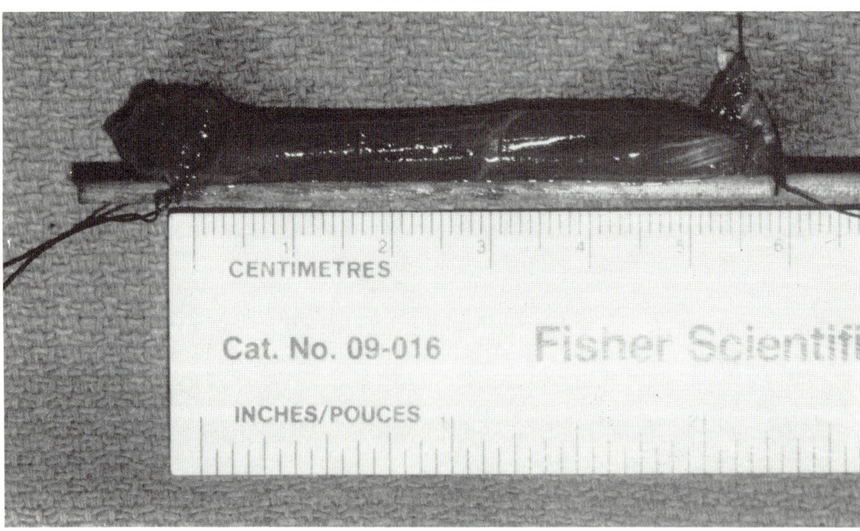

Fully dissected skeletal muscle fascicle is about 6 cm long (BRITT)

FIGURE 2c

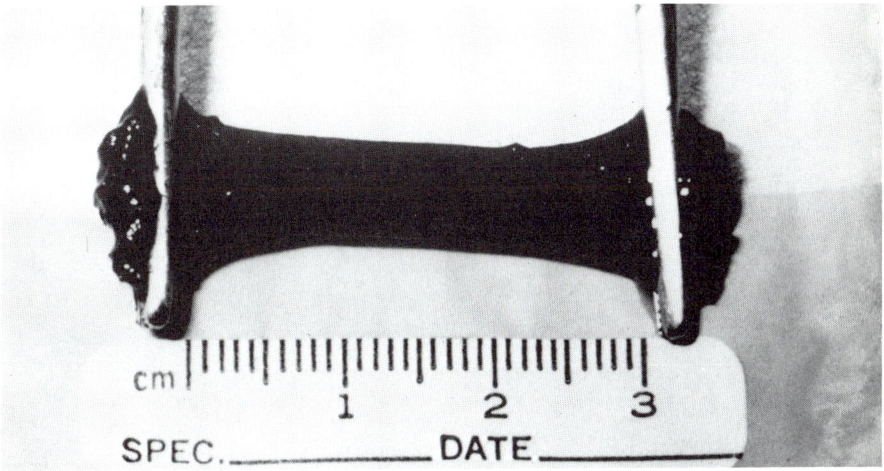

Skeletal muscle fascicle which has been dissected after being secured within a fixed length clamp. This fascicle is about 3 cm long. (REPRINTED WITH PERMISSION OF DR. SHEILA MULDOON

FIGURE 2d

Subdivided skeletal muscle fascicle in bathing chamber with tension maintained by black silk ties at each end and belly of fascicle touching 2 prongs of an electrode (REPRINTED WITH PERMISSION OF DR. TOM NELSON)

wood stick in grooves made circumferentially with scissors or a scalpel blade close to each end of the stick (Figure 2b). Securing of the sutures to the stick ought always to be done before dissection commences in order to prevent any artefactual contracture and/or stretching. The grooves should be made a little further apart from each other than are the sites of suture implantation at the ends of each fascicle. This ensures that the muscle, by being very slightly stretched when tied onto the stick, clings very closely to it, and so cannot undergo inadvertent contracture. Alternatively, the muscle may be secured within a fixed length clamp as recommended by Rosenberg, by Muldoon and by Nelson (Figures 2c) (Appendix 1-b).

To achieve reliable diagnostic information the total weight of the excised sample should be at least two grams. If the investigator wishes to additionally perform research studies, then up to fifteen grams of muscle can be safely removed from the quadriceps of an adult patient. The maximum safe yield from other muscles is considerably less. The dimensions of the sample vary considerably among different laboratories, ranging from a low of 2.0 x 0.5 cm to a high of 6.0 x 2.5 cm (Appendix 1-c).

(c) Transport of muscle to laboratory

The excised muscle samples are placed in a container (usually a thermos) filled with Krebs Ringer or Ringers Lactate

TABLE 1a

HUMAN KREBS RINGER SOLUTION

118.4 mM NaCl
3.3 mM KCl
0.9 mM $MgSO_4$
1.1 mM KH_2PO_4
11.1 mM Glucose
24.9 mM $NaHCO_3$
2.5 mM $CaCl_2$

pH of medium adjusted to 7.4 at 25°C

solution at pH 7.4 (Table Ia) and maintained at temperatures ranging from 15 to 25°C (Appendix 1-c). Colder temperatures should not be employed since cold exposure leads to unwanted and prolonged unresponsiveness of the muscle to electrical and drug stimulation. The maximum time between excision and further processing in the laboratory in no centre routinely exceeds 35 minutes (Appendix 1-c). However, on rare occasions when this time span has lasted for up to two hours, for instance, when the muscle has had to be transported from another city (usually by helicopter) we have achieved reasonable results. The muscle can be enabled to tolerate these longer transportation periods by constantly gassing the Krebs Ringer solution surrounding the muscle throughout the transportation period with carbogen (95% O_2 and 5% CO_2).

(d) Preparation of individual muscle strips

In the laboratory the specimens are usually, but not always, placed under a magnifying glass or low power microscope (Appendix 1-d) where they are trimmed free of any irregularities or remaining fat. To permit replicate measurements each is divided into several strips approximately 1.0-3.0 cm x 0.1-1.0 cm (Appendix 1-d), weighing roughly 0.1-0.3 grams. Within this relatively broad range of dimensions we have observed that alteration of the size of the specimen is associated with little difference in results.

(e) Composition of Solution in Bathing Chambers

In North American laboratories each of 2 to 20 muscles strips is placed in a bathing chamber (Figures 3a-b) filled with Krebs Ringer solution at a pH of 7.4 (Appendix 1-d). The composition of the Krebs Ringer solution in humans and pigs differs slightly. However, in pig muscle the concentrations of sodium bicarbonate, potassium chloride and potassium phosphate are higher than for human muscle (Table Ib). These differences correspond to higher serum concentrations of sodium, potassium and phosphate in pigs than in humans (23-27).

FIGURE 3a

Experimental method for isolated isometric muscle contracture study

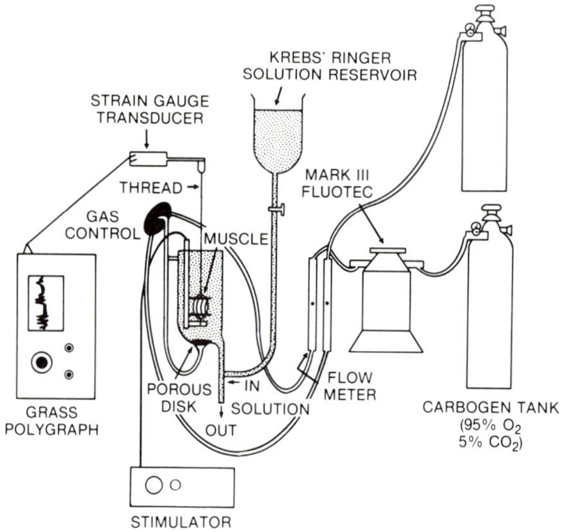

TABLE Ib

PORCINE KREBS RINGER SOLUTION

118.4	mM	NaCl
3.3	mM	KCl
0.9	mM	$MgSO_4$
1.57	mM	KH_2PO_4
11.1	mM	Glucose
35.9	mM	$NaHCO_3$
2.5	mM	$CaCl_2$

pH of medium adjusted to 7.4 at 25°C

(f) Temperature of Measurement

Although the CHC test, as originally described, was performed at 22°C, consensus at the 1982 Banff MH symposium (28) was, that in the interests of uniformity, all investigators would perform the test at 37°C. Selection of 37°C has the additional advantage of enabling the test to be completed in a shorter time

FIGURE 3b

BATHING CHAMBERS FOR THE CAFFEINE HALOTHANE CONTRACTURE TEST

There is one force displacement transducer and one micrometer screw gauge for each bathing chamber. Each chamber is served by an individual channel of an Elmed stimulator. Manifolds behind upright white board (and, therefore, not visualized) conduct gases and vapours separately to each chamber. Each chamber is served by two flowmeters and two manifolds--one for gas only and one for gas plus a halogenated vapour. Water surrounding bathing chambers is kept at $37^{o}C$ by an electronic heater (not shown). Krebs Ringer solution is conducted to each chamber through glass tubing. Flow through this tubing is controlled by a glass stopcock. Below this stopcock within the warm water bath the glass tubing is coiled to permit pre-warming of the Krebs Ringer solution. At the entrance to the bottom of the bathing chamber the Krebs Ringer solution passes through a scintered porous disc. The Krebs Ringer solution is removed from the bottom of the chamber through glass tubing which leads to a sink. Flow of Krebs Ringer solution through this tubing is controlled by a second glass stopcock situated below the water warming chamber (and so not visible in this photograph).

than at 22°C since the muscle contracts at the former temperature more rapidly, particularly in the presence of halothane. It has, however, the disadvantage that muscle, particularly MHS muscle, fatigues more rapidly at the higher temperature thereby reducing diagnostic discrimination and time available to perform the test at 37°C (29). Furthermore, considerably more equipment is required (heating units and water baths) if 37°C is chosen as the temperature of measurement. Therefore, if money is in short supply the test can be carried out quite successfully at room temperature provided the ambient temperature of the room in which the measurements are made is reasonably constant and does not fall below 21°C (Appendix 1-d).

At 37°C contractures of human skeletal muscle fascicles are are greater than at 22°C (29): in the presence of caffeine alone (MHS and normal) (Figure 4a); in the presence of halothane alone (MHS) (Figure 4b); and in the presence of caffeine plus halothane (normal) (Figure 4b). On the other hand, in MHS human skeletal muscle fascicles halothane plus caffeine initiates larger contractures at 22°C than at 37° (29) (Figure 4b). The reason for this latter difference is probably due to the greater fatiguability of MHS muscle than of normal muscle at the higher temperature in the combined presence of the two drugs.

We have found that for pig muscle fascicles 37°C is too high because at this temperature the rate of fatigue of MHS muscle, although not of normal muscle, is so rapid that even in the early stages of the experiment caffeine-induced contractures in the presence of halothane are often attenuated or absent and diagnostic discrimination is, therefore, substantially reduced (Figures 4c-d). Because the temperature of our laboratory is now usually above 22°C (due to a recent geographical move) we now measure pig muscle at 25°C, a temperature still low enough to yield good diagnostic results.

(g) Mounting of muscle strips in bathing chambers

The lower end of each muscle strip is secured by a black silk suture to the lower end of an electrode housing. The upper end of each muscle is connected via a second, and much longer

FIGURE 4a

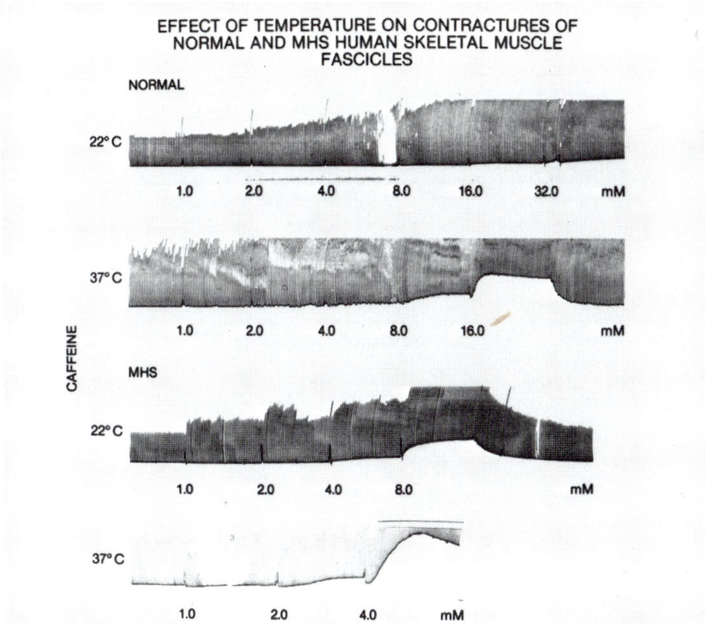

FIGURE 4b

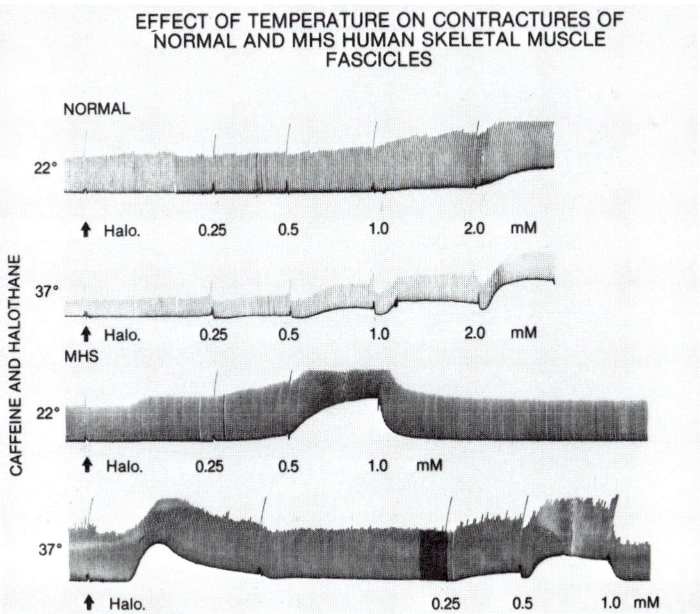

FIGURE 4c

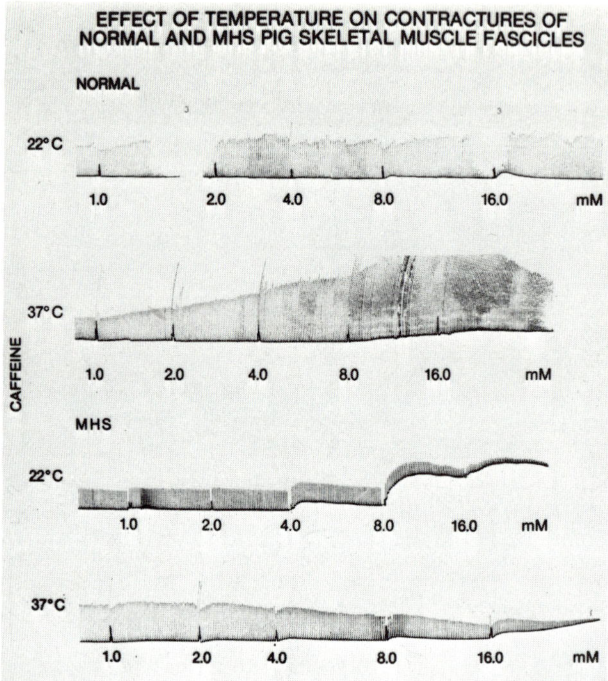

FIGURE 4d

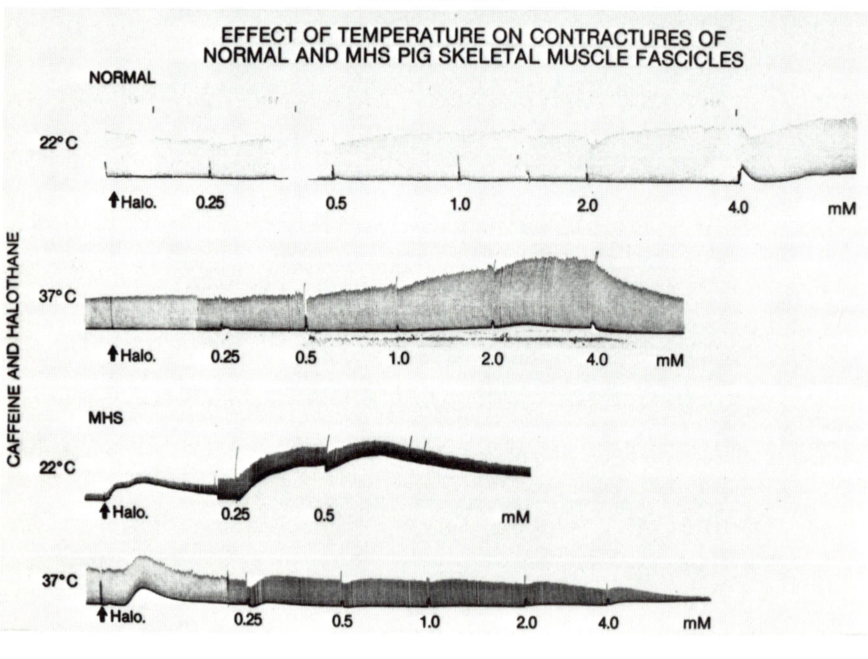

black silk suture to a force displacement transducer (Grass FT10) (Figures 2d,3a-b). Isometric tension is recorded with a polygraph. A Grass Model 7, Gould Brush Model 2200, Hewlett Packard or Beckman Model 411 are all suitable (Appendix 1-e). The initial baseline tension is set at 1.0 to 2.0 grams (Appendix 1-e). The magnitude of the contracture is not altered significantly by changing the baseline tension from 1.0 to 2.0 grams (Figure 5). However, in Gronert's and in Durbin's laboratory the initial tension is not fixed at an arbitrary level but rather is adjusted individually for each muscle strip (Appendix 1-e). The tension chosen by Gronert and by Durbin is that in the presence of which maximal contractions are produced in response to electrical stimulation.

FIGURE 5

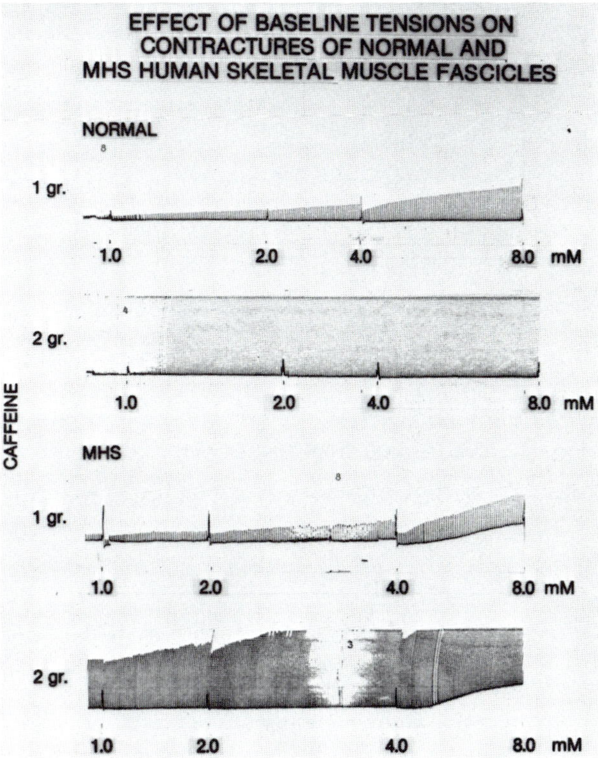

(h) Electrical stimulation of muscle strips

In order to assess viability and to simulate normal innervation the muscle is stimulated via electrodes that are made of platinum or stainless steel (Appendix 1-e, Figures 2d, 3a). The configuration of the electrodes may be of two prongs, the tips of which touch the belly of the muscle strip (Appendix 1-e). All, except the tips of the prongs may, but do not have to be electrically isolated (Appendix 1-e). The electrodes may also be in the form of a plate or a bar attached to the bottom of the muscle strip (Appendix 1-e).

The electrodes are attached to a square wave stimulator (Grass, Elmed or Custom-made) (Appendix 1-f). The controls of the stimulator (voltage, duration and frequency) should be set to induce maximal twitch tensions in each of several muscle strips (Appendix 1-f). Ideally to achieve this the stimulator, as in the case of the Elmed stimulator, ought to be able to provide individualized control of the above parameters to each strip. However, several MH investigation units use unichannel Grass stimulators that have been modified so that the outlet jack is subdivided into multiple inparallel circuits (Appendix 1-f). All these circuits are, therefore, served by the same single set of controls. Consequently the voltage leaving the outlet jack (prior to subdivision) is much higher than the voltage reaching each muscle strip. Furthermore, since all muscle strips receive the same voltage, duration and frequency of current it is unlikely that all of them are achieving maximal twitch tensions since different muscle strips attain maximal twitch tensions at different voltage inputs. If only a unichannel stimulator is available, the best that can be done is to adjust the controls to those values that produce the largest possible contractions in the largest number of muscle strips.

The frequency of stimulation varies among different laboratories between 2 and 10 seconds with most choosing either once every 5 or every 10 seconds (Appendix 1-f). The duration of stimulation ranges between 1 and 10 milliseconds. While some workers opt for a fixed duration of stimulation, others such as Rosenberg and Nelson vary the duration of stimulation so as to attain the largest possible twitch tension (Appendix 1-f). For the same reason most investigators also do not use a fixed voltage. The values actually in use range from 2 to 150 volts (Appendix 1-f). Voltages tend to be lower in those systems utilizing stimulators with individualized control of each channel (Appendix 1-f).

(i) Bubbling of carbogen and carbogen plus halothane into bathing chambers

Carbogen is bubbled into each bath through a scintered, porous disc at the bottom of each chamber (Figure 3a). The rate of flow into individual bathing chambers ranges from 75 to 500 ml and may be, but is not always, controlled by individual flow meters situated above each bathing chamber and/or by an incommon flow meter attached to the regulator of the carbogen tank and/or by an incommon anaesthetic-type rotameter (Appendix 1-g). Most investigators seem to put more faith in determining gas flow by "eyeballing" the bubbles rising upward in the bathing chamber rather than by turning the flowmeters to an arbitrarily predetermined flow rate. This is probably because slight variations in the internal diameter of the glass tubes and manifolds and in the diameter of the pores in the scintered discs at the bottoms of the bathing chambers (caused by calcium crusts) alter the flow of gas entering the bathing chambers independently of the flow meter settings. The flow rate does not seem to be critical as it varies so much from one laboratory to another, ranging from a low of 30 ml per chamber, to a high of 500 ml per chamber without significantly altering results (Appendix 1-g). In most, but not all laboratories, the carbogen line is subdivided into two lines (Figures 3a-b). One goes directly to the bathing chamber (with a branch going to a large reservoir bottle filled with Krebs Ringer solution and situated on a shelf well above the bathing chambers). The other passes through a halothane vaporizer which is most often a Fluotec-3 or a Dragerwick, but may also be a Fluotec-2, Fluomatic, Vernitrol, Copper Kettle or Ohio vaporizor (Appendix 1-g). On leaving the vaporizor the halothane laden carbogen then proceeds by a separate manifold to the bathing chambers. In a few laboratories the carbogen and carbogen plus halothane lines are not separated thus unfortunately permitting some contamination with halothane during measurement of caffeine contractures (Appendix 1-h). The length of the above lines is quite variable among laboratories. Because of problems with undesirable pressure changes in the lines due to crusting of the scintered disc with calcium and lack of control of the flow rate, some laboratories, in an effort to achieve greater constancy of gas pressure, have introduced features such as roller pumps, pressure release valves and reservoir bags into their gas lines (Appendix 1-h). These features are desirable since changes in pressure may significantly alter the concentration of halothane reaching the Krebs Ringer solution. Frequent measurement of halothane concentrations in the gas line and in the Krebs Ringer solution are, therefore, valuable. This is usually done with spectrometry, infra red analysis or chromatography (Appendix 1-i).

(j) Addition of caffeine to bathing chambers

To usually at least two duplicate fascicles (Appendix 1-i), freshly prepared caffeine in increments (from a low of 0.125 to 1.0 mM, to a high of 16.0 to 32.0 mM) is added directly to the bath (Appendix 1-j). In several laboratories each caffeine dose is left in the bath until the contracture plateau or peak has been reached (about four to five minutes) (Appendix 1-k). The caffeine is then removed and replaced cumulatively (without washing of the muscle) with another dose of caffeine double that of the previous dose (Appendix 1-j). Dr. Muldoon, however, prefers to not change the caffeine until the contracture has returned to the base line or, if no contracture has occurred, until after fifteen minutes. In the former, but not in the latter, instance, she washes the muscle with a caffeine-free bathing solution several times before the next caffeine addition. A few teams add the next dose of caffeine cumulatively after a fixed time period, usually after four to five minutes, but in the laboratory of Rosenberg, after two minutes (Appendix 1-k).

(k) Addition of halothane to bathing chambers

At least two duplicate muscle fascicles (Appendix 1-i) are equilibrated for fifteen minutes with 1.0 vol% halothane (Appendix 1-j). The amplitude of the contracture (usually baseline to plateau), in grams, induced by the halothane is measured upon completion of the fifteen minutes of halothane equilibration. The procedure for adding halothane varies considerably according to institution (Appendix 1-j). Some investigators add several concentrations, either to the same chamber cumulatively or to separate chambers (Appendix 1-j). Cattran and Gronert use 2.0% rather than 1.0% halothane (Appendix 1-j). Rosenberg adds 3% halothane to the same muscle fascicle that had been equilibrated with caffeine twenty minutes after the last caffeine addition (Appendix 1-j).

(l) Addition of halothane plus caffeine to bathing chambers

In most, but not all, institutions, caffeine is added to the bathing solution, already equilibrated with halothane, in increments from 0.1-1.0 to 32 mM (Appendix 1-j). Halothane equilibration is continued during the caffeine additions. Brownell does not perform this part of the test (Appendix 1-j). Oh and Durbin add halothane, wait fifteen minutes, wash the muscle free of halothane and then add incremental caffeine doses (Appendix 1-j). Rosenberg adds halothane and caffeine in combination only to some muscle fascicles for research purposes.

(m) Parameters measured on polygraph

The majority of investigators measure contracture as the distance from the baseline to the plateau. Since the plateau is rarely precisely horizontal, some workers arbitrarily select four or five minutes after the last drug addition as the time at which the contracture is determined (Appendix 1-k).

(n) Classification of Patients

MHS Patients - individuals who themselves have had an anaesthetic history of a MH reaction and who have had a skeletal muscle biopsy positive for MH;

TABLE 2

COMPARISON OF HALOTHANE CONTRACTURES, CSCs AND CSC-Hs - MHS PATIENTS AND THEIR RELATIVES (ACCORDING TO DEGREE OF RELATIONSHIP vs. UNRELATED NORMAL PATIENTS

		MHS PATIENTS	MHS RELATIVES 1st	2nd	3rd	UNRELATED NORMAL PATIENTS
HALOTHANE[1]	x	0.10	0.08	0.15	0.04	0.00
	SE	0.02	0.02	0.05	0.03	0.00
	N	292	362	141	83	159
	T	-2.91	-3.12	-3.01	1.78	--
	P	***	***	***	*	--
CAFFEINE[2] (CSC)	x	4.47	4.64	5.34	5.02	6.87
	SE	0.12	0.11	0.19	0.24	0.20
	N	292	363	141	83	159
	T	10.98	10.18	5.45	5.62	--
	P	***	***	***	***	--
CAFFEINE + HALOTHANE[3] (CSC-H)	x	0.89	0.97	1.24	1.08	1.96
	SE	0.03	0.04	0.08	0.07	0.09
	N	291	363	141	83	159
	T	12.82	11.82	5.78	6.29	--
	P	***	***	***	***	--

[1] Amplitude of halothane (1%)-induced contractures of skeletal muscle fascicles.

[2] Dose of caffeine required to raise the resting tension of skeletal muscle fascicles by 1.0 gram in the absence of halothane

[3] Dose of caffeine required to raise the resting tension of skeletal muscle fascicles by 1.0 gram in the presence of 1% halothane

x = arithmetic mean
SE = standard error
N = sample size
T = Student's T-test
P = probability
* $P < 0.05$
** $P < 0.01$
*** $P < 0.001$

MHS Relatives - individuals who have themselves had no anaesthetic history of MH but who have had relatives who have had a history of a MH reaction and who themselves have had a skeletal muscle biopsy positive for MH;

Negative Relatives - individuals who themselves have had no anaesthetic history of MH but who have had relatives who have had a history of a MH reaction and who themselves had a skeletal muscle biopsy negative for MH;

Normal Patients - individuals with no anaesthetic history of a MH reaction in themselves or in their relatives.

(o) Characteristics of the Halothane-, Caffeine-, and Caffeine-Plus Halothane Dose Response Curves

Typical dose-contracture curves and mean values, standard errors, sample sizes and t-tests for these curves which have been obtained in our laboratory for normal and MHS human and pig muscles are given in Figures 1a-c and in Table 2 respectively. Resting tensions (contractures) rather than twitch tensions (contractures) are measured because better diagnostic discrimination between normal and MHS muscle can be obtained with the former. This is because, while contractures are always greater in MHS than in normal human and pig muscle (Figures 6a-b), twitch tensions in human MHS muscle, although greater than normal at low caffeine doses, are lower than normal at high caffeine doses and in pig MHS muscle twitch tensions are lower than normal at all caffeine doses (Figures 6c-d).

In muscle afflicted by a severe MH defect, halothane alone induces contractures which usually disappear after a few minutes (Figure 1c). In our laboratory, this phenomenon is almost never seen in control muscle. Even when contractures do occur in normal muscle they are extremely small, being almost always less than 0.1 Gm. In human patients peak contractures occur at 3% except at the highest caffeine concentration (Table 3). In some laboratories peak contractures are only achieved at lower concentrations of halothane. These differences among laboratories are probably due to equipment differences--differences which tend to alter flow rates through and pressures within the halothane vaporizors. These

FIGURE 6a

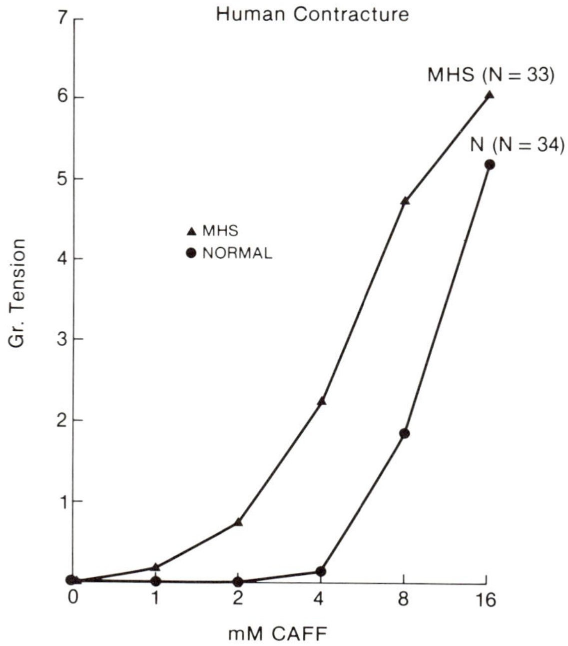

FIGURE 6b

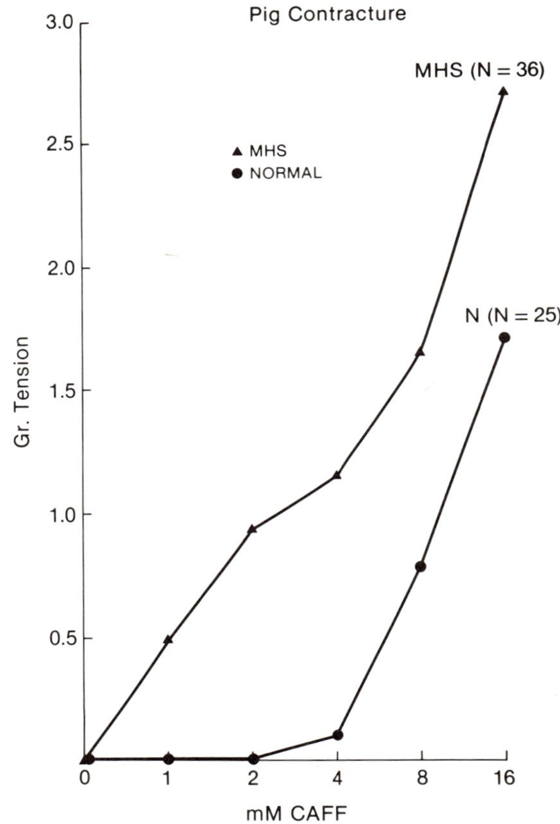

LEGEND: Caffeine dose-contracture response curves of human and pig skeletal muscle fascicles

FIGURE 6c

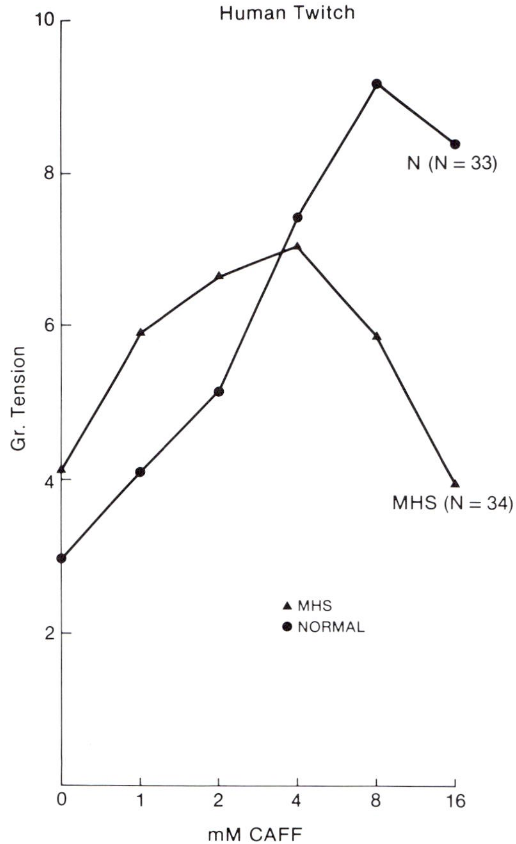

FIGURE 6d

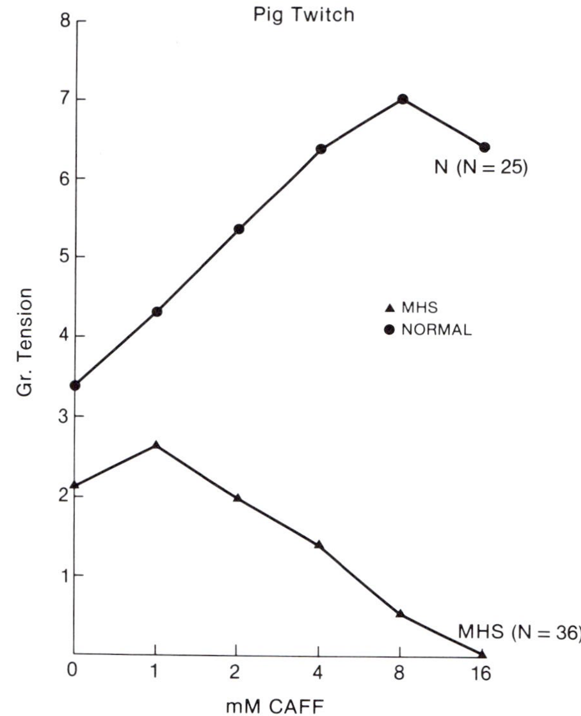

LEGEND: Caffeine dose-twitch response curves of human and pig skeletal muscle fascicles

TABLE 3

EFFECT OF HALOTHANE AND CAFFEINE ON HUMAN SKELETAL MUSCLE FASCICLES

Caff. Conc. in mM	Status	Halothane Concentration 1% N	1% MH	2% N	2% MH	3% N	3% MH	F (Halo Doses)
0.00	x	0.00	0.44	0.02	0.41	0.05	1.66	14.15 ***
	SE	0.00	0.21	0.02	0.16	0.02	0.45	
	N	17	19	17	19	14	14	
	T-NvsMHS	-2.13*		-2.32**		-3.60**		
0.25	x	0.00	0.67	0.19	1.15	0.86	1.88	15.49 ***
	SE	0.00	0.22	0.07	0.23	0.09	0.33	
	N	17	19	17	19	14	14	
	T-NvsMHS	-3.07**		-3.86***		-4.67***		
0.50	x	0.00	1.20	0.59	2.53	0.78	3.21	15.63 ***
	SE	0.00	0.32	0.13	0.39	0.16	0.40	
	N	17	19	17	19	14	13	
	T-NvsMHS	-3.78***		-4.49***		-5.82***		
1.00	x	0.21	2.61	2.05	4.54	2.98	4.24	20.55 ***
	SE	0.03	0.44	0.31	0.47	0.39	0.61	
	N	17	15	17	15	14	10	
	T-NvsMHS	-5.78***		-4.55***		-1.82*		
F(Caff Doses)		30.42***		76.42***		46.85***		

x = arithmetic mean of contracture amplitudes of skeletal muscle fascicles. SE = standard error; N = sample size; T = Student's T-test; * $P < 0.01$; ** $P < 0.05$; *** $P < 0.001$; F = analysis of variance

alterations, therefore, lead to variations in halothane concentration in the bathing chambers. Such variations tend not to correlate with dial settings on the Fluothane vaporizors or with measurements of halothane concentrations in the carbogen. Because of nonlinearity of the halothane-contracture curves a number of investigators have opted to report the contracture amplitude induced by only one dose of halothane (usually but not always 1%).

In muscle characterized by a medium degree of MH susceptibility halothane alone fails to induce contractures but caffeine alone causes contractures at lower doses than it is able to do in normal muscle. In normal muscle the contracture induced by all caffeine doses, once developed, persists for some time at a level close to its peak, often until the caffeine is washed out from the muscle (Figure 1a). In MHS muscle at high caffeine doses, on the other hand, the contracture usually declines substantially from its initial peak in an often irregular fashion (Figure 1a). This fatigue effect becomes more marked as the caffeine dose rises.

Muscle obtained from some very mildly affected MHS patients, while producing no contractures in the presence of halothane alone or in the presence of caffeine alone, does, in the presence of halothane, develop contractures at lower caffeine doses than does normal muscle.

The caffeine-contracture curve and the caffeine-halothane contracture curve are quite flat at low caffeine doses, particularly in normal muscle. This makes determination of the threshold contracture rather difficult in some cases. Just below one gram tension the caffeine-contracture curve and the caffeine-halothane contracture curve become much steeper. Therefore, a point on these curves at one gram of resting tension is much easier to determine than is a threshold contracture. At the latter level the slope may be so flat as to be almost indistinguishable from the horizontal axis.

(p) **Interpretation and Reporting of the Data to Patients and Clinical Doctors**

Measurement of the contractures produced by a single dose of halothane is relatively straight forward. However, interpretation of the caffeine-contracture curves is so complex, that simplification of the data is necessary for clinical purposes. One could choose to report: steepness of the ascending slope of the curve at low caffeine doses; the time until commencement of fatigue of the muscle as indicated by declining peak contractures in spite of increasing caffeine doses; rate of descent of individual contractures from their peaks; contractures induced by a single arbitrarily selected caffeine dose; or the dose of caffeine required to produce an arbitrarily selected rise in resting tension.

When first developing the CHC test Kalow et al (22) and Britt et al (5) decided to adopt the last of these options. Calculation of the steepness of the caffeine-contracture slope would have been cumbersome (at a time when personal computers were generally unavailable) as the slope is rarely linear but rather is usually "S" shaped. While MHS muscle on an average fatigues more rapidly than does normal muscle, the finding is not absolutely consistent and reliance on this parameter alone would have resulted in a number of false negative values. Selection of a contracture produced by an arbitrarily chosen caffeine dose has been found to yield poorer diagnostic discrimination than does calculation of the dose of caffeine which produces an arbitrarily chosen tension increase. Because of the previously mentioned difficulty in ascertaining threshold contracture Kalow et al (22), when developing the caffeine contracture test, opted for a 1.0 gram contracture--a point usually just above commencement of the steep part of the curve (5,6,21).

From the log caffeine and log caffeine plus halothane contracture curves the dose of caffeine in mM required to raise the resting tension of skeletal muscle fascicles by one gram in the absence of halothane and also in the presence of 1% halothane is determined. These values are termed the "caffeine specific concentration" (CSC) in the absence of halothane, and "caffeine specific concentration - halothane" (CSC-H) in the presence of halothane (5-7,20,21).

Among different laboratories the reference value (30) dividing normal from MHS CSC values ranges from 3.5 to 4.1 mM from MHS, while the reference value separating normal CSC-H values vary from 0.75 to 1.1 mM (Appendix 1-l). The reference value separating normal from MHS contractures observed in the presence of halothane alone ranges from 0.0 to 1.0 gram (Appendix 1-l). Those investigators reporting higher normal values for the latter parameter are usually those using higher test concentrations of halothane as in the cases of Rosenberg and Brownell (Appendix 1-l). The same or a similar method of reporting halothane contracture values is used by most but not all other institutions performing this test (Appendix 1-m).

Since MH appears to be a spectrum ranging from the very severe through the mildly afflicted, the barely normal to the very normal individuals we have subdivided the CHC test results as follows (31):

HCK - contracture in the presence of halothane alone, lower than normal CSC, and lower than normal CSC-H (the most severe variant of MH);

HC - contracture in the presence of halothane alone and normal CSC (a severe variant recorded in the laboratories of those investigators who do not measure the CSC-H):

HK - contracture in the presence of halothane alone, normal CSC, and lower than normal CSC-H (a rare intermediate variant of MH);

CK - no contracture in the presence of halothane alone, lower than normal CSC, and lower than normal CSC-H (a common intermediate variant of MH);

C - lower than normal CSC (a rare mild variant of MH in laboratories performing the CSC-H, and a common intermediate variant of MH frequently recorded in the laboratories of those investigators who do not measure the CSC-H test);

K - lower than normal CSC-H (a common mild variant of MH but not measured by a few laboratories);

N - no contracture in the presence of halothane, normal CSC and normal CSC-H (normal)

The majority of the above patients are: HCK (strongly positive for MH); CK (mediumly positive for MH); K (mildly positive for MH); and N (normal or negative for MH).

Similar subdivisions of CHC test results (although sometimes with different terminology) are used by several other investigators (Appendix 1-m). For instance, some workers divide their positive biopsy results into H, C and K. The H group contains all those patients whose muscle develops contractures in the presence of halothane alone. The C group comprises all those individuals whose muscle exhibits CSCs positive for MH but no halothane induced contractures. The K group consists of all those persons whose muscle demonstrates CSC-Hs positive for MH, but CSCs negative for MH and no halothane contractures. Rosenberg, however, classifies test values obtained in his laboratory into simply "positive" or "negative" (Appendix 1-m). Wingard sorts into "positive", "intermediate" and "negative" values (Appendix l-m). Duncan's classification of "high risk", "moderate risk" and "normal" appears to be similar in nature, although not in terminology, to that of Wingard (Appendix 1-m). Gronert's subdivision of "positive", "equivocal" and "negative" is likely different in that to him "positive" means that all parameters measured are positive for MH in all muscle fascicles, while "equivocal" means that all parameters measured are positive in one muscle fascicle only (Appendix 1-m). In summary then, we observe that the majority of investigators subdivide their patients into at least three categories that might be broadly described as strongly positive for MH, mildly positive for MH and normal.

(q) Consideration of Non-Uniformity of Test Results Among Laboratories

As the above description of CHC testing and the report in Chapter 10 of the method used by the European Group illustrate, MH investigators have not yet been able to achieve complete uniformity of performance and interpretation of this test (1-22,29,31-41). A few European investigators have claimed that only halothane is a useful agent (39,41). A few North American workers use both halothane and caffeine on separate strips of muscle but state that halothane provides more valid results (34-36) (Appendices 1-i to 1-m). Several of these latter consider that a greater than normal response to either halothane alone or to caffeine alone constitutes evidence for MH, that patients who develop contractures in the presence of halothane alone nearly always have a lower than normal threshold for caffeine induced contractures, but that those whose muscle contracts by 1.0 gram at a lower than normal dose of caffeine do not always exhibit halothane-induced contractures (1,5,10,15,16,20,37,40) (Appendices 1-i to 1-m). Most, but not all, investigators also recognize that simultaneous application of both halothane and caffeine to the muscle is necessary to ensure identification of all mild cases (1,5,10,15,16,37) (Appendices 1-i to 1-m). Thus some very mildly afflicted MHS individuals appear to these workers to react normally with either halothane or with caffeine alone but to develop greater than normal responses in the combined presence of caffeine and halothane, i.e. to have lower than normal thresholds for 1.0 gram caffeine contractures in the presence of halothane.

It is noteworthy that Oh and Durbin, who state that values, obtained with halothane alone and with caffeine alone, particularly the former, differentiate MHS from normal muscle much more reliably than does the combined use of halothane plus caffeine, do in fact add these agents to the same muscle strip but sequentially rather than simultaneously with washing out between drug additions

(Appendix 1-j). Thus Oh and Durbin (who expose the muscle first to halothane, and then after washing to caffeine) (Appendix 1-j) must be reporting caffeine contracture curves that are contaminated with halothane since no amount of washing totally removes highly lipid soluble halothane from the muscle. It would appear, therefore, that differences in technique among investigators are not so great as previously believed.

Perusal of Appendices 1-a to 1-i reveals that some of the non-uniformity in technique that does exist among the various North American laboratories performing the CHC test are necessitated by differences in equipment (vaporizors, stimulators, manifolds, etc.). Economic exigencies often prohibit repurchase of new equipment that might conform to a standardized protocol. Other types of variability of technique (resting base tension, temperature of measurement and duration of stimulation) have occurred because of relatively arbitrary decisions taken a number of years ago before any regularized consultation among the laboratories performing this test. Still others are due to deliberate choice arising out of the investigators' conviction that his or her method is superior to that of anybody elses. Furthermore, investigators are loathe to change their methods, not only for cost reasons but also because significant alteration in methodology would make difficult pooled statistical analysis of data obtained before to the change of method with that obtained after the change of method. Some differences in percentage of patients reported positive for MH are undoubtedly due to variation in criteria for selection of patients for biopsy (Appendices 1-n to 1-p). Finally, lack of simultaneous application of halothane plus caffeine tends to lower the percentage of biopsies reported positive for the MH trait thereby improving specificity but lowering sensitivity through raising the numbers of false negatives (30).

In spite of these many variables it has been surprising to the author how consistent the results have been in terms of the dividing points (reference values) between control and MHS muscle and the percentage of patients determined to have MH on the basis of the CHC test (Appendix 1-o) particularly when the parameters reported have been all three of: halothane contractures, CSC and CSC-H.

(r) Determination of Reference Value (Dividing Point) Between Normal and MHS

Since the numbers of volunteer patients who can be obtained from the general population to serve as controls will always be limited, and since it will always be impossible to deliberately challenge suspected MHS individuals with in vivo halothane, determination of the exact reliability and of the reference value, of the various parameters, of the CHC test in humans will, for the foreseeable future, remain difficult. Fortunately, however, MH also occurs in certain strains of pigs and in these animals the CHC test has proved to be a highly reliable predictor of MH reactions.

In our hands the most reliable statistical technique for choosing the reference point between the various parameters of the CHC test is based on 95% confidence limits for the CSC-H. The value obtained is then adjusted to yield the lowest number of false negative values in the probands, i.e. to provide maximum sensitivity. For halothane contractures and CSC, we have had most reliable results by using a reference value that yields the lowest number of false positive values in the control patients, i.e. to provide maximum specificity.

Setting the dividing point between normal and MHS CSCs low, and that between normal and MHS halothane contractures high produces a very specific but insensitive test that tends to yield excessive numbers of false negative values. The danger is that individuals wrongly classified as normal will be given triggering drugs during subsequent anaesthesia with potentially fatal results. On the other hand, choosing a dividing point between normal and MHS CSCs that is high, and that between normal and MHS halothane contractures that is low produces a very sensitive test that is, however, not very specific and which is, therefore, apt to yield excessive numbers of false positive values. The problem then is that individuals erroneously classified as MH will have to face lifetimes of inconvenient restrictions. In the authors opinion the

TABLE 4

COMPARISON OF SUCCINYLCHOLINE AND HALOTHANE PLUS SUCCINYLCHOLINE CONTRACTURES – MHS PATIENTS AND THEIR RELATIVES vs UNRELATED CONTROL PATIENTS: ACCORDING TO TYPE OF CHC TEST RESULT

		MHS PATIENTS AND THEIR RELATIVES				
		ALL TYPE H TESTS (HCK,HK, HC,H)	ALL TYPE C CHC TESTS (HCK,CK, HC,C)	ALL TYPE K TESTS	NEGATIVE	UNRELATED NORMAL PATIENTS
SUCCINYL-CHOLINE	x	0.35	0.25	0.02	0.00	0.00
	SE	0.18	0.10	0.01	0.00	0.00
	N	27	49	78	31	10
	T	0.89	0.82	0.84	0.00	--
	P	n.s.	n.s.	n.s.	n.s.	--
HALOTHANE + SUCCINYL-CHOLINE	x	1.53	0.14	0.39	0.08	0.16
	SE	0.26	0.16	0.04	0.14	0.12
	N	28	51	83	30	9
	T	2.21	1.90	1.31	1.08	--
	P	**	*	n.s.	n.s.	--

x = arithmetic mean of contracture amplitudes of skeletal muscle fascicles
SE = Standard error; N = sample size; T = Student's T-test; P = probability
* $P < 0.05$; ** $P < 0.01$; *** $P < 0.001$; n.s = not significant

FIGURE 7a

EFFECT OF SUCCINYLCHOLINE ON CONTRACTURES OF NORMAL AND MHS HUMAN SKELETAL MUSCLE FASCICLES IN THE ABSENCE OF HALOTHANE

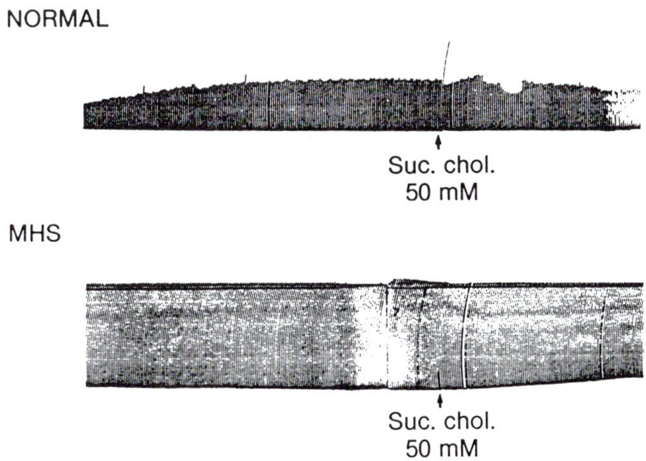

FIGURE 7b

EFFECT OF SUCCINYLCHOLINE ON CONTRACTURES OF NORMAL AND MHS HUMAN SKELETAL MUSCLE FASCICLES IN THE PRESENCE OF HALOTHANE

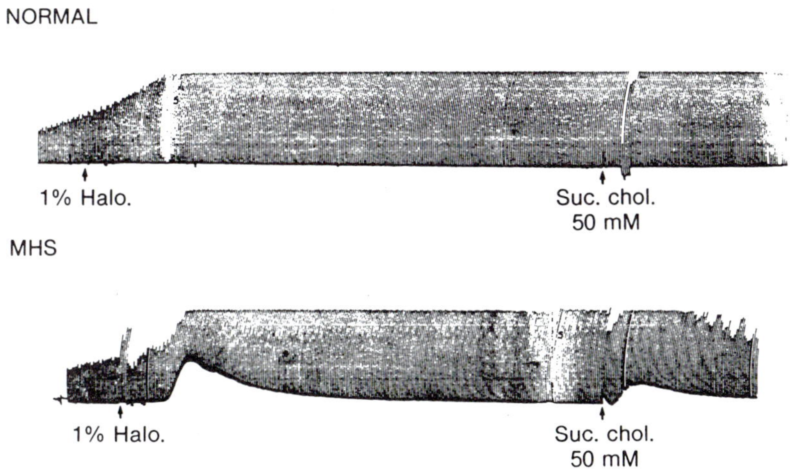

latter sin is preferable to the former since the latter may lead merely to inconvenience while the former may lead to death of the patient. Furthermore, confidence of clinical anaesthesiologists and anaesthetists would be undermined if a patient whose muscle biopsy had been falsely reported as negative subsequently developed a MH reaction during an anaesthetic with triggering drugs.

(s) Other Test Drugs

Rosenberg (42,43), has recently shown that the combined use of halothane followed by succinylcholine (50 mM) provides greater diagnostic differentiation than does the use of succinylcholine alone. For this combination of test drugs to be successful in detecting the MH trait it is essential that the bathing solution be equilibrated with halothane first. Then, while the bath is still being bubbled with halothane, succinylcholine (50 mM) is added. Values obtained in our laboratory are in essential agreement with these findings except that we have so far observed no discrimination with succinylcholine alone (Table 4, Figures 7a-b). We have also observed no positive relationship with type K CHC test results.

Removal of calcium from the bathing solution decreases halothane or caffeine induced contractures of MHS pig muscle fascicles (44) (Tables 5a-b).

Application of KCl alone to the bathing medium induces contractures but eliminates twitch tensions of MHS although not of normal muscle (45).

Equilibration of the muscle with procaine (Figures 8a-b) relieves caffeine- and halothane-induced contractures of MHS muscle (46). Equilibration with dantrolene relieves caffeine induced contractures of MHS and normal muscle (Figure 8c) (47). Differences and similarities in the effect of dantrolene on normal and MHS muscle have been discussed in detail in reference 47).

We have compared previously the effects of halothane, enflurane, isoflurane and methoxyflurane on MHS and normal muscle fascicles. Halothane shows the greatest ability to potentiate caffeine contractures, with enflurane and isoflurane following together and methoxyflurane showing the least ability. The

TABLE 5A

EFFECTS OF CALCIUM AND HALOTHANE ON CONTRACTURES OF NORMAL AND MHS PIG SKELETAL MUSCLE FASCICLES

Halo. Conc in Vol%	Status	Calcium Concentration in mM 0.0 N	0.0 MHS	1.0 N	1.0 MHS	2.0 N	2.0 MHS	4.0 N	4.0 MHS	F (Ca Doses)
0.00	x	0.00	0.00	0.00	0.00	0.00	0.04	0.00	0.02	1.57
	SE	0.00	0.00	0.00	0.00	0.00	0.03	0.00	0.02	ns
	N	9	7	9	7	9	7	9	6	
	t	0.00		0.00		-1.44		-0.99		
1.00	x	0.00	0.01	0.00	0.14	0.00	0.30	0.00	0.32	3.03
	SE	0.00	0.01	0.00	0.06	0.00	0.13	0.00	0.10	ns
	N	9	7	9	7	9	7	9	6	
	t	-1.01		-2.34*		-2.32*		-3.12**		
2.00	x	0.00	0.04	0.00	0.26	0.00	0.43	0.00	0.40	2.64
	SE	0.00	0.03	0.00	0.06	0.00	0.13	0.00	0.15	ns
	N	9	7	9	7	9	7	9	6	
	t	-1.44		-4.50***		-3.34**		-2.74**		
3.00	x	0.00	0.06	0.00	0.26	0.00	0.43	0.00	0.38	2.27
	SE	0.00	0.03	0.00	0.07	0.00	0.14	0.00	0.12	ns
	N	9	7	9	7	9	7	9	6	
	t	-1.92*		-3.75**		-3.04**		-3.13**		
	F (Halo Doses)	1.75 ns		10.97***		6.24***		6.83***		

x = arithmetic mean of contracture amplitude of skeletal muscle fascicles. SE = Standard error; N = sample size; T = Student's T-test; F = analysis of variance; P = probability; * P < 0.05; ** P < 0.01; *** P < 0.001; ns = not significant

TABLE 5B

EFFECTS OF CALCIUM AND CAFFEINE ON CONTRACTURES OF NORMAL AND MHS PIG SKELETAL MUSCLE FASCICLES

Caff. Conc in mM	Status	Calcium Concentration in mM 0.0 N	MHS	1.0 N	MHS	2.0 N	MHS	4.0 N	MHS	F (Ca Doses)
0.0	x	0.00	0.00	0.00	0.09	0.00	0.19	0.01	0.23	2.34
	SE	0.00	0.00	0.00	0.09	0.00	0.17	0.02	0.11	*
	N	7	7	7	7	7	7	7	7	
	t	0.00		-1.04		-1.12		-1.89		
2.0	x	0.01	0.20	0.01	0.61	0.06	0.87	0.04	1.53	2.89
	SE	0.02	0.07	0.02	0.18	0.04	0.29	0.03	0.45	*
	N	7	7	7	7	7	7	7	7	
	t	-2.54*		-3.34**		-2.79**		-3.33**		
8.0	x	0.64	0.53	0.47	1.00	0.56	1.17	0.62	2.27	2.89
	SE	0.56	0.19	0.11	0.31	0.19	0.29	0.24	0.61	*
	N	7	7	7	7	7	7	7	7	
	t	-0.25		-1.62		-1.77		-2.51		
	F(Caff Doses)	7.62***		57.41***		17.26***		58.66**		

x = arithmetic mean of contracture amplitude of skeletal muscle fascicles. SE = Standard error; N = sample size; T = Student's T-test; F = analysis of variance; P = probability; * P < 0.05; ** P < 0.01; *** P < 0.001; ns = not significant

FIGURE 8a

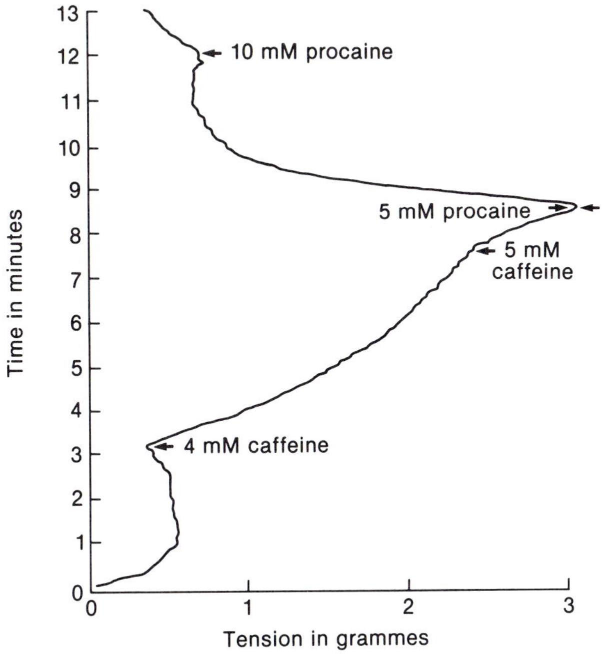

Moulds, R.F.W.; Denborough, M.A.
Britt. Med.J. 4 526 (1972)

Halothane potentiated caffeine-induced contracture of MHS skeletal muscle fascicle reversed by procaine hydrochloride.

greatest diagnostic discrimination between the MHS and the normal muscles is provided by halothane (47) (Table 6).

Ryanodine, an alkaloid extracted from Ryania Speciosa, has been suggested as an alternate to caffeine as a test drug for the diagnosis of MH (48). As ryanodine is freer than caffeine of actions on structures other than the SR and since its effect on the SR is more potent than that of caffeine, ryanodine should be preferred as a test drug to caffeine. Unfortunately, secure commercial supplies of ryanodine are not available at the present time.

To conclude, nearly two decades of experience has shown the CHC test to be a reliable means of diagnosing MH since:

FIGURE 8b

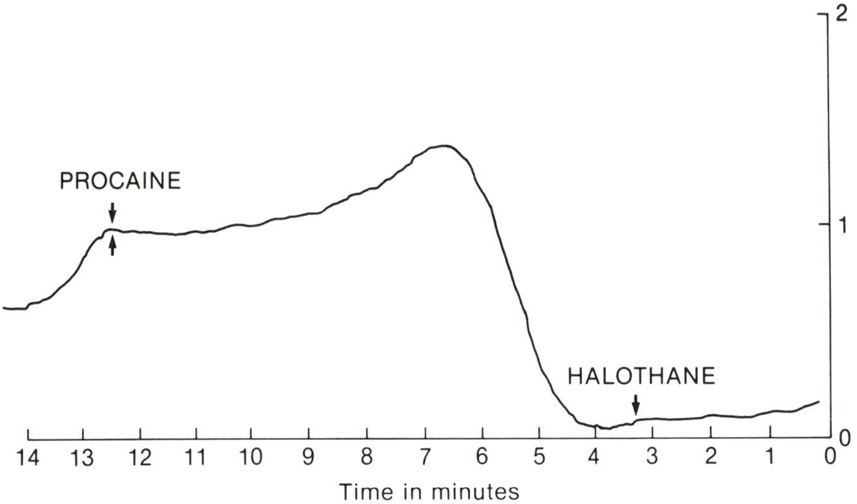

Moulds, R.F.W.; Denborough, M.A.
Britt. Med.J. 4 526 (1972)

Halothane induced contracture of MHS human skeletal muscle fascicle reversed by 5 mM procaine hydrochloride.

FIGURE 8c

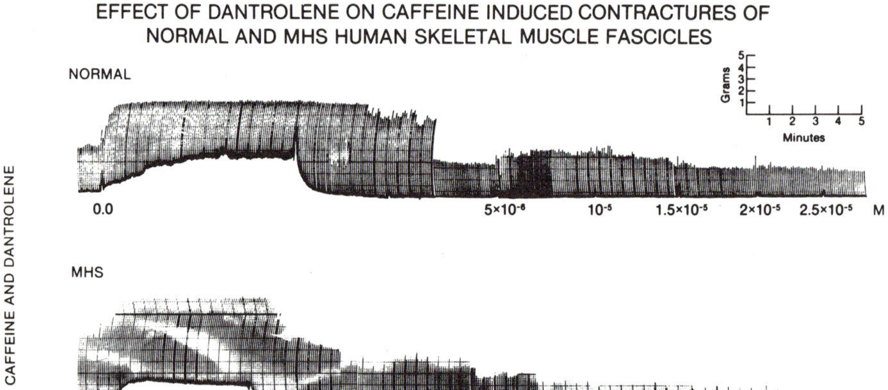

Effect of dantrolene on caffeine-induced contractures of normal and MHS human skeletal muscle fascicles

TABLE 6

CAFFEINE SPECIFIC CONCENTRATIONS[a] IN THE PRESENCE AND ABSENCE OF ANAESTHETICS

	Caffeine with				
	--	Methoxyflurane	Enflurane	Isoflurane	Halothane
Control Patients	6.85 + 0.42[b] (34)[c]	5.04 + 0.41 (19)	3.69 + 0.55 (16)	3.24 + 0.75 (10)	1.92 + 0.13 (31)
MHS Patients	2.42 + 0.25[b] (20)[c]	1.82 + 0.28 (11)	1.49 + 0.32 (9)	1.23 + 0.32 (11)	0.62 + 0.09 (20)
Ratio	2.82 + 0.32[b]	2.77 + 0.45	2.48 + 0.66	2.64 + 0.97	3.12 + 0.44
t[d]	9.70***	6.82***	3.85**	3.08*	8.72***

[a] Dose of caffeine required to raise the resting tension of skeletal muscle fascicles by 1.0 gram.

[b] Geometric mean + its standard error.

[c] Number of subjects

[d] t-statistic comparing caffeine specific concentrations of patients and controls. The levels of statistical significance are based on Duncan's multiple range test.

* P < 0.05

** P < 0.01

*** P < 0.001

REPRINTED FROM: Britt, B.A. et al. Can. Anaesth. Soc. J. 1980;27:12.

1. it provides several parameters in which MHS muscle differs from control muscle;

2. these changes are not seen in other myopathies;

3. when all parts of the test are performed, there is only modest overlapping between the MHS and the normal muscle.

CAFFEINE SKINNED FIBRE TENSION TEST (20,49-52)

INTRODUCTION

The principle of the caffeine skinned fibre tension (CSFT) test is the same as that of the CHC test except that:

1. only one fibre instead of many are isometrically mounted in a bathing chamber; and

2. each fibre has had its sarcolemma, gap junctions (excitation-contraction coupling step between the transverse tubules and sarcoplasmic reticulum (SR)) and mitochondria chemically destroyed or "skinned" (20).

METHODS (Figures 9a-c)

(a) Dissection (20)

The method of dissection is identical to that of muscle being removed for the CHC test except that the specimen is much smaller in size. A sample 1.0 cm long by 0.2 cm wide, weighing between 50 and 100 mg is ample. It is especially vital to ensure that: prior to dissection the muscle be tied with 5° black silk sutures to the stick under a very slight amount of tension so that it clings tightly to the stick and cannot artifactually contract; that as much as possible of the muscle to be dissected is secured within the sutures, leaving the least amount possible hanging free; and that the muscle be dissected meticulously and precisely along the natural lines of cleavage between fascicles, with the long cuts parallel to the long sides of the muscle cells.

(b) Skinning of Muscle Cells (20)

The muscle sample on its stick is placed immediately in a "relaxing" solution (Table 7a). Within the first hour each fascicle is transferred at least three times to a fresh "skinning" or

FIGURE 9a

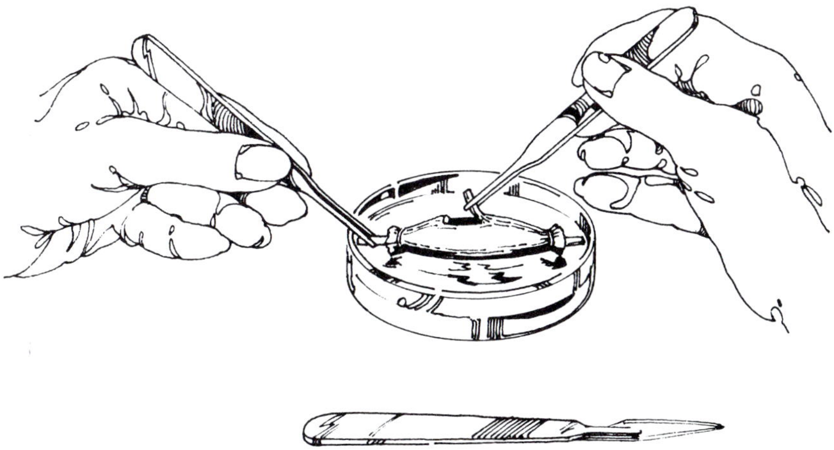

Dissection of single chemically skinned fibre from main muscle bundle under a low power microscope (not shown). Shortening of main muscle bundle is prevented by tying bundle to a balsa stick before commencement of surgical dissection.

"relaxing" solution following which the fascicles in their "skinning" solution are kept in a refrigerator at 4°C for 24 hours. The fascicles are then transferred to a "skinning-storage" solution which is identical in all ways to the "skinning" solution except for the addition of 50 per cent glycerol. Over a period of several days at -20°C this solution destroys or skins off the sarcolemma, gap junctions and mitochondria. The final preparation, ready for use after four to ten days consists, therefore, of only SR plus myofibrils.

(c) Mounting of Skinned Fibres (20)

When skinning is functionally complete, single fibres are dissected from the main fascicle (Figure 9a) under a 40 power Swift Model 31-400-00 binocular microscope. Each is then mounted

FIGURE 9b

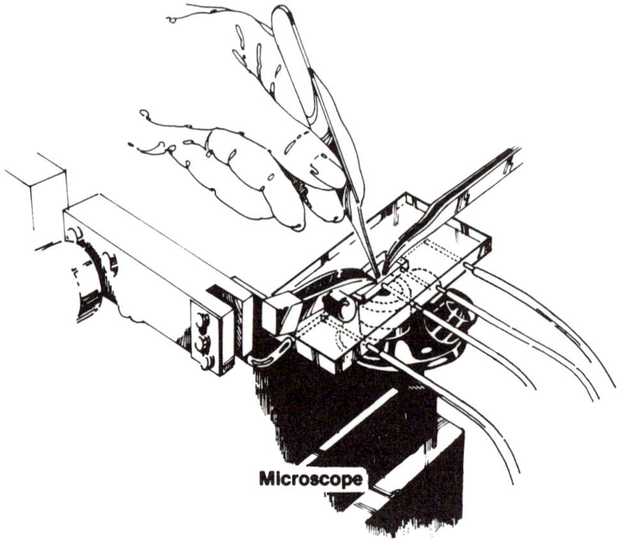

Mounting of single chemically skinned fibre within a hollow plastic bathing chamber between two clamps. One clamp is attached to a force displacement transducer. The other clamp is attached to a micromanipulator (not shown). Plastic tubes entering and leaving the well of the chamber add and remove various bathing solutions. Tubes entering and leaving the hollow wall of the bathing chamber bring and remove water maintained at the desired temperature by a water heater (not shown). The fibre is viewed during dissection by an inverted microscope (see Figure 9c).

horizontally between two clamps in a bathing chamber (Figure 9b) filled with calcium-free "skinning" solution (Table 7a). One of these clamps is attached to a micromanipulator and the other to a Grass Model FT.03C force displacement transducer. The mounted skinned fibres are each viewed through a 400 power Olympus Model KCP-TR-2 inverted microscope (Figure 9c). Changes in tension are recorded on a Gould-Brush Model 2200 square wave polygraph.

For every patient at least five fibres are assessed. The mean values of all fibres from each patient are then calculated.

SOLUTIONS FOR THE CAFFEINE SKINNED FIBRE TENSION TEST

TABLE 7a

SKINNING SOLUTION

K Propionate	170.0 mM
EGTA	5.0 mM
Imidazole	10.0 mM
Mg Propionate	2.5 mM
ATP	2.5 mM

TABLE 7b

CONTRACTING SOLUTION

K Propionate	185.0 mM
Imidazole	10.0 mM
Mg Propionate	2.5 mM
Caffeine *	40.0 mM
ATP	2.5 mM

pH = 7.0

TABLE 7c

WASHING SOLUTION

K Propionate	185.0 mM
Imidazole	10.0 mM
Mg Propionate	2.5 mM
ATP	2.5 mM

pH = 7.0

TABLE 7d

CALCIUM SOLUTIONS

pCa	Ca EGTA	EGTA	Mg	K	Imid.	ATP
6.8	1.15 mM	3.85 mM	2.46 mM	172 mM	10 mM	2.5 mM
6.4	2.10 mM	2.85 mM	2.44 mM	172 mM	10 mM	2.5 mM
6.0	3.27 mM	1.73 mM	2.42 mM	172 mM	10 mM	2.5 mM
5.6	4.14 mM	0.86 mM	2.40 mM	172 mM	10 mM	2.5 mM
5.2	4.64 mM	0.36 mM	2.40 mM	172 mM	10 mM	2.5 mM
5.0	4.78 mM	0.22 mM	2.40 mM	172 mM	10 mM	2.5 mM
4.8	4.88 mM	0.11 mM	2.40 mM	172 mM	10 mM	2.5 mM

pH = 7.02

*Larger or smaller caffeine concentrations may be used to obtain a caffeine dose-response curve.

FIGURE 9c

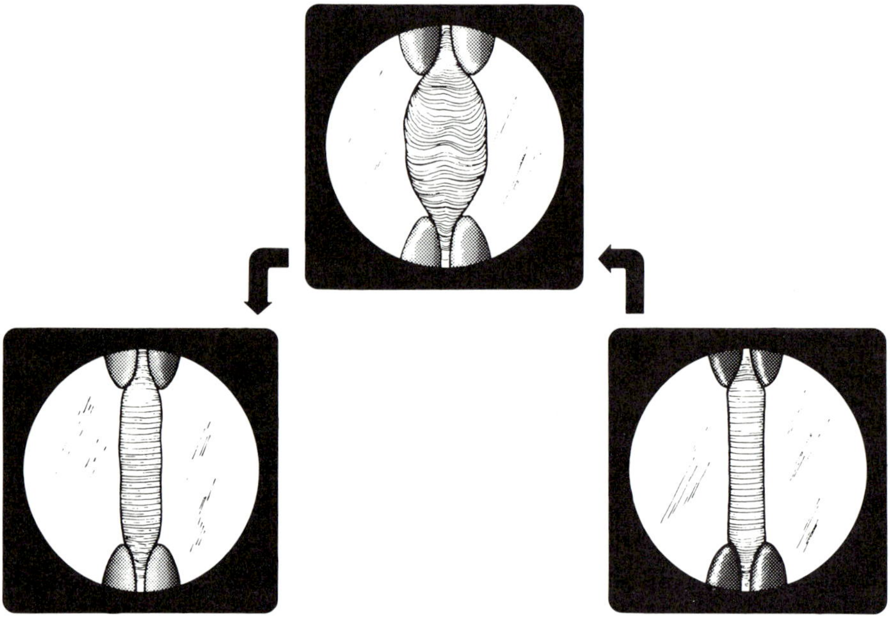

LOWER RIGHT. Single chemically skinned fibre viewed through inverted microscope in the absence of caffeine and in the presence of subcontractile dose of calcium within the sarcoplasmic reticulum.
TOP CENTRE. Same fibre after the addition of caffeine.
LOWER LEFT. After calcium released from the sarcoplasmic reticulum by the caffeine has spontaneously been reuptaken into the sarcoplasmic reticulum.

(d) Determination of Caffeine-Induced Tension Increase (Calcium Release from the Sarcoplasmic Reticulum) of Single Chemically Skinned Fibres (20)

Ca++ = pCa 6.8 is added to the "skinning" solution bathing each skinned fibre. After 30 seconds the SR having been loaded with calcium, the solution is removed from the chamber and replaced with a "contracting" solution which contains caffeine but no ionized calcium (Table 7b). The ensuing increase in tension of each muscle cell is due to release of calcium from the preloaded SR (since the exogenous solution bathing the SR-myofibrillar complex is calcium free). The procedure is repeated several times using progressively larger concentrations of caffeine so that dose-tension curves are obtained. Between every set of calcium and caffeine additions each fibre is thoroughly rinsed at least twice with a "washing" solution (Table 7c). Then any calcium still remaining in the SR is removed by adding caffeine 40 mM followed by further application of "skinning" (Table 7a) and "washing" (Table 7c) solutions. From the caffeine-tension curves the dose of caffeine required to initiate tension rise (Sk-Caff-D) and the maximum tension amplitude in kg/cm^2 (Sk-Caff-Max) is calculated.

(e) Determination of Calcium-Induced Tension Increase of Single Chemically Skinned Fibres (20)

The bathing chamber is filled with a "skinning" solution (Table 7a) which induces movement of free calcium out of the SR thus depleting it of calcium. Then after several applications of "washing" solution (Table 7c) the bathing chamber is filled with a solution that contains no caffeine but which does contain ionized calcium (Table 7d). This procedure is repeated several times using incremental concentrations of calcium so that a calcium-tension curve is obtained. Tension increases are considered to be due to the direct effect of the exogenous calcium since no calcium is available in the SR for release. From the calcium-tension curves the dose of calcium required to initiate tension rise (Sk-Ca-D) is then calculated.

(f) Determination of Adequacy of Skinning and Viability of Sarcoplasmic Reticulum (20)

Evidence for the fibres being functionally skinned and viable for the study of Ca^{++} uptake and release from the SR come from the following observations:

1. electron micrographs of the fibres show the sarcolemma to be disrupted, the transverse tubules to be separated from the SR the mitochondria to be absent or abnormal and the myofibrils and the SR to be intact;
2. after addition of caffeine to fibres preloaded with subcontractile doses of calcium or after addition of contractile doses of calcium to fibres predepleted of calcium, tensions after initial

rises always return to zero. This shows that the SR is able to take up enough free calcium from the myoplasm to allow complete relaxation to occur;

3. both with and without calcium the fibres develop rigor tension in the absence of ATP and relaxation returns only with the restoration of ATP;

4. at the end of examination of each fibre the dose of calcium which has been initially determined to induce maximum tension is reapplied. Values obtained on fibres whose maximum tension at the end of each experiment is less than 80 per cent of the maximum tension at the beginning of the experiment are excluded from statistical analysis of the calcium-tension curves.

(g) Classification of Patients

Muscle for the caffeine skinned fibre tension test has been classified as previously described above.

TABLE 8

COMPARISON OF CAFFEINE SKINNED FIBRE TENSION TEST VALUES - MHS PATIENTS AND THEIR RELATIVES (BY DEGREE OF RELATIONSHIP) vs UNRELATED NORMAL PATIENTS

	MHS PATIENTS	MHS RELATIVES			UNRELATED NORMAL PATIENTS
		1st	2nd	3rd	
x	2.88	2.71	2.98	3.26	3.43
SE	0.13	0.10	0.16	0.25	0.23
N	102	172	62	41	57
T	2.29	3.29	1.60	0.48	--
P	**	***	n.s.	n.s.	--

x = arithmetic mean of dose of caffeine required to initiate tension increase in single chemically skinned fibres
SE = Standard error; N = sample size; T = Student's T-test; P = probability
* P < 0.05; *** P < 0.001; n.s = not specific

RESULTS

The mean threshold for caffeine induced tension rise, i.e. the Sk-Caff-D is significantly lower in fibres from MHS patients and in first degree MHS relatives than in fibres from normal patients (Table 8). Furthermore, the mean caffeine induced tension rise is lower in patients with CHC test results positive for MH than in those with CHC test values negative for MH (Table 9). Thus in MHS fibres a smaller concentration of caffeine is required to initiate calcium release from the SR and thence to induce a rise in tension of the myofibrils than in the normal fibres.

Sk-Caff-Max of the MHS fibres are statistically similar to the Sk-Caff-Max of the normal fibres (20). The difference between the two groups, therefore, appears to be due to the rate at which caffeine induces release of calcium from the SR and not to the total amount of caffeine-induced calcium release from the SR.

In skinned fibres that have been depleted of calcium, the Sk-Ca-D for the MHS fibres does not differ significantly from the Sk-Ca-D for the normal fibres (20). Hence the direct responsiveness of the MHS myofibrils to exogenous calcium is, therefore, similar to that of the normal myofibrils and the MH defect cannot be accounted for by an abnormality of the myofibrillar apparatus.

INTERPRETATION OF DATA

Although substantial differences exist between Sk-Caff-Ds of MHS muscle (patients and first degree relatives) and Sk-Caff-Ds of normal muscle (Table 8) these differences are consistently less than those observed between CSCs, CSC-Hs and halothane-contracture amplitudes of MHS muscle (patients and all relatives) on the one hand and normal muscle on the other hand. Consequently, while in our hands the CSFT test may be complementary to the CHC test we do not, at present, visualize it as being a substitute for the CHC test. However, use of halothane as an additional test drug, and of separate chambers for each different bathing solution, may, as Japanese workers have indicated (see Chapter 11) improve the diagnostic discrimination of this test.

TABLE 9

COMPARISON OF CAFFEINE SKINNED FIBRE TENSION TEST VALUES - MHS PATIENTS PLUS MHS RELATIVES vs UNRELATED NORMAL PATIENTS ACCORDING TO TYPE OF CHC TEST RESULT

	MHS PATIENTS AND THEIR RELATIVES				
	ALL TYPE H TESTS (HCK, HK, HC, H)	ALL TYPE C CHC TESTS (HCK, CK, HC, C)	ALL TYPE K TESTS	NEGATIVE	UNRELATED NORMAL PATIENTS
x	2.38	2.32	2.87	3.69	3.43
SE	0.14	0.11	0.09	0.11	0.24
N	69	160	175	142	57
T	3.97	4.73	2.65	1.06	--
P	***	***	***	n.s.	--

x = arithmetic mean of contracture amplitudes of skeletal muscle fascicles
SE = Standard error; N = sample size; T = Student's T-test; P = probability
* P < 0.05; ** P < 0.01; *** P < 0.001; n.s = not significant

The CSFT test has both advantages and disadvantages when compared with the CHC test. A much smaller volume of muscle is required. The skinned muscle bundle can be stored in the freezer for long periods of time, and the test can be repeated on one sample many times (for each experiment a single cell is dissected free from the main bundle and the latter is then returned to cold storage). On the other hand, more skill is required. The equipment required is more expensive. The CSFT test cannot be begun until several days after excision of the muscle from the patient. Moreover, the CSFT test is time consuming since many fibres (at least five and preferably more) must, because of equipment limitations, be measured sequentially and not simultaneously as with the CHC test. The reason many fibres have to be measured is because in less severely afflicted MHS muscle, not all fibres are abnormal. Much greater care must be taken to prevent contamination of the solutions, chambers and instruments with unwanted calcium (water that has been through only one deionizing plus one glass distillation is usually inadequate). Even the slightest contamination of the distilled water with calcium ruins the CSFT test.

Takagi (53) and Endo and Blinks (54) have reported that the threshold for halothane contracture of single mechanically skinned fibres prepared from MHS humans is less than normal. Mechanically skinned MHS fibres have also been prepared by Hahn and Godt (55). This technique will probably not play a major role in the diagnosis of MH since the technical skill required to skin fibres mechanically is much greater than to skin fibres chemically and since very fresh specimens are essential to maintain viability of the muscle cells during mechanical skinning.

ATP DEPLETION TEST (7,56,57)

INTRODUCTION

MHS skeletal muscle exposed in vivo (21,58,59) or in vitro (7,56) to halothane loses its ATP at a faster rate than does normal muscle. This observation was made by Harrison many years ago

into a test useful in screening for MH in pigs (56). We have since slightly adapted this test to serve as a complement to the CHC test in human patients.

METHODS (Figures 10a-b)

Two muscle slices each 100 to 300 mg are rapidly removed from the patient and then each is immediately placed in a bath of Krebs Ringer solution (Table 1a) at 37°C and bubbled with carbogen. To one (but not the other) bath 4.0 vol% halothane is

FIGURE 10a

Equilibration of Skeletal Muscle Samples for ATP Depletion Test

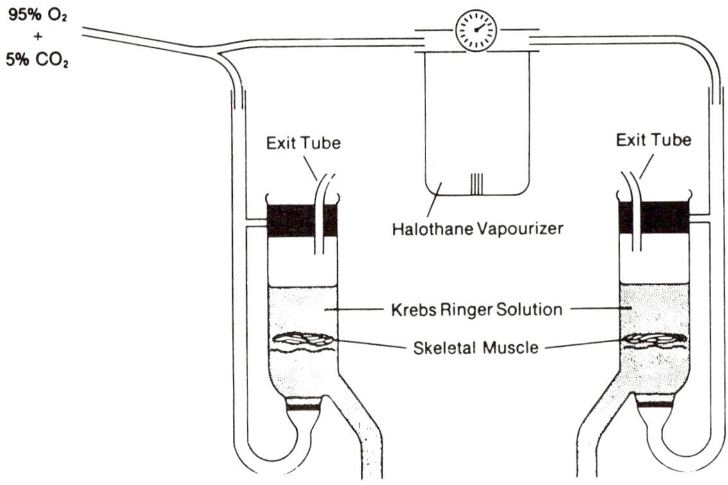

added from a Fluotec Mark 3 via the carbogen line. Equilibration of each sample is continued for 30 minutes. This time period is 15 minutes less than that recommended by Harrison (30) but we have found during preliminary experiments that this shorter equilibration has no significant effect on the final ATP depletion ratio. At the termination of the equilibration the two muscle specimens are simultaneously removed from their baths, "pancaked" in a brass clamp precooled in liquid nitrogen (Figure 10b), and then placed in liquid nitrogen where they are stored until needed for ATP measurement.

FIGURE 10b

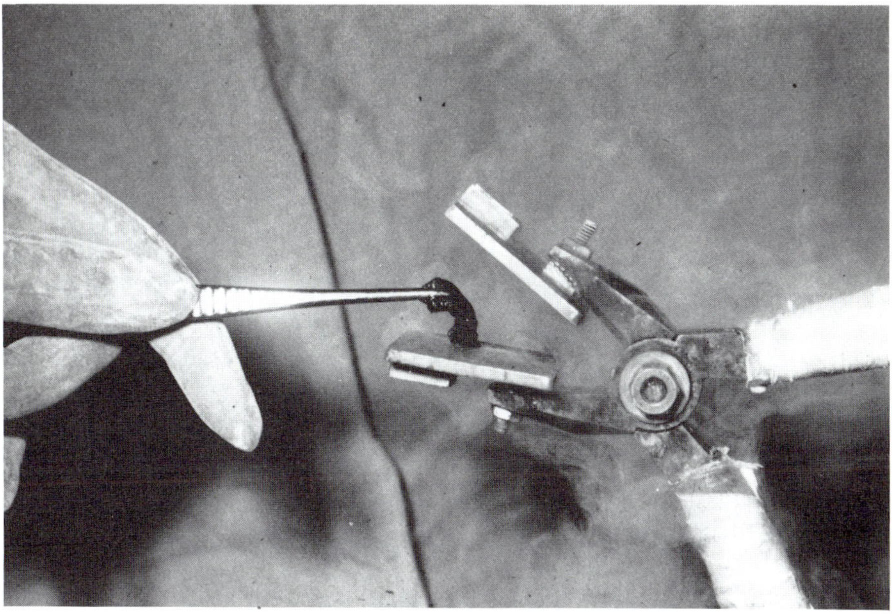

Brass clamp precooled in liquid nitrogen rapidly "pancakes" and freezes skeletal muscle sample prior to measurement of ATP and after equilibration in carbogen with or without halothane.

All the following steps are performed in a cold room at 2°C with the muscle lying on ice. Each frozen muscle specimen is sliced very thin. The still frozen slices are transferred to a preweighed and precooled Thomas flask containing 2 ml of 12% trichloroacetic acid. After weighing the flask, the weight of the muscle is calculated by subtraction. The muscle in its acid is homogenized for one minute by a motor driven Teflon pestle cooled by an ice jacket. The homogenate is centrifuged in a refrigerated International centrifuge model B20A using an 870 Fixed Angle Rotor head at 4200 x g for 20 minutes. The remaining steps are performed at 23°C. The pH of the supernatant is adjusted to 7.2 with KOH. The ATP concentration of the neutralized supernatant is determined by a Sigma kit #366-uv. This kit is based on the following equations:

	phosphoglyceric phosphokinase	
ATP + 3-phospho-glyceric acid + NADH	⟶	ATP + 2,3-diphospho-glyceric acid

	3-phosphoglyceral-dehyde dehydrogenase	
I,3-diphosphogly-ceric acid + NADH	⟶	3-phosphoglyceral-dehyde + NAD + H_3PO_4

The disappearance of NADH is observed at 340 nm in a Beckman 25 spectrophotometer. The values are expressed as uM of ATP per gram of wet frozen muscle.

The following ATP ratio is then calculated:

$$\frac{\text{ATP in muscle equilibrated with 4.0 Vol\% halothane}}{\text{ATP in muscle equilibrated without halothane}}$$

The above test may also be performed in the presence of 4.0 or 8.0 mM caffeine.

CLASSIFICATION OF MUSCLE

The muscle is classified as previously described in this paper.

RESULTS

ATP depletion ratios are significantly lower in the muscle of MHS patients than in the muscle of normal patients (Table 10). However, no differences are observed in the ATP ratio between the muscle of MHS relatives and the muscle of normal patients. ATP depletion ratios are lower than normal in patients with CHC test values positive for MH, in the presence of halothane alone or caffeine alone, but not with CHC test values positive for MH solely in the combined presence of caffeine plus halothane (Table 11).

INTERPRETATION OF DATA

For ATP ratios diagnostic discrimination between MHS and normal muscle is much poorer than for the CHC test. Significance is achieved only for muscle from MHS patients and not for muscle from MHS relatives, not even first degree relatives. It can only be considered as a modest complement to the CHC test and never as a sole diagnostic test. Its main advantages are: that no skill is required for dissection of the muscle--any shape will do; that the

TABLE 10

COMPARISON OF ATP DEPLETION RATIOS - MHS PATIENTS AND MHS RELATIVES (BY DEGREE OF RELATIONSHIP) vs UNRELATED NORMAL PATIENTS

		MHS PATIENTS	MHS RELATIVES 1st	MHS RELATIVES 2nd	MHS RELATIVES 3rd	UNRELATED NORMAL PATIENTS
ATP	x	0.89	0.93	0.94	0.94	1.00
DEPLETION	SE	0.02	0.02	0.03	0.05	0.04
RATIOS	N	209	326	132	74	102
	T	-2.75	1.39	1.09	0.82	--
	P	***	n.s.	n.s.	n.s.	--

x = arithmetic means of

$$\frac{\text{ATP equilibrated with carbogen + halothane}}{\text{ATP equilibrated with carbogen}}$$

SE = Standard error; N = sample size; T = Student's T-test; P = probability; * P < 0.05; ** P < 0.01; *** P 0.001; n.s = not significant

instrumentation and technique required are available in any routine hospital laboratory and that the reagents can be bought in kit form from companies such as Sigma Chemical Company.

RADIOACTIVE CALCIUM UPTAKE INTO THIN MUSCLE SECTIONS

INTRODUCTION

Mabuchi and Sreter have described a very simple method of measuring calcium uptake into cryostat skeletal muscle sections (60). They have suggested that such calcium uptake is less into MHS than into normal muscle (Mabuchi, K. and Sreter, F.A. Unpublished data, 1982).

METHODS

Carefully and rapidly excised skeletal muscle fascicles (about 1.0 cm long by 0.3 cm thick) are immediately frozen in liquid nitrogen for transport to the laboratory. There the muscle sections are cut to 10 u thickness in a Tissue Tek II cryostat and then are mounted on disposable plastic cover slips. The muscle on its slip is placed in a reaction medium of 0.1 M KCl, 20 mM Tris

TABLE 11

COMPARISON OF ATP DEPLETION RATIOS – MHS PATIENTS PLUS MHS RELATIVES vs UNRELATED NORMAL PATIENTS ACCORDING TO TYPE OF CHC TEST RESULT

		MHS PATIENTS AND THEIR RELATIVES				
		ALL TYPE H TESTS (HCK, HK, HC, H)	ALL TYPE C CHC TESTS (HCK, CK, HC, C)	ALL TYPE K TESTS	NEGATIVE	UNRELATED NORMAL PATIENTS
ATP DEPLETION RATIOS	x	0.87	0.82	0.94	1.00	1.00
	SE	0.03	0.01	0.02	0.02	0.04
	N	92	293	362	271	99
	T	2.40	5.04	1.44	0.03	--
	P	**	***	n.s.	n.s.	--

$x = $ arithmetic mean of $\frac{\text{ATP equilibrated with carbogen + halothane}}{\text{ATP equilibrated with carbogen}}$

SE = Standard error; N = sample size; T = Student's T-test; P = probability

* P < 0.05; ** P < 0.01; *** P < 0.001; n.s = not significant

maleate (pH = 7.0), 5 mM MgCl2, 5 mM NaAzide, 50 mM tris oxalate (pH = 7.0), 0.1 mM CaCl2 (containing about 0.5 uCuries of Ca^{45}), 0.11 mM EGTA and 3 mM ATP. After constant stirring for 15 minutes at 37°C the muscle on its slip is transferred for 30 seconds to a solution (5% acetic acid and 10% ethanol) which stops the uptake of the calcium into the muscle. Following air drying the muscle still on its slip is placed in a Beckman liquid scintillation counter where the radioactivity is counted.

CLASSIFICATION OF MUSCLE

The muscle from the MHS and normal humans is classified as previously described in this paper.

RESULTS

Mabuchi and Sreter have reported a consistently lower uptake of calcium into thin muscle sections isolated from MHS humans than into the thin muscle sections isolated from normal humans (60).

We are presently performing this test and comparing the results with the CHC test in all our MHS and normal patients using the methodology of Mabuchi and Sreter (60). Our preliminary results are given in Table 12. This data shows no significant difference in calcium uptake into thin muscle sections between normal and MHS individuals. Work in press by Muldoon (61) in radioactive calcium uptake into thin skeletal muscle sections also shows no differences between MHS and normal patients.

INTERPRETATION OF DATA

Clearly further work remains to be done to clarify whether or not this is a useful diagnostic or screening test for the MH trait. It is unfortunate that Mabuchi's and Sreter's patients have not also had CHC tests performed to confirm the validity of the clinical diagnosis of MH reactions.

TABLE 12

Ca^{45++} UPTAKE INTO THIN SKELETAL MUSCLE SECTIONS EXCISED FROM NORMAL AND MHS HUMANS

	MHS PATIENTS	UNRELATED CONTROL PATIENTS
x	0.04	0.05
SE	0.06	0.01
N	60	21
T	-1.04	
P	ns	

x = arithmetic means of Ca^{45++} uptake in uM into thin skeletal muscle sections. SE = standard error; N = sample size; T = Student's T-test; P = probability; n.s. = not significant

ACKNOWLEDGEMENTS

I wish to thank my colleagues who generously provided the information found in Appendices 1a-p. As well I gratefully acknowledge the work of my technicians Mrs. Wanda Frodis and Mrs. Elizabeth Scott whose expertise in performance of the caffeine halothane contracture test and the caffeine skinned fibre tension test has provided consistent and accurate reporting of our results.

REFERENCES

1. Anderson, I.L. and Jones, E.W. In: Aldrete, J.A. and Britt, B.A. (eds.). Second International Symposium on Malignant Hyperthermia. 1977. New York: Grune & Stratton, pp. 509-534.

2. O'Brien, P.J., Cribb, P.H., White, R.J., Olfert, E.D. and Steiss, J.E. Can. Vet. J. 1983;24:172.

3. Strobel, G.E. and Bianchi, C.P. Anesthesiology 1971;35:465.

4. Suarez-Kurtz, G. and Sudo, R.T. Muscle & Nerve 1986;9:47.

5. Britt, B.A. Can. Anaesth. Soc. J. 1973;20:431.

6. Britt, B.A. Ann. Roy. Coll. Surg. Engl. 1971;48:73.

7. Britt, B.A. and Endrenyi, L. Can. Anaesth. Soc. J. 1976;23:624.

8. Britt, B.A., Endrenyi, L., Frodis, W., Scott, E. and Kalow, W. Can. Anaesth. Soc. J. 1980;27:12.

9. Britt, B.A., McComas, A.J., Endrenyi, L. and Kalow, W. Anesthesiology 1977;47:490.

10. Brownell, A.K.W. and Szabo, M. Canad. Anaesth. Soc. J. 1982;29:218.

11. Denborough, M.A. Med. J. Australia 1982;3:204.

12. Ellis, F.R., Halsall, P.J., Ording, H., Fletcher, R., Ranklev, E., Heffron, J.J.A., Lehane, M., Mortier, W., Steinbereitner, K. and Theunynck, D. Brit. J. Anaesth. 1984;56:1267.

13. Eng, G.D., Becker, M.J. and Muldoon, S.M. Muscle & Nerve 1984;7:2185.

14. Halsall, P.J. and Ellis, F.R. Br. J. Anaesth. 1979;51:753.

15. Lambert, W. Can. Anaesth. Soc. J. 1980;27:304.

16. Nelson, T.E. Clin. Exp. Pharmacol. Physiol. 1975;2:49.

17. Ording, H., Ranklev, E. and Fletcher, R. Brit. J. Anaesth. 1984;56:1183.

18. Rosenberg, H. Can. Anaesth. Soc. J. 1980;27:303.

19. Fletcher, J.E., Rosenberg, H. and Hilf, M. Can. Anaesth. Soc. J. 1984;31:377.

20. Britt, B.A., Frodis, W., Scott, E., Clements, M-J. and Endrenyi, L. Can. Anaesth. Soc. J. 1982;29:550.

21. Britt, B.A. In: Milton, A.S. (ed.). Handbook of Experimental Pharmacology, Pyretics and Antipyretics. 1982: Heidelberg; Springer-Verlag, pp. 547-615.

22. Kalow, W., Britt, B.A., Terreau, M.E. and Haist, C. Lancet 1970;ii:895.

23. Radin, J.J., Weiser, M.G. and Fettman, M.J. Lab. Animal Sci. 1986;36:425.

24. Pon, W.G. and Houpt, K.A. The Biology of the Pig. 1978; Cornstock Publishing Associates, Cornell University Press, Ithaca, N.Y., pp. 244-275.

25. Hill, B.S. Miniature Swine. Ch. River Digest, 1983, p. 22.

26. Imeah, P. and McTaggart, H.S. In: Archer, R.K. and Jeffcott, L.B. (eds.). Comparative Clinical Haematology. Blackwell Scientific Publications, Oxford, England, 1977, pp. 271-291.

27. Schalm, O.W., Jain, N.C. and Carroll, E.J. In: Veterinary Haematology, 3rd ed. Lea and Febiger, Philadelphia, Pa., 1975, pp. 82-218.

28. Third International Workshop of Malignant Hyperthermia Investigators, Banff, Alberta, September, 1982.

29. Britt, B.A., Endrenyi, L., Scott, E. and Frodis, W. Can. Anaesth. Soc. J. 1980;27:1.

30. Galen, R.S. and Gambino, S.R. Beyond Normality: The Predictive Value and Efficiency of Medical Diagnosis. 1975. New York: John Wiley and Sons, Inc.

31. Britt, B.A. and Scott, E.A. Anesth. Analg. 1986;65:171.

32. Moulds, R.F.W. and Denborough, M.A. Br. Med. J. 1974;2:245.

33. Anderson, I.L., Rawstron, R.E. and Dunlop, D.J. NZ Med. J. 1980;91:417.

34. Rosenberg, H. and Reed, S. Anesth. Analg. 1983;62:415.

35. Fletcher, J.E. and Rosenberg, H. Anesthesiology 1985;63:190.

36. Rosenberg, H. and Gronert, G.A. Can. Anaesth. Soc. J. 1984;31:115.

37. Nelson, T.E., Flewellen, E.H. and Gloyna, D.F. Anesth. Analg. 1983;62:545.

38. Bradley, W.G., Ward, M., Murchison, D., Hall, L. and Woolf, N. Proc. Roy. Soc. Med. 1973;66:67.

39. Ellis, F.R. In: Gordon, R.A., Britt, B.A. and Kalow, W. (eds.). International Symposium on Malignant Hyperthermia. 1973. Springfield: Charles C. Thomas, p. 415.

40. Walter, F., Vulliemoz, Y., Verosky, M. and Triner, L. Anesth. Analg. 1980;59:856.

41. Verburg, M.P"., Van Bennekom, C.A., Oerlemans, F.T. and De Bruyn, C.H.M.M. In: De Bruyn, C.H.M.M., Simmonds, H.A. and Muller, M.M. (eds.). Purine metabolism in man. IV. Part B. 1984. New York: Plenum Publishing Corporation, p. 443.

42. Rosenberg, H. and Fletcher, J.E. Anesthesiology 1985;62:212.

43. Rosenberg, H. Presented at the 4th International Malignant Hyperpyrexia Workshop, York, September, 1986.

44. Nelson, T.E., Bedell, D.M. and Jones, E.W. Anesthesiology 1975;42:301.

45. Moulds, R.F.W. and Denborough, M.A. Brit. Med. J. 1974;2:241.

46. Moulds, R.F.W. and Denborough, M.A. Brit. Med. J. 1972;4:526.

47. Britt, B.A., Scott, E., Frodis, W., Clements, M-J. and Endrenyi, L. Can. Anaesth. Soc. J. 1984;31:130.

48. Casson, H. and Downes, H. In: Aldrete, J.A. and Britt, B.A. (eds.). Second International Symposium on Malignant Hyperthermia. 1978. Grune & Stratton, New York, p. 3.

49. Wood, D.S. Exper. Neurol. 1978;58:218.

50. Wood, D.S., Mozo, A. and Willner, J.H. Abstract. American Academy of Neurology, 1978.

51. Wood, D.S. IVth International Congress on Neuromuscular Diseases, Montreal, Quebec, 1978.

52. Wood, D.S., Zollman, J., Reuben, J.P. and Brandt, P.W. Science 1975;187:1075.

53. Takagi, A. Hiroshima J. Anaesth. 1977;13:155.

54. Endo, M. and Blinks, J.R. Nature New Biol. 1973;246:218.

55. Hahn, M.J. and Godt, R.E. Presented at the 2nd International Symposium on Malignant Hyperthermia, Denver, Colorado, April, 1977.

56. Harrison, G.G., Saunders, S.J., Biebuyck, J.F., Hickman, R., Dent, D.M., Weaver, V. and Terblanche, J. Brit. J. Anaesth. 1969;41:844.

57. Gronert, G.A. Anesth. Analg. 1979;58:367.

58. Berman, M.C. and Kench, J.E. In: Gordon, R.A., Britt, B.A. and Kalow, W. (eds.). International Symposium on Malignant Hyperthermia. 1973; Springfield: Thomas, p. 287.

59. Moulds, R.F.W. and Denborough, M.A. Brit. Med. J. 1974;2:241.

60. Mabuchi, K. and Sreter, F.A. Presented at the Second International Symposium on Malignant Hyperthermia, April 1-3, 1977, Denver, Colorado.

61. Muldoon, S.M. Presented at the 4th International Malignant Hyperpyrexia Workshop, York, September, 1986.

APPENDIX 1-a. TECHNIQUES OF DETERMINATION OF CAFFEINE HALOTHANE CONTRACTURE TEST IN NORTH AMERICAN LABORATORIES

INVESTIGATOR	TOTAL COST (US FUNDS)	METHOD OF PAYMENT	TIME IN HOSPITAL	# PER YEAR	TOTAL # DONE	OTHER TESTS PERFORMED	SURGICAL SITE
BRITT	$2,500	PI; Gov.	2-4 days	100	1300	H;HC;ATP;SF; RCU	VL;RA;PV
BROWNELL	$	PI; Gov.	1-2 days	50	300	H;HC	VL
CATTRAN	$2,400	PI; Gov.	3-4 days	40	200	H;HC	VL;RA
DUNCAN	$	PI; Gov.	outpatient	50	75	H;EM	VL;RA
DURBIN	$	free	variable	4-5	12	None	VL;VM
GRONERT	$1,500	PI	outpatient	30	400	H;HC	VL
MULDOON	$	Gov.	1-2 days	25	200	H;HC;RCU	VL
NELSON	$2,200	PI	1-2 days	75	220	None	VL
OH	$ 400	PI	1 day	24	4	H;HC	VL
REYNOLDS	$2,500	PI	outpatient to $3\frac{1}{2}$ days	20	60	H;HC;EM	VL;WE
ROSENBERG	$2,500	PI	1 day	75	250	H;HC	VL
SOKOLL	$	free	outpatient	20	200	None	VL
WINGARD	$ 640	free	outpatient	75	500	H;HC	VL

LEGEND: PI = private insurance; Gov. = government; H = histology; HC = histochemistry; ATP = ATP depletion test; SF = skinned fibre test; RCU = radioactive calcium uptake; EM = electron microscopy; VL = vastus lateralis; RA = rectus abdominus; PV = paravertebral; VM = vastus medialis; WE = wound edge

APPENDIX 1-b. TECHNIQUES OF DETERMINATION OF CAFFEINE HALOTHANE CONTRACTURE TEST IN NORTH AMERICAN LABORATORIES

INVESTIGATOR	ANAESTHESIA	DURATION OF SURGERY (in minutes)	LENGTH OF INCISION (in cm)	SECURED BY
BRITT	N_2O, Innovar, Diazepam	45-60	10.0-12.0	stick
BROWNELL	Femoral and lateral femoral cutaneous nerve block; Xylocaine	15	7.5-10.0	free
CATTRAN	Fentanyl, Droperidol, Thiopentone, $N_2O:O_2$; or Spinal - Tetracaine; or femoral nerve block - nesicaine	40	7.5-10.0	free
DUNCAN	Spinal - Pontocaine	30	5.0	sample 1 - stick sample 2 - free
DURBIN	N_2O, Fentanyl, Narcotic		2.5-3.5	variable length tension clamp
GRONERT	Femoral and lateral femoral cutaneous nerve block	45	10.0	free
MULDOON	Spinal or N_2O; Thiopentone, Narcotic	20-60	6.0	Cooley vascular/ fixed length clamp
NELSON	Spinal or N_2O; Thiopentone, Narcotic	30	5.0	fixed length
OH	N_2O, Fentanyl, Pancuronium	30-45	2.0-6.0	free (black silk tie at one end)
REYNOLDS	Outpatient - femoral & lateral femoral cutaneous block; Inpatient - N_2O & Narcotic	45-60	5.0	free
ROSENBERG	Femoral and lateral femoral cutaneous nerve block	40	5.0	sample 1 - clamp
SOKOLL	Adult - femoral & lateral femoral cutaneous nerve block; Child - chlorpromazine, Promazine & Demerol	20	7.5-10.0	free
WINGARD	Local or femoral nerve block	30	7.5-10.0	tongue depressor

APPENDIX I-c. TECHNIQUES OF DETERMINATION OF CAFFEINE HALOTHANE CONTRACTURE TEST IN NORTH AMERICAN LABORATORIES

INVESTIGATOR	SAMPLE SIZE (in cm)		TRANSPORT						
	LENGTH	WIDTH	CONTAINER	SOLUTION	pH	TEMP (°C)	TIME FROM OR TO LAB (in min.)	SOLUTION pre-gassed	SOLUTION per-gassed
BRITT	7.0	2.0	thermos	KR or RL	7.4	22	15	yes	no
BROWNELL	3.0-5.0	1.0	beaker	KR	7.4	22	5-15	yes	no
CATTRAN	3.5-4.0	2.5	beaker or cold plate	KR	7.4	21	0.5	yes	no
DUNCAN	4.0	1.0	beaker	KR or RL		15	15	yes	no
DURBIN	2.5-3.5	0.4-0.5	beaker	KR	7.4	20	1	yes	no
GRONERT	2.0-2.5	0.8-1.0	beaker	KR	7.4	20-22	2-3	no	no
MULDOON	4.0	1.0	beaker	KR	7.4	22	10-30	yes	no
NELSON	2.0	1.5	thermos	KR	7.4	22	3	yes	no
OH	1.0	0.5	plastic beaker	RL	7.4	20-22	10	no	no
REYNOLDS	3.0	2.0	hermeti-cally sealed glass	KR	7.45	25	10-35	yes	no
ROSENBERG	2.0-4.0	2.0	beaker	KR	7.4	20-22	5	yes	no
SOKOLL	3.0	1.0	beaker	RL	7.4	20-22	4-5	yes	no
WINGARD	2.5-3.0	0.5-1.0	beaker	KR	7.4	20-22	3	no	no

LEGEND: KR = Krebs Ringer; RL = Ringers Lactate

APPENDIX 1-d. TECHNIQUES OF DETERMINATION OF CAFFEINE HALOTHANE CONTRACTURE TEST IN NORTH AMERICAN LABORATORIES

INVESTIGATOR	MAGNIFYING GLASS OR MICROSCOPE	FASCICLE SIZE (in cm) LENGTH	WIDTH	# OF STRIPS MOUNTED	PRE-EQUIL. TIME (mins.)	BATHING SOLUTION	pH	TEMP. (ºC)
BRITT	Luxo Magnifier	2.0	0.3	8-20	15	KR	7.40	37.0
BROWNELL	No	2.0	0.3	4-5	20	KR	7.40	37.0
CATTRAN	Magnifying Glass	1.0	0.2	4		KR	7.40	37.0
DUNCAN	Loup Magnifier			4		KR		27.0; 37.0
DURBIN	12-Power Binocular Microscope	1.5	0.5-1.0	4-8	15-20	KR	7.40	37.0
GRONERT	10-Power Binocular Dissecting Microscope	1.5-2.0	0.2-0.4	6		KR	7.40	37.0
MULDOON	Microscope	2.0	0.2-0.4	4-6	15-30	KR	7.40	37.0
NELSON	No	2.0	0.2	12		KR	7.40	37.0
OH	No	0.5-1.0	0.5	8	60	KR	7.40	30.0
REYNOLDS	Loup Magnifiers	1.9	0.25-0.5	2		KR	7.45	37.5
ROSENBERG	Bipolar Microscope	1.0-2.5	0.2-0.3	4-8	30-60	KR	7.40	37.0
SOKOLL	Loup Magnifiers	3.0	0.5	2		KR	7.40	22.0
WINGARD	Magnifying Glass	2.0	0.1-0.2	2-4		KR	7.40	37.0

LEGEND: KR = Krebs Ringer

APPENDIX 1-e. TECHNIQUES OF DETERMINATION OF CAFFEINE HALOTHANE CONTRACTURE TEST IN NORTH AMERICAN LABORATORIES

INVESTIGATOR	POLYGRAPH	BASELINE TENSION (in Gms)	ELECTRODES - MATERIALS AND CONFIGURATIONS
BRITT	Grass-7	2.0	Platinum - 2 prongs touching belly of muscle
BROWNELL	Grass-7	2.0	Platinum plate
CATTRAN	Hewlett-Packard	1.5-2.0	Platinum - 2 prongs touching belly of muscle
DUNCAN			
DURBIN	Gould	that which produces maximal twitch tension	2 flat plates touching muscle close to end and near middle on opposite sides
GRONERT	4-channels Grass 4-channels Gould	that which produces maximal twitch tension	Platinum plate
MULDOON	Gould Brush	1.5-2.0 or that which produces maximal twitch tension	Platinum plate
NELSON	Gould Brush	1.0-2.0	Platinum - 2 prongs touching belly of muscle
OH	Gould 2400S	2.0	Platinum - 2 prongs touching belly of muscle
REYNOLDS		1.0-2.0	Platinum - 2 prongs touching belly of muscle
ROSENBERG	Gould 2600	2.0	Stainless steel - 2 prongs
SOKOLL	Beckmann 411	1.0	Stainless steel bar at bottom of muscle
WINGARD	Hewlett-Packard	1.0	Glass+platinum - 2 prongs touching belly of muscle

APPENDIX 1-f. TECHNIQUES OF DETERMINATION OF CAFFEINE HALOTHANE CONTRACTURE TEST IN NORTH AMERICAN LABORATORIES

INVESTIGATOR	STIMULATION				
	MANUFACTURER	# OF FASCICLES/ STIMULATOR	CHANNEL FREQUENCY (in Hz)	DURATION (in m sec)	VOLTAGE
BRITT	Elmed	1	0.5	2.0	2-20 (to maximal twitch tension)
BROWNELL	Grass	4-5	0.1	20.0	1-30
CATTRAN	Grass SD5	4	0.5	2.0	20-80 (to maximal twitch tension)
DUNCAN					
DURBIN	Grass 55C	1	0.5	10.0	100-150 (to maximal twitch tension
GRONERT	2 Grass + 2 current boosters	3	0.1	2.0	5-20 (supramaximal)
MULDOON	Grass S44	4	0.1	2.0	14-16
NELSON	2 Grass S48; 1 Grass S88	1	0.1	1.0-8.0	50 (to maximal twitch tension)
OH	Grass S88	2	0.2	4.0	40 (to maximal twitch tension)
REYNOLDS		1		1.0	To maximal twitch tension
ROSENBERG	Custom made	4	0.5	2.0-10.0	10-50 (to maximal twitch tension)
SOKOLL	Grass 54	2	0.1	1.0	40 (to maximal twitch tension)
WINGARD	Grass	1			To maximal twitch tension

APPENDIX 1-g. TECHNIQUES OF DETERMINATION OF CAFFEINE HALOTHANE CONTRACTURE TEST IN NORTH AMERICAN LABORATORIES

INVESTIGATOR	FLOW RATE FROM CARBOGEN TANK (ccs)	CARBOGEN FLOW RATE INTO INDIVIDUAL CHAMBERS (ccs)	FLOW METRES IMMEDIATELY ABOVE BATHING CHAMBERS	IS THERE A COMMON FLOW METRE CLOSE TO CARBOGEN TANK?	TYPE OF VAPORIZER
BRITT	2000	100	yes (2/chamber)	yes (on regulator)	Fluotec 3 + Dragerwick
BROWNELL	2000		yes	yes (on regulator)	Fluotec 2
CATTRAN	3000	200	yes	yes (on regulator	Fluotec 3
DUNCAN					
DURBIN			no	no (yes in line to vaporizor)	Dragerwick
GRONERT	1000	50-100	no	yes (Anaesthetic gas machine flow metre)	Vernitrol
MULDOON	300	50-100	no	yes (Anaesthetic gas machine flow metre)	Copper Kettle
NELSON		500	yes (1/chamber Brooks Sho-Rota 150 Rotometer)	no	Dragerwick
OH	2000	500	no	yes (on regulator)	Ohio
REYNOLDS	1000	500	yes (1/chamber)	yes (on regulator)	Vernitrol
ROSENBERG	1000		yes (1/chamber)	yes	Dragerwick
SOKOLL			no	no	Fluotec 2
WINGARD	500-1000	125-500	no	yes	Fluomatic

APPENDIX 1-h. TECHNIQUES OF DETERMINATION OF CAFFEINE HALOTHANE CONTRACTURE TEST IN NORTH AMERICAN LABORATORIES

INVESTIGATOR	CARB.+CARB. & HALO. MANIFOLDS SEPARATE	LENGTH OF CARB. LINES (cm)	LENGTH OF CARB.+HALO. LINES (cm)	ROLLER PUMP	PRESSURE RELEASE	RESERVOIR BAG
BRITT	yes	90	45	no	no	no
BROWNELL			90	no	no	no
CATTRAN	yes	60	120	no	no	no
DUNCAN				no	no	no
DURBIN	yes	230	230			
GRONERT	yes	60-90	120-240	yes	yes	yes
MULDOON	yes	120	210	yes	yes	yes
NELSON	yes	60	105	no	no	no
OH	no	152	30	no	yes	no
REYNOLDS	no	45	120-150	no	no	no
ROSENBERG	yes	90	120	no	no	no
SOKOLL	no	15	60	no	no	no
WINGARD	yes	45	15	no	no	no

APPENDIX 1-i. TECHNIQUES OF DETERMINATION OF CAFFEINE HALOTHANE CONTRACTURE TEST IN NORTH AMERICAN LABORATORIES

INVESTIGATOR	METHOD OF MEASURING HALOTHANE CONCENTRATION	# OF DUPLICATES		
		HALOTHANE	CAFFEINE	CAFFEINE +
BRITT	Spectrophotometry daily	2	2	2
BROWNELL	Infra-red analysis calibration	1	3	Not done
CATTRAN	Narcotest	2-4	2-4	2-4
DUNCAN		4	4	4
DURBIN	Gas chromatography calibration per six months	2	2 (strips used for halo. if no halo. contracture; on fresh strips if halo. contracture)	Not done
GRONERT	Gas chromatography mass spectrometry calibration once/yr.	3	3	Not done
MULDOON	Continuous Beckman Infra-Red Analyzer and gas chromatography daily	2	2	1-2
NELSON	Gas chromatography calibration	2	2	4
OH	Gas chromatography calibration	4	4 (strips previously used for halo.)	4 (strips previously used for halo. and/or caff.)
REYNOLDS	Mass spectrometry daily	2	2	2
ROSENBERG	Gas chromatography daily	4-8	2	2
SOKOLL	Mass spectrometer calibration once per year	2 (on same strip)	2 (on same strip)	2 (on same strip)
WINGARD	Not done	2	2	2

APPENDIX 1-j. TECHNIQUES OF DETERMINATION OF CAFFEINE HALOTHANE CONTRACTURE TEST IN NORTH AMERICAN LABORATORIES

INVESTIGATOR	OTHER TEST DRUGS	DRUG DOSES			SUCCESSIVE DRUG DOSES ADDED CUMULATIVE OR AFTER WASHING
		HALOTHANE	CAFFEINE	HALO.+CAFF.	
BRITT	Sch Sch+Halo	1.0%; 2.0%	1.0 doubling to 32.0 mM	0.25 doubling to 32.0 mM + 1.0%	Cumulative
BROWNELL	None	1.0%; 3.0%	0.25 doubling to 32.0 mM	Not done	Cumulative
CATTRAN	None	1.0%	1.0 doubling to 32.0 mM	0.25 doubling to 32.0 mM + 1.0%	Cumulative
DUNCAN	None	1.0%; 2.0%		1.0%; 2.0%	
DURBIN	None	2.0%	0.25 doubling to 16.0 mM	Not done	Cumulative + washing *
GRONERT	None	2% for 7 min; 3% for 7 min.	0.5, 1.0, 2.0, 4.0, 32.0 mM	Not done	Cumulative + + washing
MULDOON	Halo+Sch	1.0%; 2.0%; 3.0%	0.25 doubling to 16.0 mM	0.25 doubling to 16.0 mM + 1.0%	Cumulative
NELSON	None	1.0%; 3.0%	0.5 doubling to 32.0 mM	0.25 doubling to 32.0 mM + 1.0%	Cumulative
OH	None	1.0%	0.5 doubling to 16-32.0 mM	1.0 doubling to 2 mM + 1% sometimes	Cumulative + washing
REYNOLDS	KCl;Lido; Marcaine	0.5%; 1.0% 2.0%; 4.0%	0.25 doubling to 8.0 mM	0.25 doubling to 8.0 mM + 1.0%	Cumulative
ROSENBERG	Sch Halo+Sch	3.0%	0.125 doubling to 32.0 mM	0.125 doubling to 32.0 mM + 1.0%	Cumulative + washing
SOKOLL	None	1.0%	0.1 doubling to 16.0 mM	0.25 doubling to 16.0 mM + 1.0%	Cumulative
WINGARD	KCl	1.0-4.0%	0.25 doubling to 32.0 mM	0.25 doubling to 32.0 mM + 1.0%	Cumulative

* Cumulative within a dose-contracture curve with washing between sequential dose contracture curves performed on the same muscle strip

APPENDIX 1-k. TECHNIQUES OF DETERMINATION OF CAFFEINE HALOTHANE CONTRACTURE TEST IN NORTH AMERICAN LABORATORIES

INVESTIGATOR	TIME TO NEXT CAFFEINE ADDITION	PARAMETER MEASURED		
		HALOTHANE	CAFFEINE	CAFFEINE + HALOTHANE
BRITT	4 min.	contracture (B-P at 4 min)	contracture (B-P*)	contracture (B-P)
BROWNELL	4-5 min.	max.contracture	max.contracture	Not done
CATTRAN	to plateau or 4 min, whichever is less	contracture (B-P)	contracture (B-P)	contracture (B-P)
DUNCAN				
DURBIN	to plateau or 4 min, whichever is less	contracture (B-P)	contracture (B-P)	contracture (B-P)
GRONERT	to plateau or 4-5 min, whichever is greater	contracture (B-Pe**)	contracture (B-Pe)	Not done
MULDOON	to plateau or 3 min, whichever is less	contracture (B-Pe)	Not done	contracture (B-Pe)
NELSON	to plateau; if none then to twitch plateau	contracture (B-Pe)	contracture (B-Pe)	contracture (B-Pe)
OH	5 min.	contracture (B-P)	contracture (B-P)	contracture (B-P)
REYNOLDS	to plateau or 4 min, whichever is less	contracture (B-P)	contracture (B-P)	contracture (B-P)
ROSENBERG	to plateau or 2 min, whichever is greater	contracture (B-P) or 5 min, whichever less	contracture (B-P)	contracture (B-P) (Not always done)
SOKOLL	to plateau or 3 min, whichever is greater	contracture (B-P)	contracture (B-P)	contracture (B-P)
WINGARD	to plateau***	contracture (B-P)	contracture (B-P)	contracture (B-P)

LEGEND: * B-P = baseline to plateau; ** B-Pe = baseline to peak; *** plateau refers to contracture plateau. If no contracture then next dose is added without washing

APPENDIX 1-L. TECHNIQUES OF DETERMINATION OF CAFFEINE HALOTHANE CONTRACTURE TEST IN NORTH AMERICAN LABORATORIES

INVESTIGATOR	CALCULATED REPORT FOR RECORDS			NORMAL VALUES		
	Halothane	Caffeine	Caff. + Halo.	Halothane	Caffeine	Caff. + Halo.
BRITT	CA*	CSC	CSC-H	0.1 Gm	4.1 mM	1.05 mM
BROWNELL	CA	CSC	Not done	1.0 Gm	3.6 mM	Not done
CATTRAN	CA	CSC	CSC-H	0.1 Gm	4.1 mM	1.2 mM
DUNCAN	CA	CSC	CSC-H			
DURBIN	Amplitude	CSC 2**	Not done			
GRONERT	CA	Graph CCC ***	Graph CCC	N=0.4 Gm in 4 min; MH=0.5 Gm in 2 min.	4.0 mM	2.0 mM
MULDOON	CA	CSC	CSC-H	0.5 Gm	4.0 mM	1.0 mM
NELSON	CA	CSC	CSC-H	1.0 Gm	4.0 mM	0.9 mM
OH	CA	CSC	CSC-H		4.0 mM	
REYNOLDS	HSC	CSC	CSC-H	1.0 Gm	3.5 mM	1.0 mM
ROSENBERG	+ve or -ve values not reported to Pt but are kept on file			0.7 Gm (within 5 min)	0.3 Gm (at 2 mM caffeine)	Not done
SOKOLL	CA	CSC	CSC-H	0.1 Gm	4.1 mM	1.2 mM
WINGARD	CA	CSC	CSC-H	0.1 Gm	4-12 mM	0.75-1.25 mM

LEGEND: * = contracture amplitude; ** CSC 2 = the dose of caffeine which raises resting tension by 2.0 Gm; ***CCC = caffeine contracture curve

APPENDIX 1-m. TECHNIQUES OF DETERMINATION OF CAFFEINE HALOTHANE CONTRACTURE TEST IN NORTH AMERICAN LABORATORIES

INVESTIGATOR	REPORT SENT TO PATIENT	MOST SPECIFIC TEST	MOST SENSITIVE TEST
BRITT	HCK (severe); CK, HC (intermediate); K (mild); N (normal)	Halothane	Caffeine + Halothane
BROWNELL	HC; C; Indeterminate; N	Halothane	Caffeine
CATTRAN	HCK (severe); CK (intermediate); K (mild); N (normal)	Halothane	
DUNCAN	HCK, CK (high risk); K (low risk; N (all responses normal)		Caffeine
DURBIN	HC (MHS); CSC 2 < 1.0 (probably MHS); CSC 2 > 1.0 (probably negative)	Halothane	Caffeine
GRONERT	+ve - HCK (all fascicles); Equivocal - HCK (1 fascicle); N - none of the above	Halothane	
MULDOON	Child - HCK; CK; N Adult - HCK; CK; N	Halothane	
NELSON	HCK; K; N		
OH	strongly +ve; +ve; -ve	Caffeine	Caffeine
REYNOLDS	HCK; CK; MH-ve (N)	Halothane	
ROSENBERG	MH+ve; MH-ve (N)	Halothane	
SOKOLL	HCK; CK; MH-ve (N)		Caffeine
WINGARD	MH+ve; Intermediate; MH-ve (N)		

APPENDIX 1-n. TECHNIQUES OF DETERMINATION OF CAFFEINE HALOTHANE CONTRACTURE TEST IN NORTH AMERICAN LABORATORIES

INVESTIGATOR	CLINICAL CRITERIA FOR TEST
BRITT	Unexplained anaesthetic disaster; masseter spasm after Sch; unexplained cramps with or without elevated CK; sweats or fever; unexplained elevated CK; heat stroke; Neuroleptic Malignant Syndrome (NMS); SIDS
BROWNELL	Any clinical suspicion of MH; possible family history (FH) of MH
CATTRAN	Patient or FH of MH reaction; masseter spasm after Sch; cramps; NMS; heat stroke + one other abnormality, eg. FH of MH
DUNCAN	Past history or FH of MH reaction; masseter spasm + one of FH of MH, clinical muscle abnormality, myoglobinuria, unexplained anaesthetic disaster; increased CK + other muscle abnormality, cramps + elevated CK + patient or FH of MH
DURBIN	Any clinical suspicion of MH; eg. fever, rigidity, increased CK, FH of MH
GRONERT	Past history or FH of MH reaction, cramps + fatigue + sweats + elevated CK
MULDOON	FH of MH reaction; masseter spasm; unexplained rhabdomyolysis and/or muscle cramps
NELSON	FH of MH reaction; masseter spasm
OH	Combination of: 1) masseter spasm, fever and/or high CK during anaesthesia and a strong FH of MH reaction; and 2) cancelled surgery which still needs doing
REYNOLDS	Past history or FH of MH reaction; masseter spasm after Sch; rigidity after Sch; cramps + elevated CK; NMS
ROSENBERG	Past history or FH of MH reaction; masseter spasm after Sch; rigidity after Sch; cramps + elevated CK; NMS
SOKOLL	Past history or FH of MH reaction; heat stroke, cramps + elevated CK
WINGARD	Past history or FH of MH reaction; cramps + elevated CK (rarely)

APPENDIX 1-o. TECHNIQUES OF DETERMINATION OF CAFFEINE HALOTHANE CONTRACTURE TEST IN NORTH AMERICAN LABORATORIES

INVESTIGATOR	BEST CRITERIA FOR TEST	% OF POSITIVE BIOPSIES
BRITT	Masseter spasm	63.5%
BROWNELL		approximately 30%
CATTRAN		60-70%
DUNCAN	Myoglobinuria	7% - controls, relatives, probands
DURBIN	Masseter spasm	
GRONERT		20-30%
MULDOON	MH reaction with a decrease in pH and with myoglobinuria	50% (C+H measured separately) 70% (C+H measured in combination)
NELSON		
OH	Masseter spasm; fever and high CK during anaesthesia	
REYNOLDS	Combined decrease in pH, increased CK of greater than 20,000	70%
ROSENBERG	Masseter jaw rigidity after Sch + CK of greater than 20,000	45%
SOKOLL		33%
WINGARD		66%

APPENDIX 1-p. TECHNIQUES OF DETERMINATION OF CAFFEINE HALOTHANE CONTRACTURE TEST IN NORTH AMERICAN LABORATORIES

INVESTIGATOR	MINIMUM AGE FOR BIOPSY	CONCOMITTANT RESEARCH ON DIAGNOSTIC SAMPLE
BRITT	10 years	Effect of phospholipase, phospholipids, ketanserin and other dantrolene salts; SR and mitochondrial function; Ca release from SR of single chemically skinned fibres
BROWNELL	Will do a 3-year old if is an "index" case. Otherwise wait until old enough to do under regional block	Yes
CATTRAN		None
DUNCAN	$1\frac{1}{2}$ years	Effect of lidocaine; Ca metabolism O_2 consumption
DURBIN	5-6 years	None
GRONERT	25 Kg or 6 years	Contracture duration; half relaxation time; effect of dantrolene
MULDOON	5 years	Ca^{++} uptake by muscle; adenylate deaminase levels in muscle; muscle histochemistry
NELSON	23 Kg	Other drug effects; electromechanical time intervals & contraction kinetics
OH	3-4 years	None
REYNOLDS	Outpatient = midteens; Inpatient = none	Bupivacaine
ROSENBERG	4 years	Role of lipid metabolism in contracture induction; drug interaction studies
SOKOLL	6 years	None
WINGARD	10 years	None

10

THE EUROPEAN MH GROUP: PROTOCOL FOR IN VITRO DIAGNOSIS OF SUSCEPTIBILITY TO MH AND PRELIMINARY RESULTS

H. ORDING

The Danish Malignant Hyperthermia Register, Department of Anaesthesia, Rigshospitalet University Hospital, DK-2100 Copenhagen, Denmark.

ABSTRACT

Lack of International agreement on standard test procedures for the diagnosis of malignant hyperthermia susceptibility led to the formation of the European MH Group in 1983. Our group now comprises twelve countries with eighteen test centres meeting regularly. This paper outlines our protocol, specifying how the halothane and the caffeine tests are performed by members of the European MH Group, and compares results obtained in different MH centres.

INTRODUCTION

The European MH Group was established in April 1983 (1) at a meeting in Lund, Sweden. The main reason for the creation of this group was a growing concern over the lack of International agreement on standard test procedures for the diagnosis of malignant hyperthermia (MH) susceptibility. Thus it was almost impossible to compare results of in vitro tests for MH from different laboratories. One of the aims of the group was, therefore, to standardize the tests used in the screening procedure to make comparison of results between the European centres possible.

Since the first meeting in 1983 the group has met regularly. At the very beginning we were a small group of eleven people from eight countries. Since then, the numbers have steadily grown and by September, 1986, twelve countries with eighteen centres had joined the European MH Group:

Austria: Vienna

Belgium: Brussels

BRD Germany: Munchen; Wuppertal; Wurzburg

Denmark: Copenhagen

England: Leeds

France: Grenoble; Lille; Paris

Holland: Herleen; Nijmegen

Iceland: Reykjavik

Ireland: Cork

Norway: Skien

Sweden: Linkoping; Lund

Switzerland: Basel

It is expected that also Italy and Spain will join the group in the near future.

Some of the members are just now beginning the screening of patients, while others have been working in this field for many years. Common to all of us is, however, that all persons joining the group has agreed upon following a common protocol for the investigation of MH susceptibility.

When we first met, it was rather difficult to agree on a protocol, for apparently we were all following different screening procedures, chosen between those internationally suggested for this type of investigation. Based on published investigations and our own experience, we wanted to include in the protocol a caffeine test. Also, some of us had good results with a halothane test and, of course, did not want to abandon this. These two tests were, therefore, chosen to be included. Some of us had used a halothane-caffeine test previously, but since no one saw any advantage to this test and it seemed to create problems rather than solve them, it was unanimously decided that this test should not be included.

After many discussions and compromises from all participants, we have agreed on a protocol, specifying how the halothane and the caffeine tests should be performed. The protocol (1) may be changed if we feel the need for it, but an updated version can always be obtained from the secretary of the group (at present, Dr. F.R. Ellis, of Leeds, England). The latest edition is from September, 1986.

METHODS

Only quadriceps muscle should be used, preferably the vastus medialis muscle at the motor-point. The specimens should be 15-25 mm in length and 2-3 mm thick (however, specimens a little shorter are now acceptable).

Immediately after removal, the specimens should be placed in Krebs-Ringer solution at room temperature and bubbled with 95% oxygen:5% carbon dioxide. The composition of the solution is:

NaCl	118.1 mM
KCl	3.4 mM
$MgSO_4$	0.8 mM
KH_2PO_4	1.2 mM
Glucose	11.1 mM
$NaHCO_3$	25.0 mM
$CaCl_2 6H_2O$	2.5 mM
pH	7.4

The tests should be performed at 37°C. The muscle is suspended in the tissue bath, and the tension measured by a transducer connected to one end of the specimen and recorded. In order to assess viability during the test, the specimens should be electrically stimulated with 1 ms supramaximal stimulation at 0.2 Hz. Only one test should be performed on each specimen. Three tests are performed: a static caffeine test, a static halothane test and a dynamic halothane test.

For the static tests, resting tension should be gradually increased to produce good twitches. This often occurs at a tension of approximately 2 g, and is close to optimum length, as described by Gronert (2). The muscle is left at this tension for fifteen to thirty minutes after which either caffeine (0.5, 1.0, 1.5, 2.0, 3.0 and 4.0 mM) or halothane 0.5, 1.0 and 2.0%) is added. The muscle is exposed to each concentration until a contracture plateau is reached, or for three minutes if no contracture develops.

In the dynamic test (3), the muscle rests at low tension for three minutes after which it is stretched at a constant speed of 4 mm min^{-1} for 1.5 minutes to a tension of approximately 2 g. It is left resting at this length for one minute and then shortened at the same rate for 1.5 minutes. The whole cycle is thus seven minutes. After three cycles, halothane is added (same concentrations as above), each concentration for one cycle.

Threshold concentrations, i.e. the minimal concentrations of halothane or caffeine eliciting a sustained increase in tension of 0.2 g are the measured variables.

It has been discussed whether the dynamic halothane test is necessary--it is not performed in all centres. A final agreement on this point has not yet been reached. Some find it of vital importance (4), while I get the same diagnostic results with both the static and the dynamic halothane test (5). Halothane contractures are greater in the dynamic than in the static test, but in our laboratory this does not influence diagnosis (5), since threshold values are used for this purpose.

We know that halothane concentrations in the tissue baths may vary substantially and have, therefore, decided which concentrations we should obtain. These concentrations are 0.11 mM halothane corresponding to 0.5% v/v, 0.22 mM corresponding to 1% v/v and 0.44 mM corresponding to 2% v/v. These concentrations are roughly equivalent to 80% of the maximum concentrations which can be expected in the bath, with the partition coefficient for halothane in Krebs-Ringer solution at 37°C being 0.72. Our reason for choosing these seemingly odd concentrations is that these are the concentrations we found when we started to measure halothane concentrations in the tissue baths. We did not want to waste information obtained earlier by choosing 100% of the expected value, since some of us had spent years collecting normal control biopsies. Therefore, we have agreed to reach a concentration of 0.44 mM halothane in the Krebs-Ringer solution corresponding to 2% v/v halothane, with a maximum deviation of 10%.

Measuring the halothane concentrations in the bath is a very important point in the protocol. It appears that different types of tissue baths have different time constants for obtaining the desired concentrations as well as different equilibrium conditions. This fact explains a quite marked variation in halothane concentrations after three minutes at a vaporizer setting of, for example, 2%. The time stated in the protocol for halothane exposure is the time the muscle should be exposed to the desired concentration. Furthermore, it appears that normal muscle sometimes does contract with high concentrations of halothane, i.e. concentrations above 0.44 mM. Since we look at threshold concentrations for contractures of only 0.2 g, it is essential to work at the correct halothane concentrations.

RESULTS

The preliminary results of the European MH Group with this protocol were presented at the 4th International Workshop on MH in York, September, 1986 by Eva Ranklev.

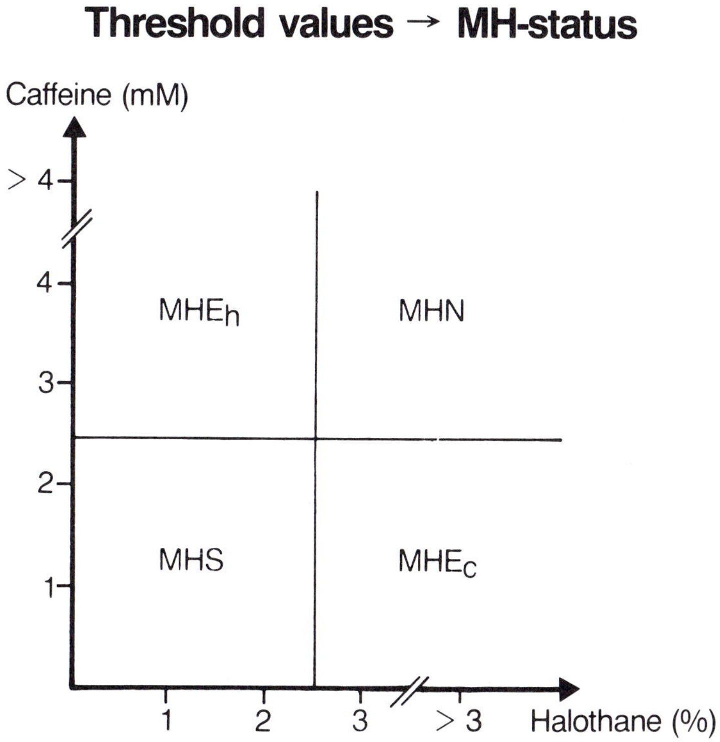

Figure 1. Diagnostic classification of patients screened for susceptibility to MH according to the European protocol (1,6).

By looking at results obtained in patients and normal controls in the actively involved centers, we decided that the threshold concentrations denoting abnormal reactions should be 0.44 mM (2% v/v) halothane or less in the halothane test, and 2 mM caffeine or less in the caffeine test. Also, we agreed upon 3 diagnostic groups (6), as can be seen in Figure 1: one is definitely MH susceptible (MHS) with abnormal results of both halothane and caffeine tests; another group is definitely non-susceptible (MHN) with normal results of both halothane and caffeine tests; the third group was created in order to minimize the

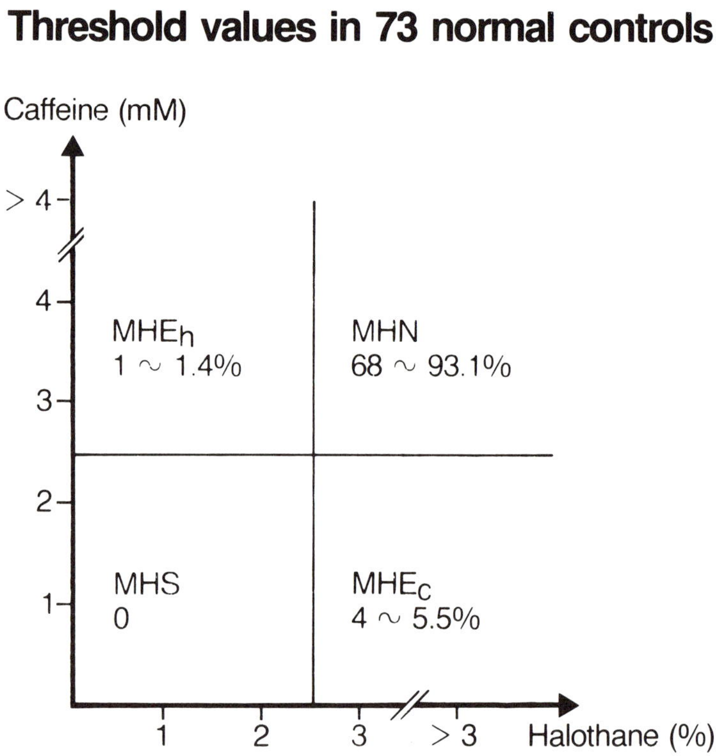

Figure 2. Diagnostic classification of 73 normal controls investigated for susceptibility to MH according to the European protocol (1,6).

risk of false positive as well as false negative diagnosis. All patients with one abnormal test result fall into this category (MHE: E for equivocal).

All centres have been asked to get their own normal control biopsies, obtained with the same type of anaesthesia as the diagnostic biopsies and from the same muscle group. Unfortunately, it is very difficult to obtain good control biopsies. Until September, 1986, we had altogether 73 normal controls (Figure 2). It is apparent from Figure 2 that none of the controls were diagnosed as MHS. Sixty-eight of the seventy-three control patients had completely normal contracture responses and were thus classified MHN. One patient, corresonding to 1.4% had an abnormal halothane response, the threshold concentration being 2%, and was classified MHE_h. Four patients, corresponding to 5.5% had abnormal caffeine responses with threshold concentrations at 2 mM and were classified MHE_c. The MHE status was thus found in 7% of the controls. But most important: the MHS status was not found in any of the 73 controls.

The distribution of results from patients investigated for MH is, of course, dependent on the selection of patients for biopsy, for example, the severity of clinical MH signs in the proband or the degree of relationship to the proband. Until September, 1986, 973 patients had been investigated following the protocol. The percentage of MHE from the investigated patients (Figure 3) is totally somewhat higher than in the control group (14 versus 7%), but not statistically significantly different. Of the 135 MHE patients, 113 were positive to halothane but not to caffeine, and 22 were positive to caffeine but not to halothane. The proportion of MHE_h is significantly greater in patients than in controls ($p < 0.01$, chi-square test), whereas the percentage of MHE results due to the caffeine test is similar for patients and controls. This finding supports the view that the halothane test is more sensitive than the caffeine test.

By now, thirty-eight probands with previous clinically fulminant malignant hyperthermia have been investigated. Of these, thirty-three were found to be MHS and five MHE. None were found to have completely normal contracture responses.

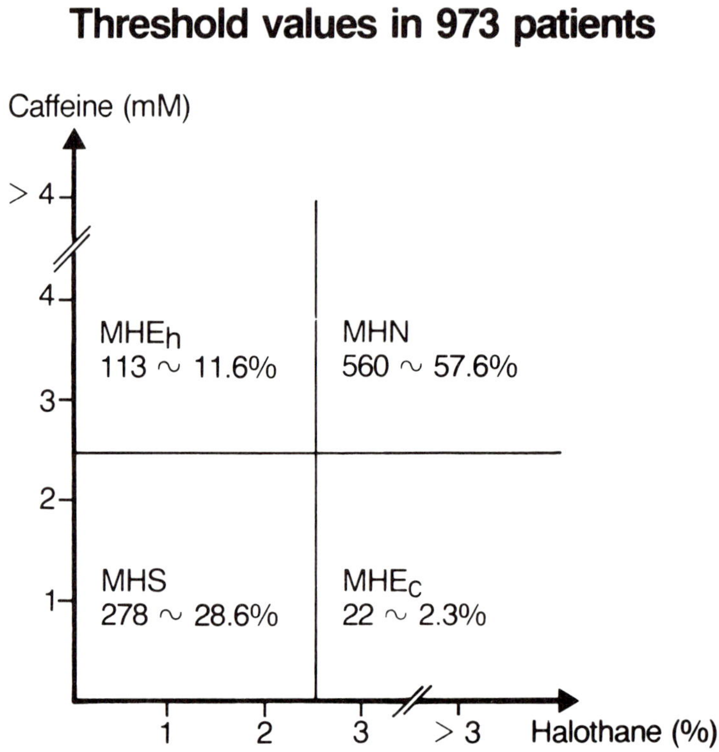

Figure 3. **Diagnostic classification of 973 patients investigated for susceptibility to MH according to the European protocol (1,6).**

Based on the presented results, we do think that this protocol fulfills the criteria mentioned earlier for a diagnostic test, although it is time-consuming and invasive. However, we still have to get more control biopsies. Also, it is to be expected that we will obtain some odd results whenever the test is set up in a new centre, since it takes quite some time and experience to get reliable results. I do hope, however, that we shall succeed in continuing this European collaboration, because I think we all gain by it. Now for the first time, we can really compare results obtained in different MH centres.

REFERENCES

1. European MH Group. Br. J. Anaesth. 1984;56:1267.

2. Gronert, G.A. Anesth. Analg. 1979;58:367.

3. Ellis, F.R., Harriman, D.G.F., Currie, S. and Cain, P. In: Aldrete, J.A. and Britt, B.A. (eds.). Second International Symposium on Malignant Hyperthermia. 1978. New York: Grune and Stratton, pp. 273-85.

4. Ranklev, E., Fletcher, R. and Blomquist, S. Br. J. Anaesth. 1986;58:646.

5. Ording, H. and Skovgaard, L.T. Acta Anaesthesiol. Scand (in press, 1987).

6. European MH Group. Br. J. Anaesth. 1985;57:1038.

11

DIAGNOSIS OF MALIGNANT HYPERTHERMIA IN JAPAN BY THE SKINNED FIBRE TEST

H. KIKUCHI, K. MATSUI AND M. MORIO

Department of Anesthesiology, Hiroshima University School of Medicine, 1-2-3 Kasumi, Minami-ku, Hiroshima City 734, Japan

ABSTRACT

As a diagnostic test of MH susceptibility, the whole muscle caffeine halothane contracture test should be employed.

In order to obtain more definitive information about MH tests, calcium uptake into the SR, calcium release from the SR and calcium related functions of the contractile system should also be utilized using single chemically skinned fibres. Both the dose of halothane or of caffeine necessary to induce threshold contractures and calcium induced calcium release should be measured.

The skill necessary to manipulate a skinned fibre is one of the great barriers to adequate performance of the skinned fibre test. Well-trained and precise hands with patience are required. There are many advantages, however, which make this technical difficulty well worth surmounting.

INTRODUCTION

What is a skinned fibre? A skinned fibre was first introduced by Natori (1) in 1954. Originally, it was a specimen of muscle fibre which was skinned in a mechanical way under mineral oil. A skinned fibre is now interpreted as a muscle fibre from which the cell membrane (sarcolemma) has been disrupted and is therefore exposed to its external environment while retaining intact the function of the subcellular organellae.

Classification

A skinned fibre is classified into two major types, according to the method of skinning, as shown in Table 1.

TABLE I. Classification of Skinned Fibres

1. Mechanically Skinned Fibres:
   - completely skinned fibres (1)
   - partially skinned fibres (2)
   - homogenized specimens (3,4)

2. Chemically Skinned Fibres:
   - EDTA treated skinned fibres (5)
   - EGTA treated skinned fibres
   - saponin treated skinned fibres (6,7)

The difference between a fibre skinned completely by mechanical means and a fibre skinned only partially by mechanical means is whether or not part of the cell membrane still remains. The partially skinned fibre is of great use in studying the sarcoplasmic reticulum (SR) because the T-system is still intact. One of the major disadvantages in preparation of mechanically skinned fibres is the technical difficulty involved. For instance, homogenized specimens are employed in heart muscle studies because of the difficulties in preparation of cardiac fibres skinned partially or completely by mechanical means.

On the other hand, chemically skinned fibres can be prepared much more easily. However, it should be always borne in mind that the chemical agent used for preparation of skinned fibres may destroy not only the cell membrane but also cell organellae such as the SR and the contractile system. For example, glycerinated fibres often show partial degeneration of the contractile system and dysfunction of the SR. EDTA (ethylene-diaminestetraacetic acid) treated skinned fibres (2) can be used from heart muscle but not from skeletal muscle (3,4). Unlike other detergents (5), certain concentrations of saponin destroy the architecture of the cell membrane while leaving the function of the SR and the contractile system intact (6,7). Saponin treated skinned fibres of skeletal muscle are used exclusively for the following experiments.

Ca^{2+} Release from the SR

Depolarization of the muscle cell membrane following stimulation generates action potentials in the T-system. Depolarization of the T-system transmits information to the SR, resulting in calcium release by a still unknown mechanism. The released Ca^{2+} activates the contractile system, causing muscle contraction. The two main functions of the SR are the release of stored calcium ions and then by active transport, uptake of calcium ions.

Factors which directly potentiate calcium release from the SR are the calcium ion itself (Ca induced Ca release, CICR) (8,9), depolarization of the SR (10,,11), decrease of magnesium concentration (12) and some agents such as caffeine (13), quinine, nicotine, chlorpromazine (14), dibucaine (15) and inhalational anesthetics (16-19). Procaine (20) and dantrolene (21) can inhibit CICR.

CHARACTERISTIC FEATURES OF A SKINNED FIBRE

1. Life of a Skinned Fibre

One of the disadvantages of a skinned fibre preparation is its short life span. The reproducible data for measurement of CICR rate can only be obtained during the first 36 hours after muscle biopsy, although it has been reported that the function of the SR can be maintained even after one to two weeks of storage at $5^{o}C$ (22). Therefore, the experiment must be commenced within twelve hours of biopsy.

2. Comparison of Skinned Fibre Tests with the Whole Muscle Caffeine Halothane Contracture Tests

There are several problems with the whole muscle caffeine halothane contracture test. The agent may not penetrate into the central part of the bundle and permeability into individual cells may be incomplete. A steep concentration gradient of the agent into a muscle bundle may affect the dose-contracture relationship. In skinned fibres, there is no barrier for the substance between the bathing medium and the SR and contractile system. By changing the solution bathing a skinned fibre, the direct action of the substance on SR and/or contractile system can be immediately observed.

3. ATP and Ca^{2+}

Unlike a muscle bundle or a skinned fibre in mineral oil (1), ATP as an energy source, cannot be produced in adequate amounts in a skinned fibre in aqueous solution (23). Therefore, for the latter preparation addition of ATP to the bathing solution is required. On the other hand, as described below, this inability to produce ATP becomes of great advantage for experiments requiring avoidance of ATP. Since spontaneous calcium storage in the SR varies between skinned fibres, preloading of the SR with fixed amounts of exogenous calcium ions is essential. This becomes, in turn, another advantage of the skinned fibre technique in that the amount of calcium storage in the SR can be externally controlled.

4. Fibre Type

A skinned fibre preparation is physiologically an ideal model of intact SR and contractile systems. The responses of the SR and the contractile system in skinned fibres are not different from those of whole muscle. No differences in the rate of CICR from the SR can be seen among muscle types. However, there are small differences between fibre types in the response of the contractile system to some substances such as calcium ions (24,25). The choice of muscle belly is limited in tests which use whole muscle bundles such as the caffeine halothane contracture test (26-28), since the ratio of type I to type II fibres varies from muscle to muscle and so influences responses to stimulation. When the rate of CICR is directly measured in skinned fibres, determination of fibre type is unnecessary, and the muscle biopsy, therefore, can be performed on any skeletal muscle site of the body.

5. Malignant Hyperthermia (MH) and Skinned Fibres

One of the most characteristic features of MH muscle is that halothane alone, caffeine alone or caffeine supplemented with halothane can enhance contracture significantly more in MHS than in normal individuals. The whole muscle caffeine halothane contracture test is generally employed for diagnosis of MH susceptibility (29). In the skinned fibre tests, halothane potentiates the CICR from the SR of MH patients more than from healthy controls (30). This caffeine and/or halothane enhancement is nothing more than an exaggeration of an already inherent predisposition to acceleration of the CICR mechanism of the SR. Using fragmented SR, the average threshold calcium concentration for CICR has been found to be markedly lower in MH than in control patients (31,32).

Using skinned fibres, abnormalities in SR function (Ca^{2+} uptake and release) and in calcium induced increase in tension of the contractile system can be detected separately. In fact, Endo's group (33-36) reported that the SR of MH exhibited a 1.8 times greater sensitivity to Ca^{2+} and had a twofold higher peak rate of CICR than normal, whereas uptake of calcium into the SR and contractile system function were normal. Therefore, it is worthwhile to investigate the function of the SR of human skeletal muscle, especially by direct measurement of the rate of CICR, for it provides more definitive information concerning MH susceptibility.

METHODS OF SKINNED FIBRE TESTING

There are two principal kinds of skinned fibre tests. One is the caffeine and/or halothane skinned fibre tension test, i.e. the test measures the dose of caffeine or halothane required to initiate an increase in tension of the contractile system (26-29) (Tables 2 and 3) and the other is the measurement of the rate of CICR (33-36) (Table 4). The apparent difference between these tests is simply due to the difference in the solutions used.

I. Principles of the Skinned Fibre Tests

Skinned fibres are enabled to take up and store a fixed amount of Ca^{2+} in the SR in the presence of sufficient ATP by dipping each fibre into a solution of appropriate pCa and appropriate ATP concentrations for a given time (Ca^{2+} loading). By activating the CICR mechanism, caffeine, halothane and certain

TABLE 2. Order of Addition and Purpose of Solutions for the Caffeine Skinned Fibre Tension (Caffeine Threshold) Test

Steps	Name of Solution*	Exposure Time (seconds)	Aims
1	G 2	Ca. 15	Relaxation
2	pCa6.7	120	Ca^{2+} loading
3	G 2	15	Wash out of Ca^{2+} from bath
4	G0.1	20	Stabilization of contractile system to Ca^{2+}
5	G0.1 Caff0.5	40	Start of challenge of caffeine
6	G0.1 Caff1.0	40	
7	G0.1 Caff2.5	40	
8	G0.1 Caff**	40	
9	G0.1 Caff40	40	Depletion of Ca^{2+} in SR
10	G 2	15	
11	pSr2.7	40	Fibre typing
12	pCa2.7	40	Contraction, maximum tension P_0 can be obtained

* Incremental doses of caff = 5.0, 7.5, 10, 15 and 20 mM until a contraction appears.

** See Table 3 for composition of solutions.

When an isometric tension (P) is generated of more than 5% of the maximum (p_0, obtained by pCa2.7), the concentration is defined as the threshold for caffeine. Halothane induced contraction may be used instead of caffeine as the test agent. When various pCa are applied instead of caffeine, sensitivity of the SR and the contractile system Ca^{2+} can be expressed as P/P_0 ratios vs pCa.

TABLE 3. Solutions for Caffeine Skinned Fibre Tension (Caffeine Threshold) Test

Name of Solutions	ATP-$MGSO_4$	EGTA	Tris-Maleate	$CaSO_4$	K-Ms*	Others
G 2	4.0	2.0	20.0	0.0	(130.0)	
G0.1	4.0	0.1	20.0	0.0	(130.0)	
G 5	4.0	5.0	20.0	0.0	(130.0)	
G0.1 Caff**	4.0	0.1	20.0	0.0	(130.0)	Caffeine
pCa6.7	4.0	10.0	20.0	1.3	(90.0)	
pCa2.7	4.0	10.0	20.0	9.1	(70.0)	
pSr2.7	4.0	10.0	20.0	0.0	(80.0)	$SrCl_2$ 0.42

Ms = Methanesulfonic acid; EGTA = Ethylene-glycol-bis-(-amino-ethylether)-N-N'-tetraacetic acid; Tris = Tris (hydroxymethyl) aminomethane

*** In all solutions the pH is adjusted to 6.8 with K-Ms* and the ionic strength is adjusted to 0.22 M with K-Ms.**

**** Incremental concentrations of caffeine. Halothane or any other agent can be used instead of caffeine as the test drug.**

Concentrations are all expressed in mM.

Free Ca ion is calculated by assuming an apparent association constant of 5×10^5 M^{-1} for Ca-EGTA and of 1.3×10^5 M^{-1} for Sr-EGTA at 20°C.

other agents can cause part or all the Ca^{2+} stored in the SR to be released. The released Ca^{2+} reacts with the contractile system, resulting in a tension rise in the presence of ATP. As the Ca^{2+} is taken back up to the SR, the contraction terminates. The amount of Ca^{2+} released is assessed as a tension-time integral of the isometric contraction generated.

2. Technique of Muscle Biopsy

The size of muscle that needs to be incised for the skinned fibre test (10-15 mm in length and 5 mm in diameter) is much smaller than that required for whole muscle caffeine halothane contracture test. The muscle is preserved in a high ATP-containing relaxing solution (G10 high ATP, see Table 4) cooled in ice (but not frozen) while being transported to the laboratory. The test ought to be started within twelve hours after biopsy to minimize the adverse effects of time related deterioration of the muscle.

3. Preparation of Skinned Fibre Preparations and Basic Procedure (Figure 1)

The biopsied muscle is dissected into thin bundles (about 0.1 mm in width) in a relaxing solution (G2) (Table 2) under a stereomicroscope. The fibres are chemically skinned by being soaked in a 0.05 mg/ml saponin containing solution for 30 minutes (7). A higher concentration such as 0.2 mg/ml of saponin applied to the muscle for more than 30 minutes may abolish all the functions of SR.

Each skinned fibre (about 2-3 mm in length) is then mounted by tying one end to a strain gauge transducer (AE80 TRANSDUCER ELEMENT, Senso Nor a.s., Norway) and the other end to a fixed site with silk threads. Isometric tension is recorded by a pen recorder and printed with tension-time integral (3390A INTEGRATOR, HEWLETT PACKARD, USA). Solution exchange is made using a hollow plastic plate containing many small wells. Each is filled with 0.5 ml of solution. The skinned fibre is placed horizontally in the central convex portion (2-3 mm) of the solution. By moving the plate manually, the fibre is transferred from solution to solution. The temperature of each of the solutions is controlled by water circulating around the wells within the plate. In our experiments, 20°C is ordinarily chosen because the fibres deteriorate rapidly at higher temperatures.

FIGURE 1. Method of Skinned Fibre Tests.

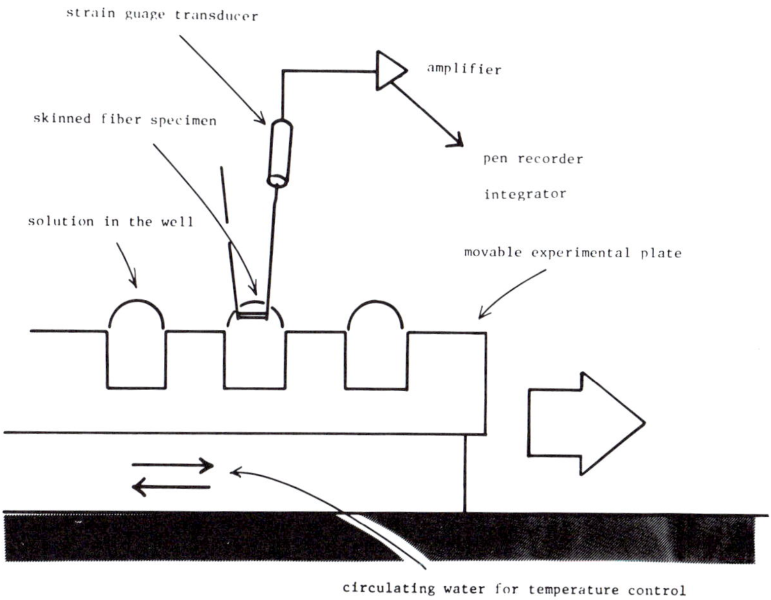

Experimental set-up for skinned fibre test. See the Methods Section, part 3, for details.

4. Caffeine and/or Halothane Skinned Fibre Tension (Threshold) Test (26-29)

This test measures the threshold dose of caffeine required to initiate a tension increase in a fibre whose SR has been preloaded with exogenous Ca^{2+}. The details of the method are given in Tables 2 and 3. Results are almost the same as those of the whole muscle caffeine halothane contracture test. As in the whole muscle test, MH fibres show enhanced sensitivity to caffeine and/or halothane which is greater in MH fibres than in normal fibres (17). Since skinned fibres have been used, the problems of the choice of muscles and the diffusion of the test agents has been overcome.

There are some drawbacks to the caffeine skinned fibre tension test. At the time of assay when a fibre reacts with caffeine and/or with halothane, both the SR function (Ca^{2+} uptake and release) and the contractile system function may be simultaneously activated, making interpretation of the data complicated. A finding of suppressed Ca^{2+} uptake into the SR with normal release of Ca^{2+} from the SR may provide the same result as that of normal Ca^{2+} uptake into the SR with enhanced Ca^{2+} release from the SR. Caffeine itself affects the sensitivity of the contractile system to Ca^{2+} (37,38).

In addition, sensitivity of the contractile system to Ca^{2+} differs among fibre types and affects the data. The fibre type can be differentiated by its response to Sr^{2+} sensitivity. Slow twitch or type I fibres are almost 10 times more sensitive to strontium ion than are fast twitch or type 2 fibre (39).

The procedures and solutions used in this test are summarized in Tables 2 and 3.

5. The CICR Skinned Fibre Test (33–36)

One of the outstanding features of this test is that it can detect an abnormality in CICR from the SR. To avoid Ca^{2+} uptake and contraction during CICR, an ATP free solution is employed. Without ATP, Ca^{2+} is released from the SR without ensuing contraction. Residual Ca^{2+} in the SR is then released by dipping the fibre in a solution containing sufficient caffeine (50 mM) and ATP to induce contraction. The smaller the ensuing contraction the greater was its CICR during the prior ATP-free period (Figure 2).

The diminution of Ca^{2+} in the SR as the result of CICR takes an exponential time course. It is expressed in the following equation:

$$CA_t = CA_i \exp(-kt)$$

where CA_t is the amount of Ca^{2+} in the SR after t minutes of stimulation with exogenous Ca^{2+}; CA_i is the initial amount of Ca^{2+} in the SR before stimulation with Ca^{2+}; k is the rate constant; and t is the duration in minutes of stimulation at a certain pCa. The rate constant, k, is taken as the rate of Ca^{2+} release at pCas of 6.5, 6.0, 5.5 and 5.0.

FIGURE 2. Skinned Fibre CICR Tension Test

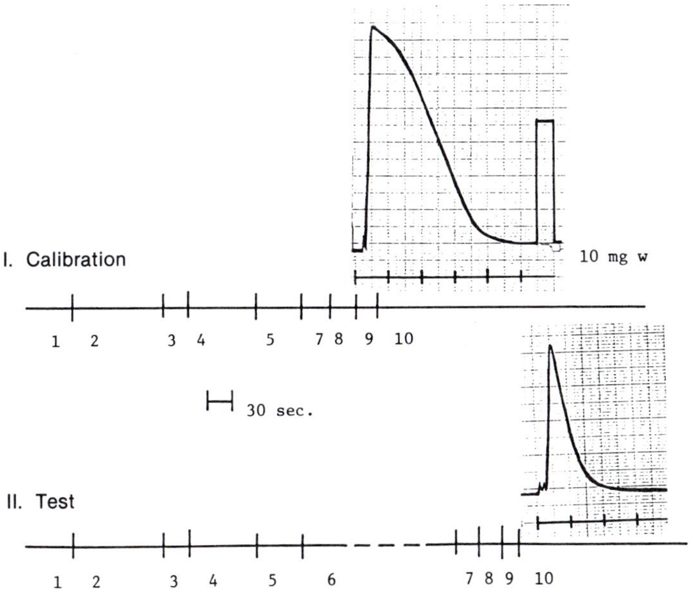

Solutions (Tables 2–4) are added at the times indicated by the following numbers.

1: 'G2', (Ca^{2+} free releasing solution);
2: 'pCa6.7', 2 minutes (for Ca^{2+} loading);
3: 'G10', (for terminating Ca^{2+} uptake);
4: 'G2 Rigor', 3 times (for washing out of ATP);
5: 'G2 Rigor Mg0', twice (for washing out of Mg^{2+};
6: test 'pCaX Rigor Mg0', t minutes when CICR takes place;
7: 'G10 Rigor Mg10 Proc10', (for stopping CICR);
8: 'G2' (for re-introducing ATP);
9: 'G0.1 Proc5', (for preassay conditioning);
10: 'Low Mg G0.1 Caff50', (for assay of Ca^{2+} in SR).

Graphic traces on the right are typical tension records in the presence of ATP and caffeine. Calibration - the fibre is examined in the absence of CICR to ensure that it is capable of maximal tension increase and complete return to baseline after tension increase. Test - then CICR is induced in the absence of ATP during which no tension increase occurs because of the lack of ATP. Finally, caffeine induced calcium release is initiated in the presence of ATP during which a tension rise does occur because of the presence of ATP. The smaller the tension rise the greater has been the CICR during the preceding ATP free period.

The bottom line is a time marker with each vertical cross hatch representing one second. The initial amount of Ca^{2+} (control) decreases as the result of previous CICR (Test). "Tests" are performed in a Ca^{2+} free ('G10 Rigor Mg0') solution with pCa (X) = 6.5, 6.0, 5.5 and 5.0 by taking appropriate times (t) to cause 30–70% decrease in Ca^{2+} concentration in the SR. Thus two parameters are varied in this test--Ca^{2+} concentrations and time.

TABLE 4. Solutions for the CICR Skinned Fibre Tension Test

Name of Solutions	ATP	$Mg-Ms_2$	$Ca-Ms_2$	EGTA	PIPES	K-Ms*	KOH**	Others
G10 high ATP	12.60	12.40	0.00	1	20	(50)	(90)	
G 2	4.25	5.08	0.00	2	20	(120)	(40)	
pCa 6.7	4.23	5.41	3.20	10	20	(90)	(70)	
G10	4.23	5.28	0.00	10	20	(100)	(60)	
G 2 Rigor	0.00	1.58	0.00	2	20	(150)	(30)	
G 2 Rigor Mg0	0.00	0.00	0.00	2	20	(150)	(30)	
G10 Rigor Mg0	0.00	0.00	0.00	10	20	(130)	(40)	
pCa6.5 Rigor Mg0	0.00	0.00	4.37	10	20	(120)	(50)	
pCa6.0 Rigor Mg0	0.00	0.00	7.11	10	20	(120)	(60)	
pCa5.5 Rigor Mg0	0.00	0.00	8.86	10	20	(110)	(60)	
pCa5.0 Rigor Mg0	0.00	0.00	9.62	10	20	(110)	(60)	
G10 Rigor Mg10 Proc 10	0.00	12.20	0.00	10	20	(100)	(50)	Procaine 10
Low Mg G0.1 Caff 50	4.11	1.00	0.00	0.1	20	(130)	(40)	Caffeine 50
G0.1 Proc 5	4.25	5.00	0.00	0.1	20	(130)	(40)	Procaine 5

See Table 3 for composition of solutions.

Ms = Methanesulfonic acid; EGTA = Ethylene-glycol-bis-(β-amino-ethylether)-N-N'-tetraacetic acid, PIPES = Piperazine-N-N'-bis-(2-ethanesulfonic acid)

* In all solutions ionic strength is adjusted to 0.2 M with K-Ms.

** In all solutions, pH is adjusted to 7.0 with KOH.

Concentrations are all expressed in terms of mM.

The association constant for Mg-ATP is $10^4 M^{-1}$. Free Ca ion is calculated by assuming an apparent association constant of $10^{6.39} M^{-1}$ for Ca-EGTA and of $10^{0.63} M^{-1}$ for Mg-EGTA at 20°C.

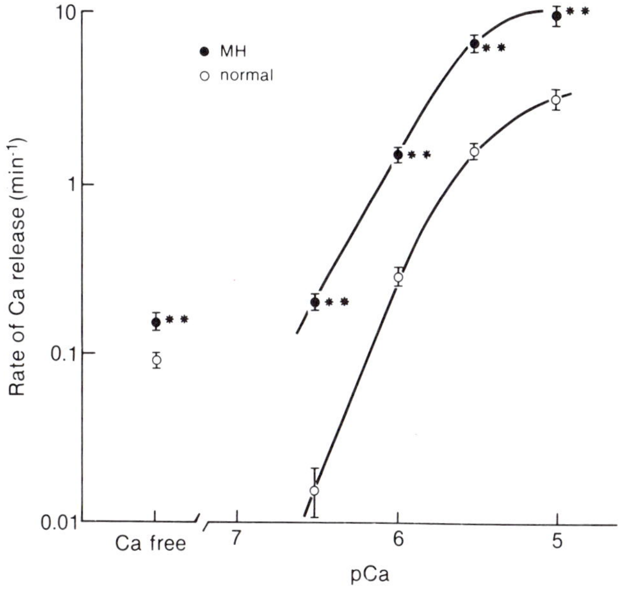

FIGURE 3. Differences in the rate of Ca release from the SR (ordinate) in the absence and in the presence of various pCa's (abscissa) between normal fibres (open circles, n=29) and typical MH fibres (closed circles, n=10). Rates are determined as described in the text (section 5). The values are expressed as Mean and SEM. ** $p < 0.01$.

Reproduced with permission from Endo, M., et al. The calcium-induced release mechanism in patients suspected of MH. Abstract of the 4th International Malignant Hyperpyrexia Workshop, York England, 1986.

The schematic diagram and the list of solutions of this test and test results are shown in Table 4 and Figure 2 respectively.

The basic composition of the standard relaxing solution 'G 2' is Mg^{2+} 1.5 mM and Mg-ATP 3.5 mM. 'Rigor' refers to an ATP free solution since the muscle fibre goes into a rigorous state in the absence of ATP. In order to terminate the CICR reaction, 'G10 Rigor Mg10 Proc 10' is used, since both Mg^{2+} and procaine are known inhibitors of CICR. By washing with pre-assay solution 'G0.1 Proc 5' just before the assay, sensitivity of the contractile system to Ca^{2+} is stabilized by lowering the concentration of EGTA.

Procaine, 5 mM, prevents Ca^{2+} leakage from the SR. At the moment of Ca^{2+} assay, the concentration of EGTA ("Assay solution, 'Low Mg G0.1 Caff 50'") is reduced to 0.1 mM in order to increase the sensitivity of the contractile system to Ca^{2+}. Mg^{2+} is also lowered to facilitate CICR.

Figure 3 demonstrates that less calcium is required to trigger a given amount of CICR in MHS fibres than in normal fibres.

ACKNOWLEDGEMENTS

The authors wish to thank Dr. Makoto Endo, Professor and Chairman, and his colleagues, Department of Pharmacology, Faculty of Medicine, the University of Tokyo, for his instruction in the basic techniques described in this paper and his courtesy in permitting use of his figure (Figure 3).

REFERENCES

1. Natori, R. Jikeikai med. J. 1954;1:119.
2. Winegrad, S. J. Gen. Physiol. 1971;58:71.
3. Fabiato, A. and Fabiato, F. Circ. Res. 1972;31:293.
4. Fabiato, A. and Fabiato, F. Eur. J. Cardiol. 1973;1:143.
5. Endo, M. Folia Pharmacol. Jap. 1976;72:9.
6. Endo, M. Heart 1974;6:1506.
7. Endo, M. and Iino, M. J. Muscle Research and Cell Motility 1980;1:89.
8. Endo, M., Tanaka, M. and Ogawa, Y. Nature 1970;228:34.
9. Ford, L.E. and Podolsky, R.J. J. Physiol. 1972;223:21.
10. Endo, M. and Nakajima, Y. Nature New Biol. 1973;246:216.
11. Kasai, M. and Miyamoto, H. FEBS Letters 1973;34:299.
12. Donaldson, S.K.B. and Kerrick, W.G. J. Gen. Physiol. 1975;66;427.

13. Endo, M. Physiol. Rev. 1977;57:71.

14. Takagi, A. Neurology 1981;73:477.

15. Yagi, S. and Endo, M. Biomedical Resarch 1980;1:269.

16. Fiehn, W. and Hasselbach, W. Eur. J. Biochem. 1969;9:574.

17. Takagi, A. and Endo, M. Folia Pharmacol. Jap. 1974;70:235.

18. Ogawa, Y. and Kurebayashi, N. J. Biochem. 1982;92:899.

19. Matsui, K., Kikuchi, H., Morio, M. et al. Abstract. Fourth International Malignant Hyperpyrexia Workshop, York, England, 1986.

20. Ford, L.E. and Podolsky, R.J. J. Physiol. 1972;223:1.

21. Ohta, T. and Endo, M. Proc. Japan Acad. 1986;62(B): 329.

22. Wood, D.S., Zollman, J. and Reuben, J.P. Science 1974;187:1075.

23. Godt, R.E. J. Gen. Physiol. 1974;63:722.

24. Kerrick, W.G.L., Secrist, D., Coby, R., et al. Nature 1976;260:440.

25. Takagi, A. and Endo, M. Exper. Neurol. 1977;55:95.

26. Nelson, T.E. and Schochet, Jr., S.S. Can. Anaesth. Soc. J. 1982;29:163.

27. Brownell, A.K.W. and Szabo, M. Can. Anaesth. Soc. J. 1982;29:218.

28. Rosenberg, H. and Reed, S. Anesth. Analg. 1983;62:415.

29. Kalow, W., Britt, B.A., Terreau, M.E., et al. Lancet 1970;2:895.

30. Takagi, A., Sugita, H., Toyokura, Y., et al. Proc. Japan Acad. 1976;52:603.

31. Nelson, T.E. J. Clin. Invest. 1983;72:862.

32. Ohnishi, S.T., Taylor S. and Gronert, G.A. FEBS Letters 1983;161:103.

33. Endo, M. Third International Workshop on Malignant Hyperthermia, Alberta, Canada, 1982.

34. Endo, M., Yagi, S., Ishizuka, T. et al. Biomed. Res. 1983;4:83.

35. Yagi, S., Ishizuka, T., Horiuchi, K. et al. Hiroshima J. Anaesth. 1983;19:71.

36. Endo, M., Horiuchi, K., Matsumura, N., Matsui, K., et al. Fourth International Malignant Hyperpyrexia Workshop, York, England, 1986.

37. Wendt, I.R. and Stephenson, D.G. Pflugers Archive 1983;398:210.

38. Horiuchi, K. J. Physiol. 1986;373:1.

39. Takagi, A., Yonemoto, K. and Sugita, H. Neurology 1978;28:497.

12

MICROSCOPY IN MALIGNANT HYPERTHERMIA INVESTIGATION

D.G.F. HARRIMAN

Neuropathology Laboratory, Department of Pathology, University of Leeds, Leeds LS2 9JT, United Kingdom

ABSTRACT

Many centres around the world offer a diagnostic service for susceptibility to Malignant Hyperthermia (MH, Denborough's Disease), but few investigate muscle histology at the same time. Microscopy plays little or no part in diagnosing susceptibility, and many biopsies are in fact structurally normal. Nevertheless, there are advantages such as defining MH myopathy when it appears, demonstrating associated disease in the susceptible and asymptomatic disease in the non-susceptible. Most associations are rare but that with central core disease is so strong that any patient and his family with that diagnosis should be considered susceptible until proved otherwise.

INTRODUCTION

Ever since in vitro tests of muscle contractility were introduced some twenty years ago to detect susceptibility to MH, this diagnostic service has been offered to suspected carriers of the MH trait in a number of centres throughout the world. But in only a few is muscle taken for histology at the time of the muscle biopsy. Some investigators may not think histology helpful or necessary as the tests on their own distinguish the susceptible from the normal. There may be other reasons, such as the difficulty of finding an interested and experienced pathologist or the mistaken idea that muscle histology has become a superspeciality.

In Leeds, England, the Malignant Hyperthermia Investigation Unit has received patients from the British Isles since 1971. The muscle biopsies and in vitro tests devised by Ellis, and recently agreed for use in Europe (1), are performed by Ellis and Halsall, and muscle is examined histologically by the writer. The biopsy is read before the results of the in vitro tests are ascertained. Over 1200 biopsies have now been performed, all on the vastus medialis at the motor point. It is hoped that the following account will convince workers in this field that it is beneficial to include microscopy in the investigation of MH.

The main way in which morphology has importance is in detecting MH myopathy (the structural myopathy associated with MH) and in uncovering cryptic muscle disorders in both susceptible and non-susceptible patients. Occasionally patients with known myopathies have to be investigated and their present muscle status can then be assessed. The non-susceptible biopsies are useful for assessing the range of normality which is influenced by age. Some older patients may show non-specific changes such as fibre splitting and single fibre atrophy. When seen in MH-positive subjects they should not be considered part of the spectrum of MH myopathy.

Finally, structural changes that sometimes follow a severe episode of MH have to be distinguished from MH myopathy. On the rare occasions on which the in vitro test results are equivocal diagnosis is sometimes helped by histology.

For the pathologist there is one unexpected benefit. As approximately half of the biopsies are MH negative and microscopically normal, a large material becomes available for quantitative analysis of the normal range.

TECHNIQUE

The biopsies are made at the motor point of the muscle permitting supravital staining of motor nerves and endplates when required. The remainder of the microscopy routine is standard: frozen section histochemistry, paraffin embedding of fixed muscle and electron microscopy. Best results are obtained when excision of specimens is made slowly and attention paid to optimum conditions for quenching and the use of small clamps for the specimen for ultrastructure (2).

MH MYOPATHY

The view that MH was an asymptomatic myopathy was suggested by two reports (3,4), one in which some members of affected families showed a rise in serum creatine kinase (CK), and another in which a muscle biopsy showed fibre necrosis, regeneration and occasional internal nuclei. It was widely assumed that a primary structural myopathy was responsible for MH. Later, when in vitro tests became established the serum CK was found to be an unreliable indicator of MH susceptibility, giving some false positive and false negative results. In the present series it was thought at first that myopathic changes were confined to MH susceptible biopsies and were, therefore, classified as MH myopathy. When the number of normal, MH non-susceptible biopsies became large it was found that many of the minor abnormalities could occur occasionally in normal muscle as well and were, therefore, non-specific. The incidence of MH myopathy dropped to approximately 30%. The revision of the bounds of normality in skeletal muscle made the contention unlikely that a primary structural myopathy could underlie MH. It was, indeed, a myopathy, but due to a defect, probably biochemical, capable of causing structural changes later in life. Ever since the discovery of MH it has been suggested that it is due to an inability to control the calcium concentration within the muscle fibre, but the defect responsible for this is still not known. It is possible that lesser disturbances of calcium flux occurring asymptomatically could lead to asynchronous myofibrillar as opposed to fibre contraction and to Z-line streaming, a prominent feature of MH myopathy as will be seen later.

There has been a true fall in the incidence of MH myopathy over the past fifteen years additional to that due to revision of early diagnoses, but it is curiously variable from year to year.

HISTOLOGY OF MH MYOPATHY

The abnormalities that comprise MH myopathy are characteristic and almost specific. They include internal nuclei, moth-eaten fibres and cores, individually or in combination, although cores and moth-eaten fibres are usually found together.

a) **Internal Nuclei**

It is not uncommon to find internal nuclei as the only abnormality in MH myopathy. In the normal muscle fibre nuclei are subsarcolemmal. Some internal nuclei are found in normal muscle, especially where fibres are inserted into the muscle fascia (terminal nuclei), or are present in focal clusters without obvious cause. To be considered abnormal and characteristic of MH myopathy in the vastus medialis they should occur in at least 2% of muscle fibres and most importantly be diffusely distributed throughout the section in at least four out of five fascicles. Often the percentage of affected fibres is high, even up to 25%. In transverse section internal nuclei are usually single, sometimes two or three together, lying anywhere within the fibre (Figures Ia and Ib). It is fibres that are counted rather than internal nuclei. In longitudinal section the nuclei are distributed in short chains within the fibre, which never reach the much greater length of nuclear rows characteristic of myotonic dystrophy. If very large numbers of internal nuclei are found the possibility of myotonic dystrophy or myotonia congenita will be considered.

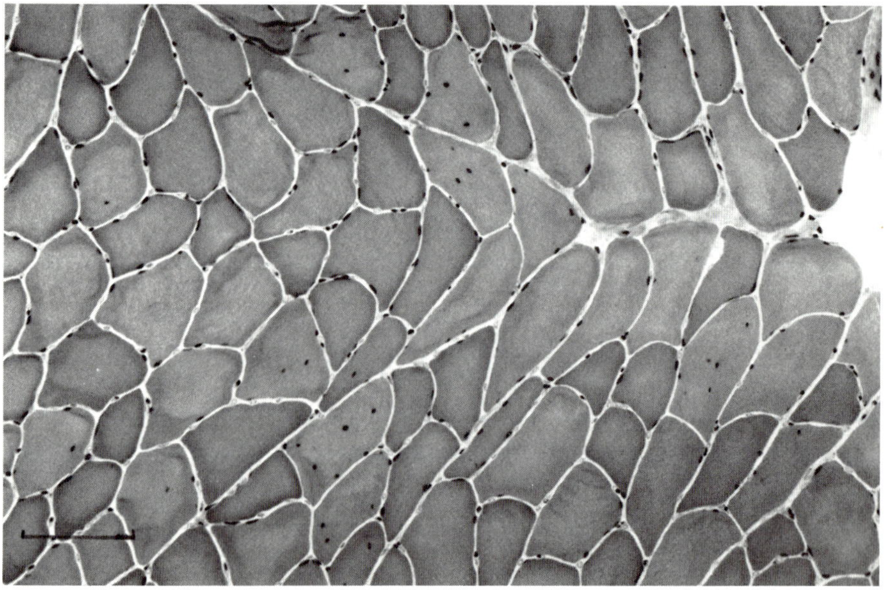

FIGURE Ia. Internal nuclei, diffuse HE. bar = 100 microns.

b) **Cores and Moth-Eaten Fibres**

These may be seen alone but more often both are present in the same biopsy, and as already mentioned are sometimes

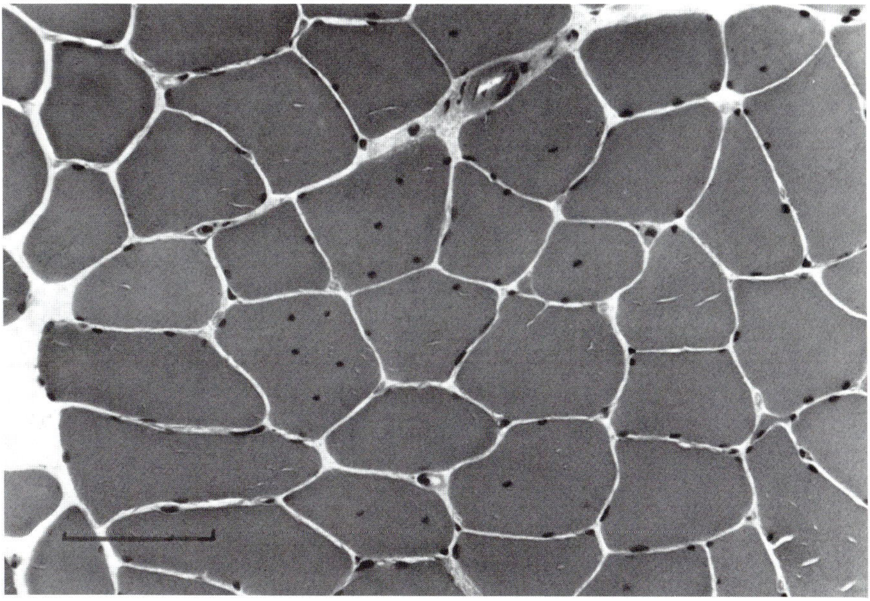

FIGURE 1b. Internal nuclei, focal. HE. bar = 100 microns.

accompanied by excess internal nuclei. Moth-eaten fibres (Figure 2) are demonstrable only in frozen sections stained for mitochondrial oxidative enzymes. They occur only in the mitochondria-rich type-I fibres, and consist of irregular areas of pallor, sometimes coalescent. Few mitochondria can be seen in the affected areas, but unstained myofibrils can be discerned. Moth-eaten fibres are rarely found in normal muscle, and never more than one or two per section. Although they occur in several myopathies, their presence as a main or exclusive feature is characteristic of MH myopathy. It is necessary to emphasize that the observer must be certain that the moth-eaten change is genuine, for it is too easily simulated by artefact whether due to poor handling of specimens or uneven sections. When they seem to be present in type 2 fibres artefact is virtually certain. It is always wise to examine serial sections, to repeat staining and to proceed to electron microscopy if there is still doubt.

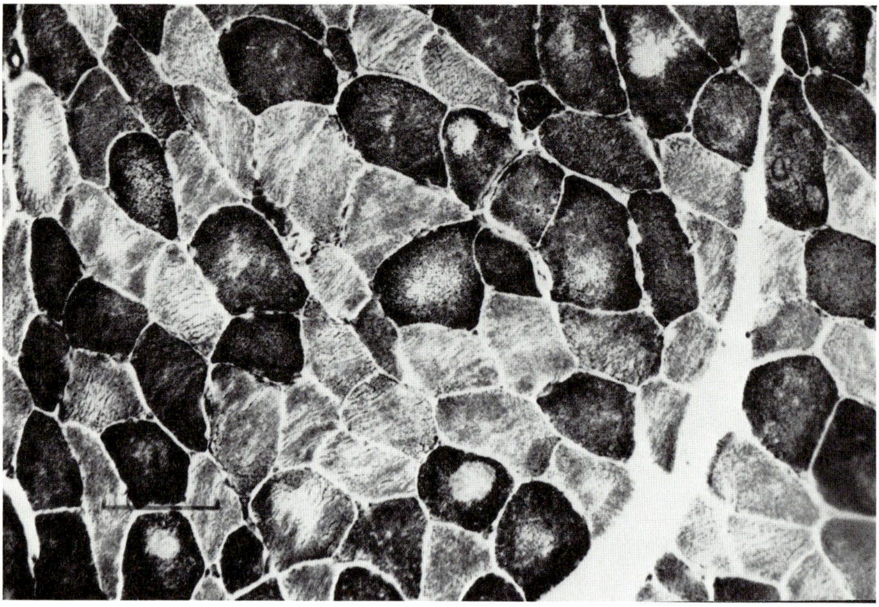

FIGURE 2. Moth-eaten and core fibres, often together in MH myopathy. NADH/TR. bar = 100 microns.

In semi-thin (resin) sections used for localization prior to ultramicrotomy it is not easy to identify moth-eaten change. In ultrathin sections in the electron microscope there is a reduction or loss of mitochondria, but in more pronounced lesions it is accompanied by some degree of Z-line streaming (Figures 3a and 3b). The frequent co-existence of moth-eaten change and cores suggests that they may have the same pathogenesis. The existence of transitional forms between the two is easy to conceive (Figure 2). It will be seen that the essential difference between them is the well-defined edge and round shape of the core.

The ultrastructure of cores is remarkably similar to that of moth-eaten fibres, the simplest core consisting of a rounded area where mitochondria have disappeared but myofibrils remain intact.

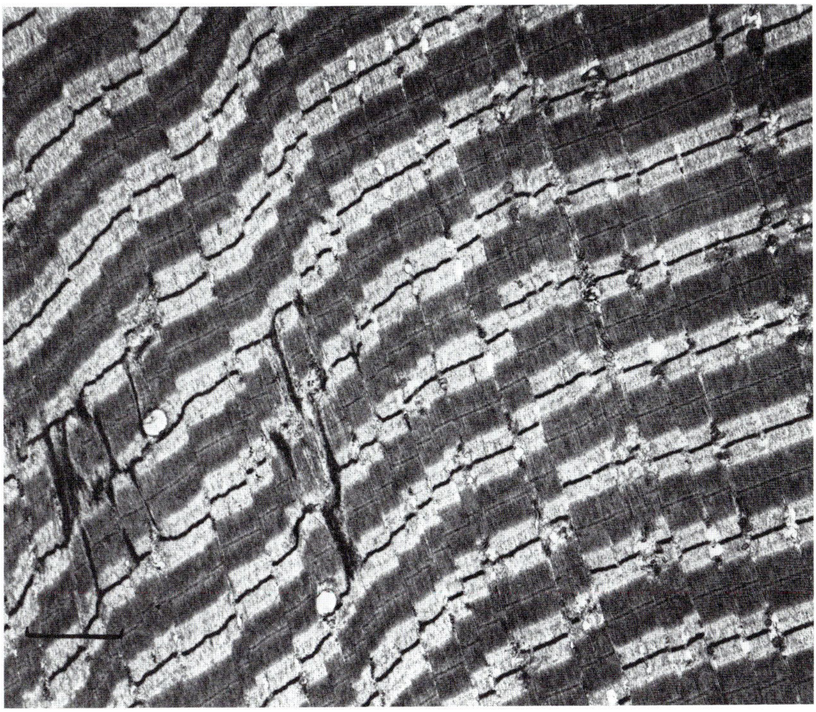

FIGURE 3a. **Scanty Z band streaming and virtual absence of mitochondria except for a few appearing at the right hand edge which are moth eaten. Electron micrograph. bar = 2 microns.**

Z-line streaming appears to develop soon, and leads eventually to large areas of fusion of Z band substance. At the early stage they can be seen by optical microscopy only in NADH sections (Figure 2) in the same manner as moth-eaten fibres. When central degeneration, in other words massive fusion of Z-band substance develops in these cores they can be seen in paraffin sections. Cores were first described in paraffin sections in a congenital myopathy later given the name of central core disease, and were said to extend the whole length of the muscle fibre, meaning that part that could be contained in a routine section. Such cores are

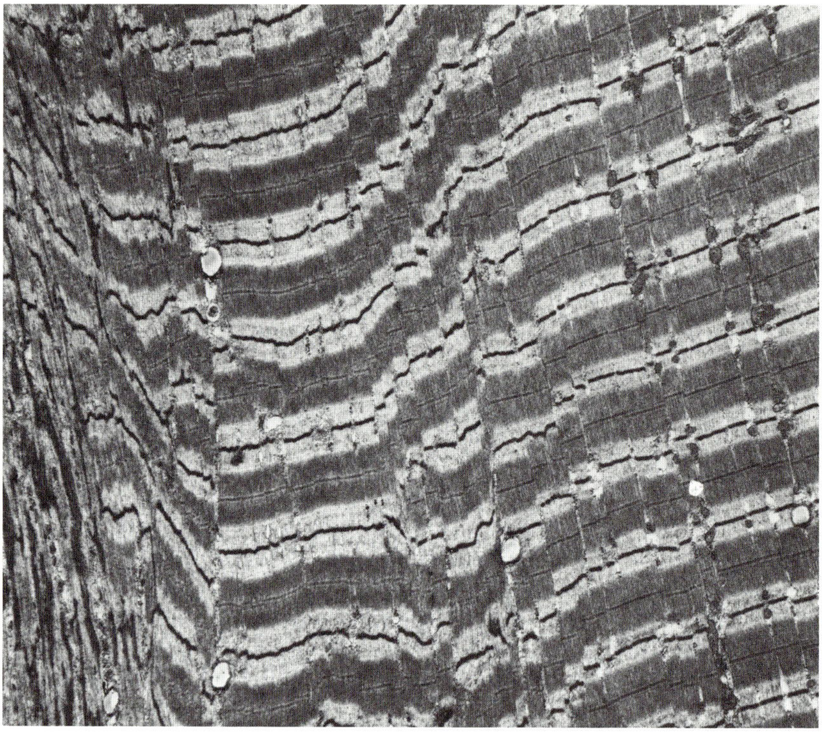

FIGURE 3b. **Extensive and severe Z band streaming and absence of mitochondria. bar = 2 microns.**

likely to be older than those of MH myopathy, explaining their greater length. Small cores sometimes occur in neurogenic disease, particularly where there is evidence of reinnervation of muscle. Here they are invariably short and if they have a dark rim are called target fibres. Terms such as core-targetoid are now considered superfluous. For further reading cores have been discussed at greater length elsewhere (2). When cores and moth-eaten fibres occur in the biopsy of a MH susceptible patient with or without internal nuclei MH myopathy can be diagnosed with confidence. The differential diagnosis of MH myopathy from central core disease is dealt with later.

NON-SPECIFIC HISTOPATHOLOGY

This refers to abnormalities that occur very infrequently in MH susceptible muscle and in "normal" muscle. Many occur in one or two fibres only, but a few are more widespread, yet without apparent effect on muscle function. If any of the following are noted in MH myopathy, it is important to realize they are non-specific and being infrequent are probably not determined by MH: single atrophic fibres, necrosis, regeneration, fibre splitting, cytoplasmic bodies, tubular aggregates, focal type 2b fibre absence, and spotted fibres. The large crescents formed of mitochondria in type-I fibres are normal in young active subjects. Many of these features have been illustrated previously (5).

THE RELATIONSHIP OF MH TO OTHER DISORDERS OF MUSCLE

Genetic or suspected genetic disorders are sometimes found together and this happens with MH from time to time, rather less commonly than claimed. Curiously, some disorders possibly related to muscle disease appear more often than by chance in the history of MH susceptible patients. They include strabismus, muscle cramps, idiopathic kyphoscoliosis and hernia (6). The proportion of these disorders actually associated with MH is small, and other evidence would be needed to justify their investigation for MH susceptibility.

Myotonia has long been known to make anaesthesia difficult especially when found unexpectedly at operation. The reaction to anaesthesia may take the form of suxamethonium (succinylcholine) spasm. In a series of 277 patients with a history of suxamethonium spasm, 147 were MHS (malignant hyperthermia susceptible) and 130 MHN (malignant hyperthermia non-susceptible), showing that about half the number of patients suffering the reaction to suxamethonium carried the MH gene[s] (7). Of interest in the MHN group was the discovery that three patients suffered from myotonia congenita on clinical and EMG examination. Although muscle histology is normal in most patients with this form of myotonia it was abnormal in the three patients mentioned, who showed an increase in fibres with internal nuclei. Such changes in the MHN biopsy should alert the pathologist to the possibility of a myotonic disorder.

Another condition sometimes making anaesthesia difficult is Duchenne dystrophy, which sooner or later involves the myocardium. Association with MH has been reported, but not usually on convincing evidence. When motor point muscle biopsies were first undertaken by the writer, general anaesthesia including halothane was provided for most muscle biopsy patients until MH became known. Some five or six Duchenne dystrophy patients per year over ten years were biopsied, without any form of anaesthetic reaction.

Congenital disorders such as the King syndrome (9,10) of short stature, small chin, antimongoloid slope of the palpebral fissures, kyphosis, webbed neck, pectus carinatum and a myopathy, were reported unequivocally to be associated with MH. Ellis and Heffron (8) have discussed links claimed with other conditions such as osteogenesis imperfecta, glycogen storage disease and thyroid disease and were not convinced by the evidence.

While a purely clinical diagnosis of MH may sometimes be acceptable, it carries more weight if supported by the in vitro tests. In the case of the King syndrome, the brief statement that three of the four affected boys died of MH may be true, but the tests were never performed.

The writer has encountered histologically a small number of other associated disorders. In MHS patients there have appeared one case each of ocular myopathy, congenital myopathy of uncertain type, myelin body myopathy and an unclassified myopathy. In the non-susceptible group ocular myopathy, Becker dystrophy and two patients with limb girdle dystrophy have presented. Neurogenic disease was discovered five times, but only in the non-susceptible.

In contrast, central core disease (CCD) and MH are very frequently associated. The present series includes seven patients with CCD of whom six were found to be susceptible. It is thus important that the diagnosis of CCD in routine biopsies should lead to MH investigation.

Central core disease was first described about thirty years ago, as a non-progressive congenital myopathy (2). The cores were easily visible in paraffin sections, quite often occupying a central position in the fibre, but some fibres contained up to five cores (Figure 4). Many cases have been described since, and when

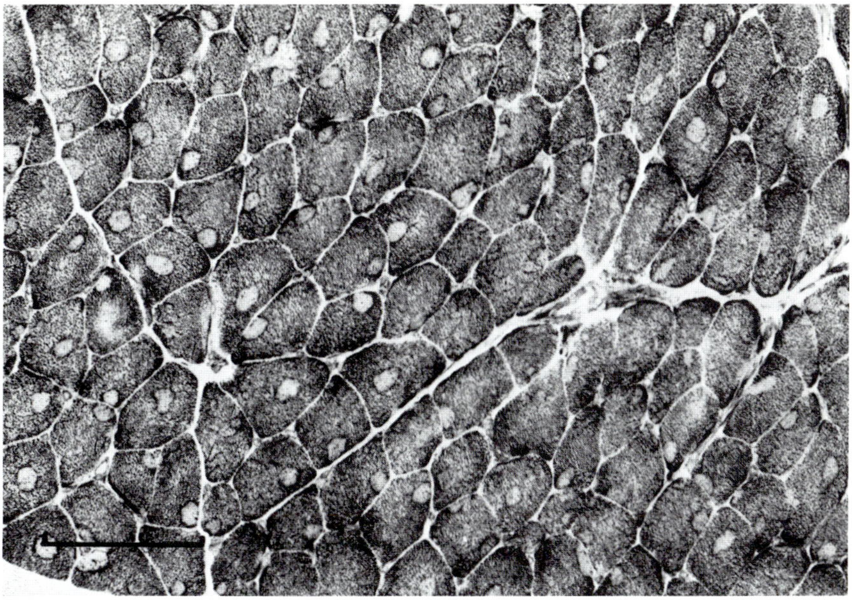

FIGURE 4. Cores in central core disease in a boy aged nine years. There is gross type-1 fibre preponderance and most fibres contain cores, both central and eccentric. Electron micrograph. NADH/TR bar = 100 microns.

frozen sections became available the cores were found to be virtually confined to type-I fibres. Their numbers and ultrastructure varied, like those in MH myopathy, but they were much longer (Figure 5).

There is one respect in which most cases of CCD differ from MH myopathy, and that is in a striking predominance of type-I fibres, not surprising in a congenital disorder. It never happens in MH myopathy. Rarely, CCD has been diagnosed in the presence of a normal ratio of fibre types, and then cannot be distinguished from MH myopathy. Which ever diagnosis is made MH investigation of the family is called for.

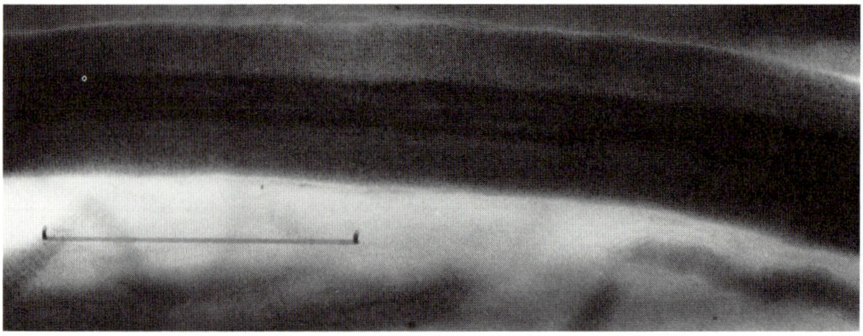

FIGURE 5. Central core disease. Single dissected muscle fibre impregnated with silver. The darkened continuous central core is evident. Block de Castro silver method. bar = 100 microns.

PUBLISHED REPORTS ON MH HISTOLOGY

Most have been made on small numbers of cases in whom the diagnosis of MH was clinical. The findings were diverse, including both neurogenic disease and primary myopathy (5). The evidence for neurogenic disease was sketchy, but it must be remembered that studies on a proximal muscle, the vastus medialis, cannot exclude its presence distally.

A recent paper by Gullotta (11) reviews forty-one patients with known or suspected MH, in whom muscle biopsies showed minimal changes in nineteen, not described in detail. Although in vitro tests were not available it is of interest that one patient had "central core disease" but without fibre type preponderance. Two other patients showed an increase in internal nuclei, in one associated with fibre splitting.

POSTMORTEM HISTOLOGY

When a patient develops hyperthermia and dies on the operating table, or dies several days later from rhabdomyolysis and renal failure, MH will be suspected. At necropsy, muscle histology (from numerous sites) can help in diagnosis. It is obviously very important to distinguish the pathology of the MH trait, if present, from that resulting from hyperthermia. If the muscle shows excess internal nuclei diffusely distributed, and moth-eaten fibres and cores, then MH myopathy unrelated to the MH episode can be diagnosed, but their absence does not exclude MH. Muscle should be quenched as soon as possible after death, otherwise the demonstration of cores may not be possible. Internal nuclei, on the other hand can be shown many days after death. If widespread necrosis is found (which is always focal and will obliterate any evidence of MH myopathy in affected areas), the effect of heat is being observed whether due to MH or rarely to other causes.

POST MH EPISODE HISTOPATHOLOGY

If a patient has survived an MH episode, and some months later is investigated, the muscle tissue may show sequelae. Some fibres are heavily stained with NADH and indicate a late stage of regeneration. One unique case showed partial type-I fibre atrophy at the same stage.

CONCLUSIONS

Although malignant hyperthermia is now well known, the number of centres around the world in which it can be thoroughly investigated are still few. Diagnosis is too often clinical and occasionally by a muscle biopsy without in vitro tests. This practice is never justified as up to 70% of patients found to be susceptible to MH by the in vitro tests have normal muscle histology. But histology is useful occasionally in diagnosis when in vitro tests are equivocal, in defining MH myopathy and above all in detecting associated disease in the susceptible and unsuspected diseases in the non-susceptible. So long as an invasive test has to be performed to detect MH susceptibility, patients should not be deprived of the additional information provided by muscle histology.

ACKNOWLEDGEMENTS

My thanks are due to Dr. F.R. Ellis and to Dr. Jane Halsall for the prompt transmission of the results of the in vitro tests in every case. Senior Technician Peter Walsh discovered that long cores block stain by silver.

The grants for research in malignant hyperthermia and neuromuscular disease from the Yorkshire Regional Health Authority and the Muscular Dystrophy Group of Great Britain are gratefully acknowledged.

REFERENCES

1. Ellis, F.R., Halsall, P.J., Ording, H., Fletcher, R., Ranklev, E., Heffron, J.J.A., Lehane, M., Mortier, W., Steinbereitner, K., Sporn, P., Theunynck, D. and Verburg, R. Br. J. Anaesth. 1984;56:1267.

2. Harriman, D.G.F. In: Adams, J.H., Corsellis, J.A.N. and Duchen, L.W. (eds.). Greenfield's Neuropathology, Fourth Edition. 1984: London, Edward Arnold.

3. Isaacs, H. and Barlow, M.B. Brit. Med. J. 1970;1:275.

4. Steers, A.J.W., Tallack, J.A. and Thompson, D.E.A. Brit. Med. J. 1970;2:341.

5. Harriman, D.G.F. In: Mastaglia, F.L. and Walton, Sir J. (eds.). Skeletal Muscle Pathology. 1982. Edinburgh, London, Melbourne, New York: Churchill Livingstone.

6. Britt, B.A. In: Britt, B.A. (ed.). Malignant Hyperthermia. International Anesthesiology Clinics 1979;17:63.

7. Ellis, F.R. and Halsall, P.J. Br. J. Anaesth. 1984;56:381.

8. Ellis, F.R. and Heffron, J.J.A. In: Atkinson, R.S. and Adams, A.P. (ed.). Recent Advances in Anaesthesia and Analgesia, Volume 15. 1985. Edinburgh, London, Melbourne, New York: Churchill Livingstone.

9. King, J.O. and Denborough, M.A. J. Pediat. 1973;83:37.

10. McPherson, E.W. and Taylor, C.A. Jr. Amer. J. Med. Genet. 1981;8:159.

11. Gullotta, F. and Spiesz-Keifer, C. Anasth. Intensiver Notfallmed. 1983;18:21.

13

COUNSELLING OF MALIGNANT HYPERTHERMIC SUSCEPTIBLE INDIVIDUALS

A.K.W. BROWNELL

Professor of Clinical Neurosciences and Medicine, The University of Calgary; The Malignant Hyperthermic Detection Program, Foothills Hospital, Calgary, Alberta T2N 2T9

ABSTRACT

The counsellor for malignant hyperthermia susceptible (MHS) individuals must be concerned with a broad range of issues. Initially, there must be a review of the evidence on which the diagnosis of malignant hyperthermia (MH) susceptibility was made. If the diagnosis of MHS appears valid, then the counsellor must be prepared to:

a) review the genetics of MH;
b) discuss the drugs that should be avoided;
c) comment on the safety of general anaesthesia with non-triggering agents; and
d) comment on lifestyle.

During the counselling session, it is valuable to point out areas of controversy and review unknowns. All MHS individuals should be given information as to how they may become involved with lay organizations devoted to MH. In those instances where contracture studies have indicated patients are free of the MH trait, the counsellor may be asked to advise concerning the validity of the test result.

INTRODUCTION

When asked to counsel MHS individuals, one can emphasize to them the tremendous advances that have been made since attention was first focused on this disease by Denborough's letter to Lancet readers in 1960 (1). MHS individuals can be reassured that the disease is widely recognized by anaesthetists, general anaesthesia can be provided safely and effectively for them using non-triggering drugs, effective treatment is available for MH reactions if they occur and family physicians, obstetricians and gynaecologists, general internists, paediatricians, surgeons and dentists are becoming better acquainted with the implications of MH susceptibility in their patients. In spite of this progress, uncertainty and controversy still exists in many areas, which makes counselling difficult, and can lead to confusion for MHS individuals and their families.

Any approach to counselling the MHS individual must consider several issues.

1. Is the diagnosis of MHS based on firm evidence?
2. Should counselling be confined to the subject of general anaesthesia and genetics only, or should it be much broader and include items such as lifestyle, use of drugs other than general anaesthetics, disease associations, etc.?
3. How confident can we be that patients with negative studies are not at risk for MH?

In this chapter these issues will be addressed.

IS THE PATIENT MHS?

To answer this question implies a review, by the counsellor, of the evidence for MH in the patient. Is the diagnosis based on a clinical suspicion or as a result of a laboratory test for susceptibility?

a) Clinical Diagnosis

If, during the course of a general anaesthetic using halothane and succinylcholine, metabolic acidosis, hyperthermia and muscle rigidity develop and post-anaesthesia, muscle pain and weakness occur accompanied by marked elevations in serum creatine kinase (CK) and myoglobinuria, there will be no doubt that the patient had MH and must be considered MHS.

Fortunately, nowadays, it is rare for a full blown MH reaction to occur as anaesthetists are likely to institute appropriate therapeutic steps at a stage when the diagnosis of clinical MH is suspected but unproven and, therefore, it remains unclear on clinical grounds if the problem really is MH. This situation makes counselling, in relation to MH susceptibility, extremely difficult.

In addition, the clinical suspicion threshold for MH varies significantly from one anaesthetist to another. Thus, while one anaesthetist may judge masseter spasm occurring after succinylcholine as within normal limits and proceed with an anaesthetic that proves to be totally unremarkable, another anaesthetist may conclude that a similar degree of spasm is suggestive of early MH and either abort the anaesthetic or switch to a "safe" anaesthetic technique. If the diagnosis of MHS is made in this clinical situation, again, counselling is most difficult since it is recognized that up to 50% of such individuals will have normal contracture responses (2-5) and, therefore, are not MHS.

In conclusion, today, except in very rare instances, it is not going to be possible to determine with certainty on clinical grounds alone, if most patients have had a MH reaction and, therefore, need counselling.

b) <u>Tests for MH Susceptibility</u>

Since the report by Kalow et al in 1970 (6) of abnormal contracture responses in biopsied muscles from individuals who had survived episodes of MH, contracture tests have been adopted as the primary laboratory test for determining MH susceptibility. During the test, a strip of biopsied muscle is exposed to:

1. halothane in varied concentrations;
2. caffeine in varied concentrations; or
3. a fixed concentration of halothane and varied concentrations of caffeine.

Many other tests have been reported that may prove to be useful for the determination of MH susceptibility but since they have not gained widespread use in many laboratories, they will not be considered further in this chapter (7-10) but rather will be evaluated in Chapter 8.

In spite of the worldwide acceptance of the value of contracture tests in determining MH susceptibility, there is still lack of agreement as to what constitutes an abnormal test. As a result of this, the counsellor must rely on the interpretation of the test results by the laboratory where the test is performed. Thus, it is conceivable that a patient could be tested in two different laboratories and receive contradictory advice as to their MH status. This creates an additional problem when members of the same pedigree are investigated by different laboratories. The European Malignant Hyperpyrexic Group (11) is attempting to resolve some of these problems by standardizing their test procedures and interpretation of test responses. The formidible problems associated with obtaining sufficient control specimens of muscle to introduce a new contracture test in a diagnostic laboratory, where controls have already been established for another type of contracture test, is the major road block preventing the introduction of a standardized contracture test that could potentially produce results that would be interpreted the same worldwide.

To properly evaluate a diagnostic test, it is necessary to have information on parameters such as sensitivity and specificity (12). Although solid published data is not available, all people who operate diagnostic laboratories agree that contracture tests have a high degree of sensitivity. What is much less clear is their specificity. There is very little published in this area. Mortier and Biesel (13) reported abnormal contracture responses in control patients with neuromuscular disease but no history of MH. Isaacs (as reported by Gronert) (14) and Heimann-Paterson (15) have made similar unpublished observations. Takagi et al (16), using skinned muscle fibers, reported abnormal contracture responses in some fibers from patients with Duchenne Muscular Dystrophy and asymptomatic patients with elevated serum CK.

Our own studies (17) correlating the results of contracture tests, clinical MH reactions and the presence of a clinically or pathologically well-defined myopathy do, however, suggest a high degree of test specificity.

Clearly, there is still much to be done in relation to standardizing the methods used in contracture testing and the interpretation of the results and until such a time as this happens, the counsellor has to depend on the diagnostic laboratory for interpretation of the test results.

THE PATIENT PROVEN TO BE MHS

a) Genetics

Prior to the advent of the effective preventative measures for MH, it had been observed on the basis of occurrence of full blown MH reactions that MH was a genetic disease and susceptibility followed an autosomal dominant pattern of inheritance (18,19). This observation of autosomal dominant inheritance has been confirmed by contracture studies.

Thus, when a new case of MH is identified, the patient must be informed that he/she has a genetic disease, that it follows an autosomal dominant pattern of inheritance, and the counsellor must ensure that the patient recognizes what this means for their relatives. The counsellor must stress to the patient the importance of contacting relatives to inform them of the existence of the MH gene in their pedigree and to inform relatives of the necessity of discussing this problem with their family physician or other appropriate medical personnel. The need for an appropriately labelled Medic-Alert bracelet should be emphasized as well as giving them information on how to receive lay publications dealing with MH, such as "The Communicator" or "The Hot Line".

Kalow et al (20) stressed that an autosomal recessive pattern of inheritance does occur and McPherson et al (21) suggested that sporadic and multifactorial determined cases are common. Future studies carried out with more standardized protocols at a time when there is more information relating to such items as test specificity and sensitivity may eventually clarify the significance of these suggestions.

b) **What Drugs Should be Avoided?**

It should not be necessary to burden the MHS patient with the pharmacological knowledge that the anaesthetist who provides regional or general anaesthesia for the MHS patient must possess. All the patient should need to know is that there are drugs used during anaesthesia that are safe, and others that are MH triggers. If the patient requires anaesthesia, he/she should consult the anaesthetist prior to the planned anaesthetic.

Unfortunately, many MHS patients, as a result of their own personal experience or because accounts of other MHS patients' experiences, are reluctant to accept on good faith alone that an anaesthetist who is not known to them will have the knowledge, skill and experience to deal with their particular anaesthetic problem safely. The patient is then faced with the task of trying to assess the anaesthetist's competence without having a sound base from which to do this. During the preoperative consultation, the opportunity to discuss their upcoming anaesthetic fully should go a long way toward reassuring the patient that the anaesthetist is capable of handling their special anaesthetic problems. If, on the other hand, following this type of encounter the patient has not had their apprehensions satisfactorily allayed, the patient should feel free to request referral to a centre with recognized expertise in dealing with MH patients.

It is very important to counsel MHS patients about the use of both prescription and non-prescription drug use as this may be a very confusing area for them and may lead to patients being denied or being afraid to use definitely required drugs because of their being considered potential triggers. Because most of the recommendations against the use of various drugs are based on non-published, anecdotal reports, it is impossible for a counsellor to make an informed judgment concerning the validity of these recommendations.

Much of the information about and many of the recommendations concerning drugs that MHS patients should avoid, were made at a time when very little was known about MH and any drug was considered a potential trigger. It is this author's belief that now is the time to reappraise many of these recommendations to find out if there is any scientific justification in making them. This author strongly agrees with Gronert's comment that "the whole area of triggering MH episodes in humans in the absence of general anaesthesia, is loaded with prejudice, emotional responses and personal opinions and there is almost no factual data to rely on" (22).

A recent addition of "The Hot Line" (23) listed a total of seventy-one non-prescription drugs which were described as "drugs which may induce or aggravate MH reactions". It appears that the common factor all these drugs share is that they contain caffeine. Yet, Flewellen and Nelson (24) found no evidence to support claims that caffeine had any negative effect on their patients with MH and they calculated that massive doses of caffeine would be required to produce the equivalent serum levels in humans that can produce contracture responses in the in-vitro situation. This author's own personal experience with MHS patients would support the findings of Nelson and Flewellyn in relation to caffeine. There is no published evidence to support the claim that drugs containing caffeine are contraindicated in MHS patients.

The use of local anaesthetics in MHS patients must be discussed with every patient. Recommendations about the avoidance of amide local anaesthetics in MHS patients has lead to very significant problems in regard to dental care for many patients.

On the basis of a retrospective review of patient records from episodes of MH (25), it seemed that all patients who received lidocaine as one of the drugs during the treatment of an episode of MH, fared less well than those who did not. As well, it has been observed (26) than at physiological pH, amide local anaesthetic augmented caffeine contractures, whereas ester local anaesthetics either depressed or did not influence caffeine contractures in either human or rat skeletal muscle. These two observations seem to have been the basis for the often repeated warning to avoid amide local anaesthetics in MHS individuals and to use ester type local anaesthetics instead.

Evidence is now accumulating to suggest that this recommendation is no longer valid for the following reasons.

1. There are no reports in the literature clearly implicating amide local anaesthetics as the trigger for MH reactions (27).

2. Berkowitz and Rosenberg (28) reported use of the amide local anaesthetic mepivicaine in 103 patients, of which 52 were biopsy positive for MH without any suggestion of triggering an episode of MH.

3. Several experts have suggested in "The Communicator" (29-31) that amide local anaesthetics are safe for dental anaesthesia.

4. Lidocaine failed to trigger MH reactions in susceptible pigs (32,33).

This author routinely questions all MHS patients investigated about prior adverse reactions to local anaesthetics and has never found any history of it, thus confirming the experience of many others. In Calgary, lidocaine is now routinely used as the local anaesthetic for induction of regional anaesthesia during MH biopsies. No complications in relation to triggering of MH have been encountered (34).

There is no evidence to support the recommendation that amide local anaesthetics can trigger MH in susceptible individuals and, therefore, there can be no justification to continue to recommend against the use of amide local anaesthetics in MHS persons (35). Controversy surrounds the issue as to whether or not epinephrine can or cannot be used safely in combination with amide local anaesthetics.

In conclusion, this author believes that on the basis of published evidence, the majority of MHS patients do not need to be concerned about avoiding drugs other than those which are administered during general anaesthesia. In making such a recommendation, this author recognizes that this is only a generalization and therefore must be interpreted in the context of a particular patient's complete medical history. Obviously if a known MHS individual has had an unfavourable reaction to a particular compound, then good medical practice dictates that the substance should be avoided in that particular individual. However, much broader evidence than this would be required before one would want to extrapolate this type of information into a general recommendation for all MHS patients.

IS GENERAL ANAESTHESIA SAFE WITH NON-TRIGGERING AGENTS?

The widespread use of standardized protocols for the management of MHS patients combined with a marked decrease in the number of deaths from MH, suggests that the special protocols for general anaesthesia in MHS patients are effective in preventing MH. Recent case reports of MH occurring in susceptible patients who received general anaesthesia without triggering agents and with and without dantrolene pretreatment (36,37) suggest the need to observe all MHS patients very closely both during the anaesthetic and in the immediate post-operative period for signs of MH. It is reassuring that each one of these patients responded when appropriate treatment was instituted for what was thought to be MH.

Although no data is available on the number of general anaesthetics given to MHS individuals each year, the numbers must be very large when one considers the worldwide occurrence of MH and, therefore, in spite of the few case reports mentioned above, we are still justified in reassuring MHS patients that they can face a general anaesthetic confidently.

LIFESTYLE

Occasionally, reports surface suggesting that in certain pedigrees, sometimes fatal MH-like reactions may be triggered in MHS persons by stress. At the Wausau Symposium, Wingard (38) reported a large pedigree from which he concluded that the life-threatening events which occurred outside the operating room were related to the MH trait. He stated that stress was the precipitant for many of these events and that many of the individuals who experienced these reactions were highstrung and had been treated for emotional problems. Unfortunately, because the documentation in the report is often quite incomplete, it is difficult for the reader to draw firm conclusions.

In a review article on MH, Britt (39) states that "MH crises may be induced in susceptible individuals by stress, for instance, by high environmental temperatures, sudden changes from very high to very low environmental temperatures (diving into cold water on a very hot day), mild infections, extreme emotional excitement, and extensive muscle injury or violent muscle exercise". There is no documentation of the basis for these statements.

The case reported by Gronert et al (40) is the only published example of what may be modified MH reactions triggered by extreme physical or emotional stress or fatigue. In this instance, the authors concluded that their patient was MHS on the basis of an abnormal contracture test. Thus, for the purposes of counselling, one must conclude that the vast majority of MHS patients are not at risk for awake episodes of MH precipitated by stress and there is no need to alter their lifestyle following establishment of a diagnosis of being MHS. If, in the future, well documented case reports and family studies are reported, it may be necessary to revise this recommendation for selected families.

Recently, two additional case reports have been described relating to this subject. In the first case Krisovic-Horber et al (41) reported a child who presented with recurrent reactions associated with bradycardia and decreased respirations followed by an elevated temperature. A muscle biopsy indicated malignant hyperthermia susceptibility. Family history revealed that two older sisters had recently died during similar reactions (42). In the second, Britt (43) has described a family containing four individuals who had sustained hyperthermic anaesthetic reactions--one fatal and three non-fatal. Muscle biopsies in two of the surviving patients were positive for MH. One of these latter patients, and the one patient who had a reaction but no biopsy, subsequently died at age 21 years and 5 years respectively, during hyperthermic reactions. In the former, a combination of emotional excitement, cocaine and alcohol appeared to be the precipitating factors. In the latter case, the fever of 108°F was preceded by "flu-like" symptoms.

IS THE PATIENT WITH NORMAL CONTRACTURE STUDIES FREE OF THE MH TRAIT?

This is one of the basic assumptions underlying the application of diagnostic testing for MH susceptibility. Is there evidence in the literature to support this assumption?

Unfortunately, there is not! Fortunately, despite the lack of literature on the subject there is a large unreported experience from the various test centres world wide which support the claim that anaesthetics with known triggering agents can be given safely to patients who have been proven to be free of MH by contracture testing. The solution to this dilemma will not be forthcoming until the various test centres world wide report follow ups on the anaesthetic experiences of patients who have been diagnosed as non-MHS, and have had subsequent general anaesthetics using known triggering agents. At the present time, we are carrying out just such a survey. Although we have tested some 300 patients, the number with negative tests who have undergone subsequent general anaesthesia appears to date to be quite small. Therefore, unless all test centres systematically follow up and report on their patient outcomes, it will still be a long time before there will be evidence to support this claim.

UNKNOWNS

The counsellor for MHS individuals may be faced with additional questions for which no definite answers are available. The following are examples.

1. Is there a relationship between the Sudden Infant Death Syndrome and MH as suggested by Denborough (44)?
2. Are patients with Duchenne Muscular Dystrophy (DMD) at greater risk for a syndrome of MH or does the occurrence of MH in patients with DMD just represent a chance occurrence, as had been observed in the pedigree our group reported (45)?
3. Should all patients with Muscular Dystrophy, congenital myopathies such as Central Core Disease or metabolic myopathies such as Periodic Paralysis be considered at risk for MH?
4. Can MH be a syndrome as well as a disease?
5. Is there a relationship and, if so, what is it between MH and the neuroleptic malignant syndrome?
6. Is there a relationship between MH and heat stroke?

Obviously, there is much that still has to be learned and this information will only be forthcoming if there is careful and complete clinical and laboratory evaluation of all MHS patients on an ongoing basis and publication of full reports of any unusual reactions or associations.

LAY ORGANIZATIONS DEVOTED TO MH

Three such organizations exist: two in North America and one in Great Britain. The largest one is the Malignant Hyperthermia Association of the United States (P.O. Box 3231, Darien, CT. 06820). The Malignant Hyperthermia Association (Box 144, 2 Bloor St. W., Toronto, Ontario M4W 2E2) is a smaller, Canadian-based association. The third, the Malignant Hyperpyrexia Association of Great Britain is based in Leeds (11 Gorse Close, Newthorpe, Nottingham, U.K.). Both the North American organizations publish regular newsletters which provide up-to-date information for patients and their families. Counsellors should be familiar with both of these publications and should recommend MHS patients to join these associations and subscribe to their publications.

CONCLUSION

Any physician who provides counselling for MHS individuals and families faces a difficult task because the many unknowns that remain create uncertainty, and uncertainty is particularly troublesome when one is dealing with a potentially life-threatening condition like MH. As well, both the scientific and lay literature relating to MH contains many recommendations for patients that cannot be substantiated on the basis of published evidence. In this chapter, this author's approach to counselling has been reviewed. Some of the recommendations may be at variance with accepted "current wisdom" but they are made on the basis of a review of the best information available today. It appears that apart from the need to avoid "triggering drugs" during general anaesthesia, the vast majority of MHS individuals can look forward to an otherwise unrestricted lifestyle.

REFERENCES

1. Denborough, M.A. and Lovell, R.R.H. Lancet 1960;2:45.

2. Schwartz, L., Koka, B.V. and Rockoff, M.A. Anesthesiology 1983;59:A438.

3. Rosenberg, H. and Reed, S. Anesth. Analg. 1983;62:415.

4. Ellis, F.R. and Halsall, P.J. Br. J. Anaesth. 1984;56:381.

5. Flewellyn, E.H. and Nelson, T.E. Canad. Anaesth. Soc. J. 1982;29:42.

6. Kalow, W., Britt, B.A., Terreau, M.E. and Haist, C. Lancet 1970;2:895.

7. Willner, J.H., Wood, D.S., Cerri, C. and Britt, B.A. New Engl. J. Med. 1980;303:138.

8. Endo, M., Yagi, S., Ishizuka, T., Horiuti, K., Koga, Y. and Amaha, K. Biomed. Res. 1983;4:83.

9. Eng, G.D., Becker, M.J. and Muldoon, S.M. Muscle and Nerve 1984;7:618.

10. Lopez, J.R., Alama, L., Caputo, C., Wikinsi, J. and Ladezma, D. Muscle and Nerve 1985;8:355.

11. The European Malignant Hyperpyrexic Group. Br. J. Anaesth. 1984;56:1267.

12. Department of Clinical Epidemiology and Biostatistics, McMaster University Health Sciences Centre. Can. Med. Assoc. J. 1981;124:703.

13. Mortier, W. and Biesel, C. Brain Dev. 1982;4:347.

14. Gronert, G.A. Anesthesiology 1983;59:273.

15. Heimann-Patterson, T. Personal Communication.

16. Takagi, A., Sunohara, N., Ishihara, T., Nonaka, I. and Sugita, H. Muscle and Nerve 1983;6:510.

17. Brownell, A.K.W., Paasuke, R.T. and Fowlow, S.B. Presented at the 4th International MH Workshop, York, England, September 3-5, 1986.

18. Denborough, M.A., Forster, J.F.A., Lovell, R.R.H., Maplestone, P.A. and Villiers, J.C. Br. J. Anaesth. 1962;34:395.

19. Britt, B.A., Locher, W.G. and Kalow, W. Canad. Anaesth. Soc. J. 1969;16:89.

20. Kalow, W., Britt, B.A. and Chan, F-Y. Int. Anesthesiol. Clin. 199;17:119.

21. McPherson, E. and Taylor, C.A. Am. J. Hum. Genetics 1982;11:273.

22. Gronert, G.A. The Communicator 1984;2:3.

23. The Hotline. 1984;2:3.

24. Flewellen, E.H. and Nelson, T.E. Anesth. Analg. 1983;62:115.

25. Britt, B.A. and Kalow, W. Canad. Anaesth. Soc. J. 1970;17:293.

26. Bianchi, C.P. Fed. Proc. 1968;27:126.

27. Adragna, M.G. Anesthesiology 1985;62:99.

28. Berkowitz, A. and Rosenberg, H. Anesthesiology 1985;62:651.

29. Gronert, G.A. The Communicator 1982;1:1.

30. Flewellen, E.H. The Communicator 1983;2:2.

31. Yagiela, J. The Communicator 1984;2:2.

32. Harrison, G.G. and Morrell, D.F. Br. J. Anaesth. 1980;52:385.

33. Wingard, D.W. and Bobko, S. Anesth. Analg. 1979;58:99.

34. Paasuke, R.T. Personal Communication.

35. Paasuke, R.T. and Brownell, A.K.W. Canad. Anaesth. Soc. J. 1986;33:126.

36. Ruhland, G. and Hinkle, A.J. Anesthesiology 1984;60:159.

37. Grinberg, R. and Gordon, A. Canad. Anaesth. Soc. J. 1983;30:273.

38. Wingard, D.W. In: Henschel, E.O. (ed.). Malignant Hyperthermia, Current Concepts. 1977. New York: Appleton-Century-Crofts, pp. 79-95.

39. Britt, B.A. In: Milton, A.S. (ed.). Pyretics and Antipyretics, Handbook of Experimental Pharmacology. 1982. New York: Springer-Verlag, pp. 547-615.

40. Gronert, G.A., Thompson, R.L. and Onofrio, B.M. Anesth. Analg. 1980;59:377.

41. Krisovic-Horber, R. Presented at the 4th International MH Workshop, York, England, September 3-5, 1986.

42. Britt, B.A. Personal communication, York England, 1986.

43. Britt, B.A. Malignant hyperthermia--tragedy of an American family 1968-1986. Submitted for publication to Anesth. Analg. 1986.

44. Denborough, M.A. Med. J. Aust. 1981;1:649.

45. Brownell, A.K.W., Fowlow, S.B. and Paasuke, R.T. Neurology 1985;35(Suppl.):195.

14

DANTROLENE--AN UPDATE

B.A. Britt

Departments of Anaesthesia and Pharmacology, University of Toronto, c/o Toronto General Hospital, Toronto, Ontario M5G 2C4

ABSTRACT

Dantrolene sodium, a hydantoin derivative, relaxes skeletal muscle by a peripheral rather than a central action by inhibiting release of calcium from the sarcoplasmic reticulum (SR). Both resting and twitch tensions are decreased in the presence of dantrolene. The action of dantrolene on the SR may be indirect rather than direct. Thus in both isolated SR preparations and in skinned fibre preparations dantrolene does not greatly alter caffeine, calcium or chloride induced calcium release from the SR.

Dantrolene prolongs the action potential duration and effective refractory period of normal polarized Purkinje muscle. Dantrolene also blocks calcium release from vascular smooth muscle. Dantrolene depresses canine urethral and bladder pressures and reduces the amplitude of calcium induced contractions of rabbit jejunum.

Dantrolene is formulated in 70 cc bottles which contain 20 mg of dantrolene, 3.0 grams of mannitol and enough sodium hydroxide to raise the pH to 9.5 when the material is reconstituted with 60 ml of water. This rather unusual formulation is necessary because of the relative water insolubility of dantrolene. Care must be taken not to allow extravasation of dantrolene into the tissues because the high pH of the solution would be injurious to the tissues. The vein site should be changed every few hours to avoid serious thrombophlebitis.

Since the introduction of dantrolene a marked improvement in survival from malignant hyperthermic (MH) reactions has occurred. While some of this improvement has probably been due to other factors (better monitoring and better education of anaesthetists about the diagnosis and management of MH reactions) a considerable portion of the improvement must be attributed to the use of dantrolene.

The dose of dantrolene required to successfully resuscitate patients from MH reactions has varied from less than 1.0 to 17 mg/kg. All patients who have received, early in the reaction, at least 6.0 mg/kg of the official formulation have survived. The drug is administered as a bolus at a maximal rate of 1 mg/kg/minute up to a maximum of 10 mg/kg under EKG control. This bolus may be repeated every 15 minutes. Once temperature, heart rate and muscle stiffness have begun to decline, an IV drip containing 1-2 mg/kg of dantrolene may be administered over a three to four hour period.

In patients dying in spite of the use of dantrolene, one or more of the following has usually been present: use of a non-official formulation; delay in commencing dantrolene; use of an insufficient amount of dantrolene; concomitant use of contraindicated drugs; true diagnosis not malignant hyperthermia.

In pigs dantrolene has been effectively used prophylactically, to prevent halothane induced MH reactions, in doses ranging from 5 to 24 mg/kg in up to four divided doses. Dantrolene has also been used in MH humans prior to elective anaesthesia. Because deliberate halothane challenge is impossible in human patients its efficacy in this species is difficult to determine.

In some, but not all, MHS patients, dantrolene has reduced the severity of the skeletal muscle aches, pains and cramps from which these individuals not infrequently suffer. Dantrolene is also useful in certain other situations involving non MH patients. For instance, dantrolene is valuable in the management of heat stroke, neuroleptic malignant syndrome, cerebral palsy, Parkinsons disease, chorea, athetosis, spinal cord injury, multiple sclerosis, myotonia and tetanus. Dantrolene also attenuates succinylcholine induced muscle fasciculations, postoperative muscle pains and rises in serum potassium.

Early undesirable but transient effects of dantrolene include dizziness, diplopia, dysarthria, sensations of muscle weakness and swelling of the eyeballs and tongue. These symptoms disappear a few hours after discontinuation of the drug. Liver damage, chronic pleural effusion, acute pericarditis and lymphocytis have rarely been reported but their true relationship to dantrolene ingestion has not been well documented. Long term high dose dantrolene therapy rather consistently causes a fine cuneiform rash over the face and back.

Dantrolene is metabolized through reductive and oxidative pathways. The nitro group of dantrolene is reduced to amino-dantrolene. The amine group is acetylated. Oxidation of dantrolene produces 5-OH dantrolene.

INTRODUCTION

Dantrolene sodium (dantrium, Norwich-Eaton Pharmaceuticals) is a lipid soluble hydantoin derivative (1,2) (Figure 1). It relaxes skeletal muscles by acting directly on the muscle. The site of action of dantrolene within the muscle is not yet clearly

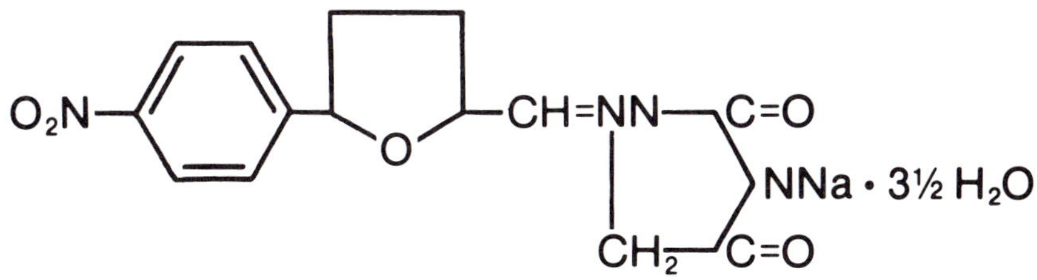

Dantrolene
(Dantrium)

FIGURE 1: Dantrolene Na (Dantrium) is a benzathiazepine derivative which, by inhibiting release of calcium from the sarcoplasmic reticulum, decreases contraction and heat production in skeletal muscle during MH reactions.

understood. It is believed, however, that it either directly or indirectly, inhibits calcium release from the sarcoplasmic reticulum (SR) (3). Dantrolene is presently marketed as dantrolene sodium but other dantrolene salts and analogues of dantrolene also have muscle relaxant properties (4).

ABSORPTION, BINDING AND METABOLISM OF DANTROLENE

About 70% of a single oral dose of dantrolene is absorbed, with peak plasma concentrations of about 1 mg/L occurring 4 to 6 hours after administration of a 100 mg dose (5). Dantrolene is bound to human serum albumin, plasma proteins and to skeletal

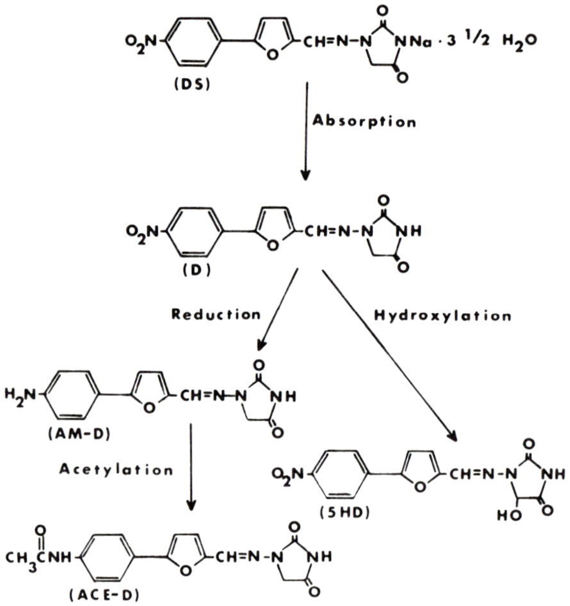

FIGURE 2: Metabolism of dantrolene sodium (DS) and dantrolene (D), based on identified metabolites: 5-hydroxy-dantrolene (5HD), aminodantrolene (AM-D), and acetylated dantrolene (ACE-D).

Reprinted with the permission of Dr. K.O. Ellis, Scientific Affairs Dept., Norwich-Eaton Pharmaceuticals, Norwich, New York.

muscle SR (5). Metabolism of dantrolene sodium (as measured by high performance liquid chromatography (6) or spectrofluorometry (7,8)) occurs primarily in the liver through both reductive and oxidative pathways (8-10). The nitro group of dantrolene is reduced to aminodantrolene. In some animals, including man, the amine is acetylated (9,10). Oxidation of dantrolene produces 5-OH dantrolene (9,10) (Figure 2). Ellis and Wessels have shown that 5-OH dantrolene is able to relax skeletal muscle as well as dantrolene (9). Upon repeated administration of dantrolene the plasma concentration of 5-OH dantrolene may exceed that of dantrolene (5). On the other hand these workers have discovered that aminodantrolene and acetylated dantrolene have minimal relaxing properties (9). Fifteen to 25% of dantrolene is excreted in the urine (primarily as the hydroxy metabolite) with negligible faecal excretion (5). The elimination half-lives of dantrolene and 5-hydroxydantrolene are about 6 to 9 hours and 15.5 hours, respectively (5). There appears to be a direct correlation between the plasma concentrations of the drug and its therapeutic effects (5).

MODE AND SITE OF ACTION OF DANTROLENE SODIUM

I. SKELETAL MUSCLE

A. Non Malignant Hyperthermic Muscle

1. Sarcolemma

Dantrolene attenuates the twitch tensions of normal skeletal muscle that is being stimulated either directly or indirectly through its nerve (11,12) and of normal human and porcine skeletal muscle fascicles (13). The ability of dantrolene to inhibit twitch tensions in normal humans and pigs is antagonized by increasing the concentration of calcium in the bathing medium (14-17). Microelectrode studies show that a dose of dantrolene sodium which reduces the twitch responses of electrically stimulated skeletal muscle has no effect on either resting or action potentials of the muscle membrane of cats (1,18). Dantrolene does not change total membrane capacitance or membrane resistance (19). Electromyograms (EMGs) are not altered by dantrolene. Since, therefore, dantrolene does not affect electrical excitability of the muscle membrane and does not disrupt the transverse tubules, it cannot act by decreasing electrical conduction between the post-synaptic junction and the SR.

Dantrolene has no influence on the resting influx of Ca^{++45} from the extracellular fluid to the myoplasm of normal giant barnacle muscle cells (20). Dantrolene, therefore, probably does not relax barnacle muscle cells by preventing influx of extracellular fluid calcium. Similar studies have not yet been reported on normal mammalian muscle.

Dantrolene does inhibit potassium induced contractures of normal intact skeletal muscle fascicles (19-26). On the other hand, Brocklehurst (27) has shown that dantrolene does not alter any aspect of potassium induced contractures of frog muscle fibres regionally (but not completely) skinned by mechanical means. These experiments suggest that dantrolene prevents calcium release from the SR by an indirect action which requires an intact sarcolemma. For instance, dantrolene might inhibit release of regenerative or "trigger" calcium from the sarcolemma. This postulation is supported by the observation of Oba (28) that extracellular dantrolene always depresses twitch tension of frog skeletal muscle while intracellularly injected dantrolene sodium transiently increases and then decreases frog muscle twitch tension. These effects are more potent at high than at low temperatures.

Dantrolene raises the minimum electric current needed to produce stimulation in a rat skeletal muscle preparation (29). This effect is completely antagonized by the calcium ionophore A23187 which elevates cytoplasmic calcium by increasing its release from the SR (29). Dantrolene, therefore, is not able to prevent A23187 induced calcium release from the SR. The above work suggests that dantrolene might have a primary, although not necessarily exclusive, action on the sarcolemma. Dantrolene might act by displacing regenerative ("trigger") calcium from binding sites on the transverse tubular membrane to the extracellular fluid (28). Then the released calcium might accelerate further release of calcium from the SR.

It is of interest that in chicks succinylcholine produces rigid paralysis, hyperpyrexia and death (30). While dantrolene provides marked protection against these untoward events. The muscle cells of fowls have multiple motor end plates whereas the muscle of normal mammals have only one motor end plate per cell (30). The adverse response of chicks to succinylcholine might be due to the drug acting on many, rather than one, cholinergic neuromuscular receptor, thereby creating an excessively rapid rise in myoplasmic calcium (30). Could it be that the sarcolemma of MHS mammals also contains multiple rather than single cholinergic neuromuscular receptors, and perhaps also a greater than normal complement of adrenergic receptors?

2. Sarcoplasmic Reticulum (SR)

a) Calcium Release

The effect of dantrolene on calcium release from the SR has been studied by several investigators using in vivo and in vitro, and direct and indirect techniques.

Ellis and Carpenter (31) and Krarup (12) have reported that in the rat diaphragm, dantrolene inhibits direct and indirect electrically induced twitch responses, and antagonizes caffeine potentiation of twitch responses. They have also found that the addition of EDTA increases, while the addition of calcium decreases these various dantrolene effects. Leslie and Part (32) have shown that in rat extensor digitorum longus, soleus and segmental tail muscles dantrolene also inhibits twitch tensions with the inhibition being significantly greater in fast than slow muscle (32). These in vitro responses, at least in normal muscle, are unaltered by curare (19,22,23,33). Desmedt and Hainaut (20,34) have observed that in giant barnacle muscle fibres micro-injected with the photoprotein aequorin, the resting rate of light emission reversibly decreases upon exposure to dantrolene sodium. Since aequorin emits light only in the presence of calcium this experiment suggests that dantrolene lowers the concentration of calcium in the myoplasm.

They postulate that dantrolene sodium reduces the number of Ca^{++} releasing sites on the SR membranes activated by a given depolarization.

Other work has demonstrated that dantrolene inhibits electrically (22) induced Ca^{++45} release from the SR to the myoplasm of normal frog skeletal muscle cells. The ability of dantrolene to depress twitch tensions is greater in rat extensor digitorum brevis (a fast muscle) than in rat soleus (a slow muscle) (35). Dantrolene also suppresses twitch tensions of: cat gastronemius muscle (36), rat skeletal muscle (37) and the adductor pulleris muscle of paraplegic humans (38).

These various in vivo and in vitro findings suggest that dantrolene, in some way prevents release of calcium from the SR to the myoplasm, either by a direct action, by inhibition of release of "trigger" or "regenerative" calcium from the sarcolemma or by inhibition of the excitation-contraction (EC) step between the transverse tubules and the SR. Dantrolene, however, has not been found to alter calcium induced calcium release (CICR) from the SR of normal rabbit skeletal muscle (39).

Confusingly, Flewellen et al (40) have claimed that dantrolene augments in normal pig and human skeletal muscle fascicles caffeine potentiated twitch tensions at 2.0 but not at 0.5 or 1.0 mM concentration of caffeine. However, Flewellen explains this apparent contradiction of other workers' results by postulating that dantrolene, by improving calcium uptake into the SR has, therefore, made more calcium available for release and so ensuing twitch tensions become correspondingly greater.

b) <u>Calcium Uptake</u>

Evidence for acceleration by dantrolene of calcium uptake into the SR is not unequivocal. Some investigators (13,14,26,41-44) have observed that dantrolene does inhibit caffeine and/or halothane induced contractures of normal swine or human skeletal muscle. Dantrolene also inhibits succinylcholine induced contractures in caffeinated or halogenated skeletal muscle fascicles isolated from normal humans (24,45,46).

Similarly, Fairhurst et al (25) have found that dantrolene inhibits ryanodine induced contractures both in the presence and in the absence of halothane potentiation. As ryanodine induces release of calcium from the SR this antagonism by dantrolene may reflect opposing influences of these two agents on cytoplasmic concentrations of free calcium. Other workers, however, have claimed that dantrolene sodium has no influence on ryanodine induced contractures of frog rectus muscle (47) or caffeine induced contractures of normal rat phrenic and frog sartorius muscles (19,22). These apparent discrepancies might in part be due to non-uniform experimental techniques.

Thus, some but by no means all studies suggest that dantrolene increases calcium uptake into the SR. Since, however, these studies have been done on intact skeletal muscle fascicles, they do not elucidate whether the action of dantrolene is directly on the SR or is mediated indirectly via the sarcolemma or the E-C coupling step.

Furthermore, these caffeine induced contractures may not be due to increased calcium uptake into the SR but rather may be due to continued excessive calcium release from the SR during the intervals between twitches. The extent of this caffeine induced calcium release may be so great that the muscle becomes incapable of complete relaxation between twitches. The influence of the dantrolene on so called "resting tension", therefore, may in reality be an effect on calcium release from the SR rather than calcium uptake into the SR.

c) Calcium Binding

Studies with (C^{14}) dantrolene sodium have shown that the drug binds to the mitochondria and SR of pig heart and skeletal muscle with low affinity. In addition, skeletal muscle SR possess a separate class of sites which bind dantrolene sodium with high affinity. These high affinity sites suggest a specific site of action for this drug on skeletal muscle SR (48). However, the isolation technique employed for the SR do not remove the excitation-contraction (E-C) coupling step at the gap junction. The high affinity binding sites, therefore, may be in this location rather than in the SR.

B. Malignant Hyperthermia Susceptible Muscle

1. In Vivo Studies

The efficacy of intravenous dantrolene sodium in the therapy of malignant hyperthermic (MH) crises was first observed by Harrison (49,50). He noted that dantrolene relieved halothane induced MH reactions of susceptible Landrace swine. Other investigators soon reported that intravenous dantrolene sodium was beneficial in the therapy of MH reactions induced by halothane and/or succinylcholine in other strains of pigs, for instance, Pietrain (5,51-53), Poland China (54,55) and Landrace pigs (56). However, although some indices of increased muscle metabolism in porcine MH such as venous hypercarbia, lactacidaemia and hypoxaemia and glucose metabolism respond rapidly, while others (such as muscle temperature and plasma potassium) only respond slowly to dantrolene (5,57,58).

In addition, pretreatment with large single oral doses of dantrolene prior to induction of anaesthesia completely prevents the development of symptoms of MH in Landrace pigs (49,59) and in Poland China pigs (54,60). Nelson and Flewellen (40,41,61,62) have noted that dantrolene, but not procainamide, inhibits indirectly evoked foretoe twitches of normal and malignant hyperthermia susceptible (MHS) pigs. The addition of halothane augments this dantrolene induced depression of foretoe twitches in MHS but not in normal pigs.

In MHS swine resting cytoplasmic calcium levels are about four times greater than normal (57). These levels are increased by a further 20-fold after administration of halothane or succinylcholine (57). Dantrolene reduces both resting baseline levels of cytoplasmic calcium and prevents or reverses the effects of halothane and succinylcholine on cytoplasmic calcium concentrations (63,64). Thus dantrolene appears to act by lowering, in some way, the cytoplasmic concentration of calcium.

We are aware of at least four MHS patients who have had two skeletal muscle biopsies each--one without and one with in vivo dantrolene pretreatment. The skeletal muscle fascicles of one patient reported by Rosenberg (65) developed caffeine and also halothane induced contractures which were not inhibited by prior in vivo dantrolene sodium administration. However, the skeletal muscle fascicles of three other patients, one reported by Lambert (66) and two others measured by ourselves (unpublished data), developed caffeine and halothane induced contractures which were substantially less after than before in vivo dantrolene sodium therapy.

2. **In Vitro Studies**

In vitro investigations in preparations from muscle tissue samples from malignant hyperthermia susceptible (MHS) pigs have provided clues to the mode of action of dantrolene on MHS muscle (5,67).

Halothane 2% depolarizes the sarcolemma of skeletal muscle from MHS swine, but does not affect the resting membrane potential of normal pig muscle. This halothane induced depolarization of MHS sarcolemma is reduced by dantrolene (68) which suggests that the major site of action of dantrolene on MHS muscle may be at the sarcolemma/sarcotubular system.

As in normal skeletal muscle fascicles, contractures induced in MHS skeletal muscle fascicles by halothane, caffeine, succinylcholine, potassium chloride and thymol are inhibited by dantrolene (13,26,69). Similarly in vitro dantrolene inhibits twitch amplitudes of MHS as well as of normal pig muscle fascicles in the absence and in the presence of halothane (14) and in the presence and absence of caffeine (13) . On the other hand direct calcium uptake into isolated porcine MHS SR (70) or the SR of a single skinned fibre (71) is unaffected by dantrolene. While release of calcium from the SR isolated from MHS porcine skeletal muscle occurs at a much lower concentration of exogenous or "trigger" calcium than in normal pig SR (17,72), this accelerated CICR from MHS pig SR is not altered by dantrolene (73) or is only lowered by a small

ratios, and bind similar amounts of dantrolene (5). This evidence also indicates that dantrolene does not directly affect the uptake of calcium into the SR (5,74).

Muscle mitochondria from MHS pigs accumulate myoplasmic calcium during halothane challenge with resultant damage to the mitochondria. Dantrolene prevents this accumulation (75). Thus dantrolene may also act by inhibiting calcium uptake into the mitochondria. This in turn would prevent overactivity of phospholipase A_2 and, therefore, excessive release of long chain fatty acids which would, without the suppressant effect of dantrolene, increase the sensitivity of the SR to CICR (76).

II. HEART MUSCLE

In the past, dantrolene has been stated to have no significant action on vascular smooth muscle or on heart muscle (77-85). More recent work, however, has shown that dantrolene does alter the function of both vascular and heart muscle (86-94).

Salata and Jalife (95) have reported that dantrolene decreases the strength of canine myocardial contraction. Plizga and Hall (92) have also observed that dantrolene reduces the contractile force of isolated rabbit left atria without altering the effective refractory period, suggesting to them that the inhibitory effect of dantrolene on cardiac muscle is on the SR and not on sarcolemmal excitability. However, in anaesthetized dogs dantrolene does prolong the refractory period of the right atrium and ventricle (96). Moreover, Salata and Jalife (95) have found that dantrolene prolongs the effective refractory period of normally polarized Purkinje fibres.

Dantrolene sodium has no significant effect on resting membrane potentials (89-91,94,95,97), upstroke velocity of phase 0 (95), atrio ventricular conduction velocity (95,98,99) or pacemaker activity of Purkinje fibres (95). Bailey and his colleagues (100) have observed that dantrolene sodium inhibits the positive inotropic effects of Anthopleurin-A (a potent cardiostimulant polypeptide) on rat and rabbit atria.

On the one hand, Salata and Jalife have noted that dantrolene depresses the plateau phase of the action potentials of canine Purkinje fibres (95). On the other hand, in guinea pig and frog myocardia dantrolene prolongs action potential duration while increasing twitch tension amplitudes (95,101-104). Lascault et al (97) have also shown that dantrolene prolongs the duration of action potentials in isolated sheep Purkinje fibres. These workers, however, have not found a similar effect in myocardial fibres. Hatae et al (101) have, therefore, concluded that dantrolene sodium increases the slow inward calcium current. They have also observed that in guinea pigs, but not in frogs, the dantrolene-induced increase in twitch tension is preceded by a transient reduction in myocardial twitch tension. They have attributed this to inhibition of calcium release from the SR by dantrolene similar to that postulated for skeletal muscle.

In isolated rabbit hearts subjected to cardioplegia and reflow, dantrolene induces ultrastructural preservation, lowers ventricular stiffness and end diastolic pressure and does not alter contractility (5,105). These results suggest that dantrolene may improve post ischaemic diastolic contracture and compliance (105). In rats, with ligated coronary arteries, dantrolene decreases the number of episodes of ventricular tachycardia and raises the ventricular fibrillation threshold (5,106). Dantrolene also suppresses ventricular arhythmias induced by chloroform in mice or by coronary artery occlusion in dogs (5,107).

In anaesthetized dogs, dantrolene accelerates the heart rate (5,106), elevates systemic blood pressure and decreases the cardiac index by increasing systemic vascular resistance (5,98,102). In anaesthetized normal swine dantrolene accelerates the mean arterial blood pressure but does not influence cardiac output, pulmonary artery occlusion pressure or pulmonary vascular resistance index (5,99,108).

III. VASCULAR SMOOTH MUSCLE

Ally et al (109) have reported that dantrolene blocks intracellular calcium release from rat superior mesenteric artery smooth muscle by competitively antagonizing thromboxane A_2. Dantrolene also inhibits pressor responses to noradrenaline (norepinephrine) and angiotensin (but not those to potassium) in man (109).

IV. OTHER TYPES OF SMOOTH MUSCLE

Dantrolene depresses spontaneous smooth muscle contractions in guinea-pig uterus (5,110), and in isolated guinea-pig trachea, uterus and ileum, rat ileum, and rabbit jejunum in response to potassium, various agonist agents, and electrical stimulation (5,79,110-113). Dantrolene also reversibly reduces the amplitude of calcium induced contractions of rabbit jejunum in vitro (113). Thus dantrolene sodium might act on intestinal smooth muscle by inhibiting calcium movement across the cell membranes (113). However, dantrolene in cumulative doses of approximately 10 mg/kg has no consistent effect on intralumenal duodenum pressure response in anaesthetized dogs (5,79). This may not be surprising, as in vitro tests have shown that skeletal muscle is approximately 14 times more sensitive to relaxation induced by dantrolene than is intestinal smooth muscle (5,79).

Dantrolene has been reported by Harris and Benson (114) to have no significant inhibitory effect on acetylcholine-induced canine detrusor muscle contractions. Furthermore, dantrolene has no effect on acetylcholine-induced or electrically induced contractions in smooth muscle strips from rat, rabbit or human urinary bladder (115). In rabbits, dantrolene does not alter urinary bladder pressure or capacity (115), and doses up to 14 mg/kg in normal and chronic T-10 paraplegic dogs only minimally (15 to 20%) reduces peak urethral pressure during pudendal nerve stimulation (5,116). However, Khalaf et al (117) have observed that dantrolene causes dose-dependent depression of maximal canine urethral and bladder pressures in anaesthetized dogs. In some spinal cord injured human patients dantrolene sodium decreases urethral pressures and lowers residual urine volumes (118). Urethral resistance is also reduced by dantrolene in such patients (119). That dantrolene can increase the effectiveness of voiding in many patients with neuromuscular disorders is probably due more to a decrease in tone of the striated sphincter than to any alteration in urinary bladder wall smooth muscle (115).

V. MOTOR NERVES

Several workers have attempted to find an effect of dantrolene sodium on motor nerves. Thus Ellis et al (11) have shown in crossover circulation experiments of vascularly isolated but neurally intact hind limbs of normal dogs that dantrolene has no effect on the excitability or on the conduction velocity of the sciatic motor nerve. The action of d-tubocurarine (a non-depolarizing neuromuscular blocker) or of edrophonium (an anticholinesterase agent) on twitch tension is not altered by dantrolene (11). In vivo studies on spinal motoneurons in rats show no modification of membrane potential, cell firing or membrane resistance after administration of dantrolene (120). Ohta et al (121) have observed that dantrolene does not alter action potentials in presynaptic nerve cells but does reduce the amplitude of excitatory post-synaptic cells. Finally, Colton and Colton (122) have shown that, at lobster neuromuscular junctions, dantrolene sodium does not appear to affect either spontaneous or evoked transmitter release nor excitatory or inhibitory end plate function.

To the contrary other workers have observed a variety of actions of dantrolene on the motor nerves. A report by Durant et al (123) suggests that dantrolene decreases neuromuscular transmitter mobilization which in turn causes a reduction in the available store of acetylcholine. They suggest that this effect is due to impairment by dantrolene of calcium release from bound stores within the nerve terminal (5,123). These workers feel, however, that this effect of dantrolene is unlikely to be of physiological consequence in vivo since it is only observed at high frequencies of nerve stimulation and does not cause a significant reduction of end plate potential amplitude during trains of stimuli

(5,123). However, in the rat phrenic nerve-diaphragm preparation, dantrolene has been found to reverse the block to indirect stimulation caused by caffeine (5,124). This suggests that presynaptic influence on stimulus-secretion coupling is affected by intraterminal calcium ion balance (5,124). Additionally, dantrolene decreases rhythmic hyperpolarizations during depolarization of sensory nerve terminals in frog muscle spindle, whereas methylxanthines increase the threshold and sensitivity of hyperpolarizations (5). Electronmicroscopic and x-ray analysis of the spindles suggests that dantrolene may act by modulation of nerve terminal responses by inhibiting calcium release, and increasing potassium efflux, whereas methylxanthines have the opposite effect (5,125).

VI. BRAIN

In rat brain synaptosomes, neither dantrolene, nor nifedipine inhibit calcium uptake into neurons bathed in potassium, and dantrolene is much less powerful an antagonist of stimulated calcium uptake than is nifedipine (5,126). On the other hand, dantrolene has been demonstrated to increase glutamic acid decarboxylase activity at the corpus striata, substantia nigra and cortical levels, and to have a marked central GABAergic effect (5,127). Despite structural similarities to hydantoin and some local anaesthetics dantrolene appears to be devoid of anticonvulsant or anaesthetic properties (5,128). Dantrolene decreases K^+ evoked and electrically evoked serotonin release from the hypothalamus in a dose-dependent manner (129). This suppression effect is negatively correlated to the concentration of calcium.

VII. RESPIRATORY SYSTEM

The respiratory effects of dantrolene appear to be negligible in spite of its pronounced muscle relaxant effects (5,130). It has been proposed (5,53) that the sparing effect on respiration is due either to a reflex modification of the nervous drive to the respiratory muscles or to resistance of the respiratory muscles to the depressant effect of dantrolene. In anaesthetized cats or rats, intravenous dantrolene depresses twitch responses in various muscle groups but has little effect on respiration (5,81,131).

VIII. ENDOCRINE SYSTEM

Insulin release stimulated by glucose, mannose and glyceraldehyde is inhibited by dantrolene in isolated rat pancreatic islets (5). However, that stimulated by leucine, arginine, ouabain, potassium or 3-iso-l-methylxanthine is not inhibited (5). Calcium utilization is unaffected but glucose oxidation is decreased by dantrolene (5). At normal calcium concentrations dantrolene inhibits glucose-stimulated calcium efflux, but this is unaffected by dantrolene in the absence of calcium (5). These results suggest that dantrolene interferes with mobilization of B-cell calcium stores (5,132), although in another in vitro study dantrolene describes the opposite effect, potentiates glucose-stimulated insulin release (5,133).

In rats large intraperitoneal doses of dantrolene 25 to 50 mg/kg cause significant adrenal enlargement, reduced serum glucocorticoids and decreased hepatic cytochrome P-450 after 5 days (5,134). In porcine adrenal glomerulosa cells dantrolene greatly inhibits angiotensin II-induced increases in fractional efflux of calcium and delays the onset and rate of aldosterone secretion (5,135-137). Dantrolene does not alter the potency of gonadotrophin releasing hormone stimulation of luteinizing hormone release in pituitary cell cultures (5,138).

IX. LIVER

Dantrolene depresses state 3 oxidation of L-glutamate in intact mitochondria and the oxidation of NADH in submitochondrial particles but only slightly reduces succinate oxidation in intact mitochondria even at very high concentrations (139).

THERAPY WITH DANTROLENE

I. INTRAVENOUS DANTROLENE FORMULATIONS

Dantrolene sodium began to be used in the treatment of human MH reactions in 1975 (unpublished data). As the intravenous lyophillized formulation was not officially approved at that time these early patients were treated with a variety of non-official formulations, in most of which only a small fraction of the added dantrolene actually dissolved and became available to the muscle cells. Furthermore, because preparation of these homemade dantrolene sodium solutions usually began only after diagnosis of the MH reactions, the majority of the early patients began to receive their dantrolene only after prolonged delays--often of several hours (67).

The first intravenous dantrolene sodium formulation consisted of dantrolene sodium 300 mg, Mannitol 26.64 grams, sodium hydroxide 48 mg and water ad 600 ml (50). Its disadvantage was that several hours of stirring was required to achieve solubilization of the dantrolene sodium--and once solubilized the dantrolene remained effective for only about six hours.

A more simple technique was to place five 100 mg capsules of dantrolene in a one litre bottle of warm 15 or 20 percent mannitol and shake. The resulting dantrolene solution was run through a millipore blood filter placed in the intravenous line. Since no sodium hydroxide was used in this formulation the common ion effect of the first formulation was avoided and so the amount of dantrolene dissolved was greater (K.O. Ellis, Personal Communication, Norwich-Eaton Pharmaceuticals, Norwich, N.Y., 1981).

A rather different formulation was that of Fukuchi et al (140,141) who prepared dantrolene by sealing 1400 mg of crystalline dantrolene in an ampoule which was then heat sterilized at 105°C for 60 minutes. Three hundred and fifty ml of aqueous solution of 1.0 M nicotinamide was placed in a separate vial and subjected to autoclave sterilization. When needed these two vials were mixed to give 0.4 W/V percent of dantrolene at pH 9.0-10.3. Another Japanese formulation was described by Tomonaga et al (142). This group dissolved dantrolene by using cyclodextrin.

In order to achieve a preparation that would dissolve rapidly and which would also remove the starch, talc, magnesium, stearate and lactose present as binding agents in dantrolene sodium capsules, Gronert and his colleagues (143) recommended placing thirty-two 100 mg capsules in litres of sterile water and stirring for 30 minutes. They then filtered the resulting liquid through a 201 grade filter paper. The clear orange filtrate was sterilized by passing it through a 0.45 micron millipore filter. Eighty ml of 10 percent citric acid was added to the sterile filtrate to lower its pH to 3.0, thereby inducing precipitation of dantrolene crystals. The crystals were collected by vacuum filtration and dried overnight in a laminar flow hood and then weighed asceptically into 500 ml lots and placed into sterile 100 ml vials. When desired the crystals were reconstituted with 100 ml of a sterile aqueous solution containing 88 mg of sodium hydroxide. The suspension was immediately added to a sterile 900 ml solution containing 44.5 grams of mannitol to make the solution isotonic. The dantrolene dissolved within two to three minutes. This preparation, like that of Harrison (50) had the disadvantage of a common ion effect with the sodium hydroxide.

Fortunately, these and other (144) tedious "moonshine" efforts are now past history. An official intravenous dantrolene sodium formulation was finally approved for use in North American hospitals in 1979 (145). It is lyophillized in combination with sodium hydroxide and mannitol. Thus a 70 ml vial contains 20 mg of dantrolene, 3.0 grams of mannitol and enough sodium hydroxide to raise the pH to 9.5 when the material is reconstituted with 60 ml of water.

II. EPIDEMIOLOGY OF IN VIVO DANTROLENE USE DURING MH REACTIONS IN HUMANS

Due to the infrequency of human MH reactions, investigation of dantrolene during such crises has been limited to retrospective studies of individual MH case reports of individual patients (5). There have been no controlled trials with dantrolene and no comparisons with other drugs because no other efficacious drugs are currently available for the management of MH (5). Even if several patients were available for treatment within one institution over a short time span, the use of placebo controls would have been unethical due to the severity of the reactions, and since the range of pharmacological and physical supportive measures varies enormously according to the range of signs and laboratory abnormalities that the patient exhibits (5). Historically, however, the mortality and morbidity associated with MH despite supportive therapy is well documented. Therefore, it is not difficult to conclude that dantrolene has made a very significant contribution to the treatment of this very potentially lethal condition (5,99,146).

The efficacy of dantrolene sodium therapy for MH reactions in humans has now been demonstrated by a large number of published case reports (147-164), personal communications and reviews (67,82,162-166). This has been done by comparing simultaneous survival rates in countries in which dantrolene has been approved for use with survival rates in countries with similar medical standards in which dantrolene has not been approved for use.

We are now aware of 162 patients who have received dantrolene sodium between 1975 and 1987 (Table 1). One hundred and forty-seven of these have received the official formulation (Table 1). One hundred and thirty-one of this latter group have survived (Table 2) giving a mortality rate of 10.88% among patients treated with the official formulation. The mortality rate among the 761 patients who have had MH crises and have not received dantrolene sodium during the same time span is 39.29%. Dantrolene sodium has, therefore, been associated with a 28.41% reduction in mortality ($x^2 = 42.63$, d.f. = 1, $p < 0.001$) (Table 2). This improvement is remarkable in that the data include a number of patients who received inadequate amounts of dantrolene and/or dantrolene long after the onset of the reaction. Some of the patients who died, therefore, might have expired, not because of true failure of control of MH by dantrolene, but rather because the dantrolene reaching the muscle cells was inadequate in amount and/or arrived too late in the course of the reaction to prevent death (Table 3).

TABLE 1: Relationship between formulation of dantrolene therapy and survival from MH reactions.

Dantrolene formulation	No. of Patients Died	Survived	Totals	x^2
Official	16	131	147	11.52
Homemade	7	8	15	
Totals	23	139	162	

TABLE 2: Relationship between dantrolene therapy (official formulation, begun pre arrest) and survival from MH reactions

Dantrolene*	No. of Patients Died	Survived	Totals	x^2
Used	16	131	147	42.63
Not Used	299	462	761	
Totals	315	593	908	

* Patients who received homemade formulations of dantrolene or who received first dose of dantrolene post arrest are excluded.

TABLE 3: Relationship between giving dantrolene pre vs post cardiac arrest and survival from MH reactions

Time Given	No. of Patients Died	Survived	Totals	x^2
Pre arrest	15	86	101	22.42
Post arrest	10	3	13	
Totals	25	89	114	

The dose of dantrolene employed has ranged from less than 1.0 mg/kg to 17 mg/kg (1,5,67,146,167). The majority of cases have received less than 5.99 mg/kg (Table 4). All patients who have received early an initial dose of at least 6.0 mg/kg have survived (67) (Table 4). In patients not receiving dantrolene, deaths have occurred following maximum temperatures as low as 37°C, while in patients receiving dantrolene no patient has died whose maximum temperature was under 38.8°C (67). In patients who have not received dantrolene, maximum temperatures tend to be higher than in those patients who have received dantrolene (67) (Table 5). The highest temperature reached in the absence of dantrolene has been 46°C while the highest temperature recorded in a dantrolene treated patient has been 43°C. As is to be expected, survival is higher when diagnosis is made early and dantrolene infusions are promptly instituted (67).

TABLE 4: Relationship between dose of dantrolene and survival from MH reaction

Dose of Dantrolene mg/kg^{-1}	No. of Patients*	
	Died	Survived
<1.0	2	11
1.0-1.99	4	31
2.0-2.99	2	21
3.0-5.99	3	25
6.0-8.99	0	8
>9.0	0	10

*** Data include only patients receiving official formulation prior to cardiac arrest.**

In patients dying in spite of dantrolene sodium administration, one or more of the following has nearly always been present:

i) use of a homemade formulation (which on later analysis nearly always has been found to contain much less dantrolene sodium than its creators had imagined) (67) (Table 2);

TABLE 5: Relationship between dantrolene therapy and maximum temperature attained during MH reaction.

	No. of Patients			
	Died		Survived	
Dantrolene	Yes	No	Yes	No
Max.temp. °C				
<37.9	0	2	20	69
38.0-39.9	6	22	72	161
40.0-41.9	12	85	34	152
42.0-43.9	5	159	12	57
>44.0	0	13	0	0

ii) inordinate delay in starting dantrolene sodium therapy (often due to failure to have the official lyophillized IV dantrolene sodium formulation on hand in the operating theatre) (67) (Table 6);

iii) treatment with an insufficient amount of dantrolene sodium (62);

iv) concomitant use of contraindicated remedies, for instance, calcium salts, cardiac glycosides, isoproterenol and sympathomimetic vasopressors; and

v) true diagnosis not MH but rather some other fever causing condition, for instance, hyperthyroidism, pheochromocytoma, primary hypoxia, infection of embolus.

TABLE 6: Relationship between time after induction of commencing dantrolene therapy and survival from MH reaction.

Time after Induction (minutes)	No of Patients	
	Died	Survived
0-29	1	5
30-59	0	4
60-119	1	5
120-179	2	3
>180	8	4

III. RECOMMENDED ROUTE AND DOSAGE OF DANTROLENE SODIUM DURING MH REACTIONS

Reconstituted lyophillized dantrolene sodium ideally should be administered at the rate of 1.0 mg (2.5 ml) per kg per minute under EKG control until the temperature begins to fall, muscle stiffness starts to subside and/or the heart rate begins to decline. A maximum of 10 mg/kg may be given over a fifteen minute period. This loading dosage may be repeated every fifteen minutes if the reaction recurs. Even in the absence of recurrence of the reaction, the loading dose(s) of dantrolene ought to be followed by a maintenance infusion of 1.0-2.0 mg/kg during each three to four hour period until all evidence of an active MH crisis has disappeared. As soon as possible intravenous dantrolene should be replaced with oral dantrolene as intravenous dantrolene induces thrombophlebitis of peripheral veins. Further details of emergency dantrolene therapy are given in Chapter 16.

IV. DANTROLENE BEFORE ELECTIVE ANAESTHESIA OF MHS INDIVIDUALS

In the management of porcine MH reactions dantrolene sodium has proved effective not only in the therapy of established crises but also prophylactically before elective anaesthesia (49,50,54,55,59,60,168). The total preanaesthetic dose of dantrolene sodium in porcine studies has ranged from less than 1 mg/kg to 24 mg/kg (54,55,59,67). The time of start of treatment extends from two days to two hours before anaesthesia. Kerr (55) has found that a single oral dose of 8 mg/kg four hours before halothane challenge is as effective as smaller but multiple doses of dantrolene sodium. Gronert (54) has shown that pretreatment with 5 mg/kg intravenously prevents, 1-3 mg/kg attenuates, and 0.1 mg/kg has no effect on induction of reactions in MHS swine by halothane and succinylcholine. Flewellen and Nelson (60,61) have reported that the intravenous dose of dantrolene sodium which produces 95% of maximal muscle relaxation (i.e. reduction of toe twitch tension) is sufficient to prevent MH crises in swine. They have also determined that this dose is 3.5 mg/kg. They also have discovered that this same dose corrects already established MH reactions. Further details on the use of prophylactic dantrolene for known and suspected MHS patients is given in Chapter 15.

V. USE OF DANTROLENE IN DISORDERS OTHER THAN MALIGNANT HYPERTHERMIA

A. The Neuroleptic Malignant Syndrome (NMS)

Neuroleptic malignant syndrome (NMS) is a rare but sometimes lethal adverse response to antipsychotic therapy (5). It usually develops over a period of 24 to 72 hours after drug exposure and is characterized by akinesia, muscle rigidity, hyperthermia, tachycardia, cyanosis, autonomic dysfunction, altered consciousness, diaphoresis and raised serum creatine kinase (CK) concentrations (5,169). Estimates of mortality from NMS range from 20 to 30% and the risk appears to increase when depot forms of a psychotropic drug are administered (5,170,171). Haloperidol (169,172-175) and fluphenazine (169) are the most commonly implicated drugs, but occasional episodes with other butyrophenones, phenothiazines (5), tricyclics (176), monoamine oxidase inhibitors and other antipsychotic drugs have been reported (5,175-179). The syndrome usually but not always occurs in psychiatrically ill patients at the start of treatment with neuroleptic drugs (5).

Certain events in the clinical course of the disease, and some in vivo muscle tests, suggest some similarities in the pathogenesis between NMS and MH (5). However, other clinical and experimental evidence indicates that the underlying defect is central (presynaptic) in NMS while it is known to be peripheral (postsynaptic) in MHS (5). Thus in NMS the defect is thought to be related to dopamine-receptor blockade in the hypothalamus and basal ganglia (5). Additionally, no apparent genetic link exists in NMS (5). Unfortunately, as with MH there are no controlled studies of treatment of humans in this syndrome, only case reports (5).

Caffeine halothane contracture studies in skeletal muscle obtained from survivors of NMS reactions have yielded variable results. Merry (180), Tolletson (181) and Scarlett (182) have reported normal values while Caroff and Rosenberg (183) has reported that 5 out of 7 NMS survivors have had biopsy values like those observed in NMS patients. Similarly, Araki et al (184) have found that single skinned fibres from NMS survivors develop tension increases at lower than normal concentrations of caffeine.

Treatment traditionally has consisted of drug withdrawal and symptomatic measures such as cooling, correction of dehydration and treatment of cardiovascular, respiratory or renal complications (5,185–187). Recently the use of dopamine antagonists (amantadine and bromocriptine) and dantrolene have produced promising results (5,172,174,175,178). Whether the efficacy of dantrolene is due to similarities in the aetiology of NMS and MHS or to the fact that dantrolene may be a nonspecific anti-hypercatabolic agent is not yet known.

In NMS, as in MH, thermogenesis is ultimately due to calcium induced contracture of the skeletal muscles (5). Thus, in both MH and NMS dantrolene acts by preventing an abnormal rise in cytoplasmic calcium (5).

Because of the somewhat experimental nature of the use of dantrolene in treating patients with NMS dosage regimens have varied enormously (5). Single intravenous bolus doses (less than 3 to 10 mg/kg) and repeated oral doses (25 to 600 mg/day) have been used, often in combination with other therapies (5). Although the use of dantrolene has been limited to a small number of patients, in all cases to date, dantrolene therapy has been associated with normalization or significant reduction of temperature, usually within twelve hours of commencing intravenous therapy, and over a somewhat longer period with oral dantrolene therapy (5). Most patients also experience an improved clinical condition, though this occurs at a later state (5). In some cases withdrawal or transfer from intravenous to oral dantrolene therapy has been associated with a deterioration of the patient's condition and an increase in body temperature (5).

B. Heat Strokes

Heat stroke is a relatively ill-defined condition, but does have many clinical similarities to those of MH (5). Patients usually present with severe mental confusion or even coma with markedly elevated rectal temperature (5). The excessive heat production is derived from skeletal muscle when heat stroke occurs with exercise (5). Treatment traditionally involves rehydration with intravenous fluids, cold baths, and other symptomatic measures (5).

However, due to the similarities with MH dantrolene has been used in a few heat stroke patients (5). A 19-year-old male who, following strenuous exercise, collapsed and was unconscious, unresponsive to pain, and with a rectal temperature of 42°C, was treated with intravenous fluids, intravenous chlorpromazine and immersed in a cold bath (5). Upon investigation the patient also had metabolic acidosis, myoglobinuria and disseminated intravascular coagulation (5). Intravenous dantrolene 20 mg produced rapid (within 1 hour) and sustained clinical improvement (5). The patient was subsequently found on skeletal muscle biopsy to show increased sensitivity to caffeine and to halothane, findings which are diagnostic of MH (5,188). Another male, 41 years old, experienced similar complications and continued to deteriorate despite being treated with intravenous fluids and cold immersion (5,189). Administration of intravenous dantrolene 4 mg/kg brought about a rapid improvement and his temperature dropped from 42°C to 37°C in thirty minutes (5,189).

C. Use of Dantrolene to Control Skeletal Muscle Pain Associated with the MH Trait

Oral dantrolene has been used to alleviate skeletal muscle cramps, aches, pains and spasms which occasionally afflict MHS patients. For instance, Gronert et al (190) have reported their experience with a muscular patient who during periods of extreme physical or emotional stress suffered from episodes of fever, sweats and aching joints. A skeletal muscle biopsy was positive for MH. The serum creatine kinase (CK) was elevated. All his distressing symptoms were relieved by giving him 2.2 mg/kg of oral dantrolene sodium over a period of 15 hours. We have had similar experiences with a number of our own patients (unpublished data).

We have carried out a double blind study to test the value of dantrolene sodium for this purpose. We have found that dantrolene appears to lower serum CKs and to attenuate partially, but not completely, muscle discomfort in some, but not all MHS humans. Patients who have chronic elevation of the CK and whose skeletal muscle fascicles exhibit contractures in the presence of halothane alone are the most likely to obtain relief from skeletal muscle cramps through dantrolene therapy (Britt, unpublished data).

D. Use of Dantrolene in Neurological, Neuromuscular and Skeletal Muscle Diseases Other than MH

Oral dantrolene has been used for a number of years to relieve muscle spasms associated with such conditions as cerebral palsy, Parkinson's disease, chorea, athetosis, stroke, spinal cord injury, multiple sclerosis, myotonia, the continuous muscle fibre activity syndrome and tetanus (191-212). The advantage of dantrolene for these conditions is that even at very high doses it causes only slight reduction in skeletal muscle power. Moreover, it does not adversely affect the frequency of epileptic seizures which afflict many patients suffering from spastic disorders (213). Meyler et al (210) have, however, found that high doses (above 200 mg/day) are not more effective than low doses (less than 200 mg/day) in the long-term management of chronic muscle spasticity.

E. Use of Dantrolene Before Elective Anaesthesia of Normal Patients with Succinylcholine

Dantrolene (3.0 mg/kg) pretreatment has been used successfully to attenuate succinylcholine induced fasciculations, muscle pains and rises in serum potassium (214). Furthermore, Plotz et al (215) have shown that preanaesthetic treatment of twenty children with oral dantrolene sodium not only significantly reduces the incidence of post-succinylcholine skeletal muscle fasciculations but also significantly inhibits pre- and post-operative serum CK levels.

SIDE EFFECTS OF DANTROLENE SODIUM

Consistent side effects of dantrolene sodium therapy, which largely subside after a few days, include: dizziness, diplopia, dysarthria, a sensation of swelling of eyeballs and tongue, a feeling that limb muscles are rubbery, spaghetti-like and weak (not confirmed on objective management), uncertainty of location of feet when ascending or descending stairs, nausea, epigastric discomfort and diarrhoea (53,99,216). Dantrolene sodium elevates serum potassium (217).

Rarely after prolonged very high dosage hepatic dysfunction has been reported (199,201,218). In a report of a large population of patients receiving dantrolene for long periods, biochemical abnormalities of liver function (including moderate elevations of serum transaminases, alkaline phosphatase and bilirubin) have been observed to occur in about 1.8% of patients. Symptomatic hepatitis occurs in 0.6% and fatal hepatitis in 0.3% (5). The risk of hepatic injury appears to be about 2-fold greater in females, in patients receiving doses over 300 mg/day and in those treated for over 60 days (5,219). The major histological pathology observed is subacute hepatic necrosis or chronic active hepatitis, although cholestasis has been reported (5). Subsequent to these findings, the dosage recommendations for long term use of dantrolene have been reduced, resulting in a marked decrease in the number of severe hepatic disease reported (data on file, Norwich Eaton). In all those surviving, liver function studies have returned to normal after withdrawal of the drug.

Experiments on rats have shown that:

i) the action of dantrolene is on the bile acid-independent rather than the bile acid-dependent fraction of the bile (220);

ii) dantrolene inhibits the hepatic mixed function oxidase (MFO) system activity and cytochrome P_{450} content (221,222). The dose required to induce these changes is in the range of 25 to 400 mg/kg for five days. This dose range is, however, far in excess of the known effective dose required for humans. Nevertheless, if therapy of patients continues over several days, liver function studies ought to be performed.

In a recent report Petusevsky et al (223) have recorded that long-term dantrolene therapy was followed by chronic pleural effusion in three patients, one of whom also developed acute pericarditis. A fourth patient manifested both pleural and pericardial effusions. After the dantrolene sodium was discontinued, resolution of the pleural process was prolonged. This report does not prove a causal relationship between dantrolene therapy and pleural and pericardial reaction, but it should alert one to search for such a possibility in all patients on chronic dantrolene sodium treatment. Pleural-pericardial reaction has also been reported (224).

Lymphocytic lymphoma has been observed in one patient on long term high dose dantrolene (225). Whether the lymphoma was a fortuitous finding or was induced by the dantrolene therapy cannot of course be determined on a clinical sampling of one. Nevertheless, the report should make one cautious about the prolonged administration of large amounts of dantrolene sodium.

Other rare effects of dantrolene include hallucinations (226), exacerbation of respiratory depression (227) and leucopenia (228). Long-term high dose dantrolene sodium therapy rather consistently causes a fine cuneiform rash over the face and back (229) which over time progresses to a boil-like rash (Britt, unpublished data). A claim that dantrolene sodium has induced asystole during MH reactions in two susceptible swine cannot be accepted as the arrest in each case was likely due to the MH crisis itself rather than to the effects of dantrolene (230).

Dantrolene sodium solution is highly alkaline (pH 9.6) and intravenous injection may cause extravasation and tissue necrosis. Additionally, even if confined within the veins, dantrolene infusion continued for more than a few hours leads to severe local thrombophlebitis. The site of infusion should, therefore, be changed every few hours.

In spite of these several undesirable actions of dantrolene, it nevertheless remains the only agent which is effective in the management of malignant hyperthermia.

REFERENCES

1. Norwich-Eaton Pharmaceuticals (Division of Norwich-Eaton, Ltd.). 1979. Dantrium Intravenous (Dantrolene Sodium Intravenous). Product Monograph.

2. Snyder, Jr., H.R., Davis, C.S., Bickerton, R.K. and Halliday, R.P. J. Med. Chem. 1967;10:807.

3. Lee, C., Durant, N.N., Au, E. and Katz, R.L. Anesthesiology 1981;54:61.

4. Ellis, K.O. White, Jr., R.L., Wright, G.C. and Wessels, F.L. J. Pharm. Sci. 1980;69;327.

5. Ward, A., Chaffman, M.O. and Sorkin, E.M. Drugs 1986;32:130.

6. Hackett, L.P. and Dusci, L.J. J. Chromatography 1979;179:222.

7. Hollifield, R.D. and Conklin, J.D. Arch. int. Pharmacodyn. 1968;174:333.

8. Hollifield, R.D. and Conklin, J.D. J. Pharm. Sci. 1973;62:271.

9. Ellis, K.O. and Wessels, F.L. Nauyn-Schmiedeberg's Arch. Pharmacol. 1978;301:237.

10. Cox, P.L., Heotis, J.P., Polin, D. and Rose, G.M. J. Pharmaceutical Sci. 1969;58:987.

11. Ellis, K.O., Castellion, A.W., Honkomp, L.J., Wessels, F.L., Carpenter, J.F. and Halliday, R.P. J. Pharm. Sci. 1973;62:948.

12. Krarup, C. J. Physiol. 1981;311:389.

13. Britt, B.A., Scott, E., Frodis, W., Clements, M-J. and Endrenyi, L. Can. Anaesth. Soc. J. 1984;31:130.

14. Anderson, I.L., Lipicky, R.J. and Jones, E.W. In: Aldrete, J.A. and Britt, B.A. (eds.). Second International Symposium on Malignant Hyperthermia. 1978. New York: Grune & Stratton, p. 509.

15. Ellis, K.O. and Wessels, F.L. In: Wasserman, R.H. et al. (eds.). Calcium-binding proteins and calcium function: Proceedings of the 2nd International Symposium on Calcium-Binding Proteins and Calcium Function in Health and Disease, Cornell University, Ithaca, N.YU., June 5-90, 1977, pp. 273-274, North-Holland, Amsterdam, 1977.

16. Nelson, T.E. Anesthesiology 1983;59:A225.

17. Yamamoto, Y., Suzuki, A. and Hotta, K. Jap. J. Physiol. 1977;27:95.

18. Nott, M.W. and Bowman, W.C. Clin. Exp. Pharmacol. Physiol. 1974;1:113.

19. Ellis, K.O. and Bryant, S.H. Naunyn-Schmiedeberg's Arch. Pharmacol. 1972;274:107.

20. Desmedt, J.E. and Hainaut, K. J. Physiol. 1977;265:565.

21. Ellis, K.O. and Carpenter, J.F. Naunyn-Schmiedeberg's Arch. Pharmacol. 1972;275:83.

22. Putney, Jr., J.W. and Bianchi, C.P. FEBS 1973;32:772.

23. Moulds, R. Brit. J. Pharmacol. 1977;59:129.

24. Halsall, P.J. and Ellis, F.R. Br. J. Anaesth. 1979;51:753.

25. Fairhurst, A.S., Hamamoto, V. and Macri, J. Anesthesiology 1980;53:199.

26. Okumura, F., Crocker, B.D. and Denborough, M.A. Br. J. Anaesth. 1980;52:377.

27. Brocklehurst, L. Nature 1975;254:364.

28. Oba, T. Can. J. Physiol. Pharmacol. 1981;59:358.

29. Morgan, K.G. and Bryant, S.H. J. Pharmacol. Exp. Therap. 1977;201:138.

30. Korczyn, A.D., Shavit, S. and Schlosberg, I. Eur. J. Pharmacol. 1980;61:187.

31. Ellis, K.O. and Carpenter, J.F. Fed. Proc. 1971;30:670.

32. Leslie, G.C. and Part, N.J. Br. J. Pharmac. 1984;81:351.

33. Flewellen, E.H., Nelson, T.E. and Bee, D.E. Anesthesiology 1980;52:126.

34. Desmedt, J.E. and Hainaut, K. Biochem. Pharmacol. 1979;28:957.

35. Leslie, G.C. and Part, N.J. Br. J. Pharmac. 1981;72:665.

36. Tamai, Y. Electroencephalotraphy and Clinical Neurophysiology (Suppl.) 1982;36:221.

37. Kotsias, B.A. and Muchnik, S. Arch. Neurol. 1978;35:734.

38. Miglietta, O. Am. J. Phys. Med. 1977;56:293.

39. Miyamoto, H. and Racker, E. J. Membrane Biol. 1982;66:193.

40. Flewellen, E.H., Nelson, T.E. and Bee, D.E. Canad. Anaesth. Soc. J. 1980;27:16.

41. Nelson, T.E. and Flewellen, E.H. Canad. Anaesth. Soc. J. 1979;26:484.

42. Anderson, I.L. and Jones, E.W. Anesthesiology 1976;44:57.

43. Anderson, I.L., Rawstron, R.E. and Dunlop, D.J. N.Z. Med. J. 1980;91:417.

44. Nelson, T.E. In: Aldrete, J.A. and Britt, B.A. (eds.). Second International Symposium on Malignant Hyperthermia. 1978. New York: Grune & Stratton, p. 23.

45. Fletcher, J.E. and Rosenberg, H. Anesthesiology 1985;63:190.

46. Nelson, T.E. and Denborough, M.A. Clin. Exp. Pharmacol. Physiol. 1977;4:315.

47. Casson, H. and Downes, H. In: Aldrete, J.A. and Britt, B.A. (eds.). Second International Symposium on Malignant Hyperthermia. 1978. New York: Grune & Stratton, p. 3.

48. Sengupta, C., Meyer, U.A. and Carafoli, E. FEBS 1980;117:37.

49. Harrison, G. In: Huelsz et al. (eds.). Anaesthesiology. 1977. Amsterdam: Exerpta Medica, p. 452.

50. Harrison, G.G. Br. J. Anaesth. 1975;47:62.

51. Hall, G.M., Lucke, J.N. and Lister, D. Anaesthesia 1977;32:472.

52. Hall, G.M., Lucke, J.N. and Lister, D. Br. J. Anaesth. 1980;52:165.

53. Hall, G.M., Lucke, J.N., Orchard, C., Lovell, R. and Lister, D. Anaesthesia 1982;37:1167.

54. Gronert, G.A., Milde, J.H. and Theye, R.A. Anesthesiology 1976;44:488.

55. Kerr, D.D., Wingard, D.W. and Gatz, E.E. In: Aldrete, J.A. and Britt, B.A. (eds.). Second International Symposium on Malignant Hyperthermia. 1978. New York: Grune & Stratton, p. 499.

56. Hall, L.W., Trim, C.M. and Woolfe, N. Br. Med. J. 1972;2:145.

57. Allen, P., Lopez, J.R., Jones, D. Alamo, L., Papp, L. et al. Anesthesiology 1985;63 (Suppl. 3A):268.

58. Verburg, M.P., Oerlemans, F.T.J.J., van Bennekom, C.A., Gielen, M.J.M., de Bruyn, C.H.M.M. and Crul, J.R. Acta Anaesthesiol. Scand. 1984;28:1.

59. Harrison, G.G. Br. J. Anaesth. 1977;49:315.

60. Flewellen, E.H. and Nelson, T.E. Anesthesiology 1980;52:303.

61. Nelson, T.E. and Flewellen, E.H. Anesthesiology 1979;50:118.

62. Flewellen, E.H. and Nelson, T.E. FEBS 1978;37:331.

63. Lopez, J.R., Alamo, L., Jones, D., Allen, P., Lapp, L., et al. Anesthesiology 1985;63(Suppl. 3A):273.

64. Lopez, J.R., Alamo, L., Jones, D., Allen, P.D., Papp, L., et al. Biophys. J. 1985;47(Part 2):313A.

65. Rosenberg, H. (Letter). Can. Anaesth. Soc. J. 1980;27:304.

66. Lambert, W. (Letter). Can. Anaesth. Soc. J. 1980;27:304.

67. Britt, B.A. In: Milton, A.S. (ed.). Handbook of Experimental Pharmacology, Pyretics and Antipyretics. 1982. Heidelberg: Springer-Verlag, p. 547.

68. Gallant, E.M., Godt, R.D. and Gronert, G.A. Muscle and Nerve 1979;2:491.

69. Sullivan, J.S. and Denborough, M.A. Br. J. Anaesth. 1981;53:1217.

70. White, M.D. and Denborough, M.A. J. Gen. Pharmacol. 1984;15:129.

71. Araki, M., Takagi, A., Fujita, T. and Matsubara, T. Biomed. Res. 1985;6:73.

72. Nelson, T.E. FEBS Letters 1984;167:123.

73. Ohnishi, S.T., Taylor, S. and Gronert, G.A. FEBS Letters 1983;261:103.

74. White, M.D., Collins, J.G. and Denborough, M.A. Biochem. J. 1983;212:399.

75. Stadhouders, A.M., Viering, W.A.L., Verburg, M.P., Ruitenbeek, W. and Sengers, R.C.A. Acta Anaesth. Scand. 1984;28:14.

76. Cheah, K.S. Presented at the 4th International Malignant Hyperthermia Workshop, York, 1986.

77. Butterfield, J.L. and Ellis, K.O. Fed. Proc. 1973;32:772.

78. Ellis, K.O., Wessels, F.L. and Carpenter, J.F. J. Pharmaceutical Sci. 1976;65:1359.

79. Ellis, K.O., Butterfield, J.L., Wessels, F.L. and Carpenter, J.F. Arch. Int. Pharmacodyn. Ther. 1976;224:118.

80. Gollan, F. and McDermott, J. Proc. Soc. Exp. Biol. Med. 1979;160:42.

81. Bowman, W.C., Houston, J., Khan, H.H. and Rodger, I.W. Eur. J. Pharmacol. 1979;55:293.

82. Gronert, G.A. Anesthesiology 1980;53:395.

83. Ellis, R.H., Simpson, P., Tatham, P., Leighton, M. and Williams, J. Anaesthesia 1975;30;318.

84. Lullmann, H. and Schramm, U. Pharmacol. Res. Comm. 1982;14:967.

85. Wendt, I.R. Proceedings of the Australian Physiological and Pharmacological Society 1979;10:132.

86. Bers, D.M., Philipson, K.D. and Langer, G.A. Am. J. Physiol. 1981;240:576.

87. Hiraoka, M., Kinoshita, H. and Kawano, S. Abstract. Proceedings of the IUPHAR 9th International Congress of Pharmacology, 92, 1984.

88. Hiraoka, M., Kinoshita, H. and Kawano, S. Jap. J. Physiol. 1985;35:123.

89. Honerjager, P. and Alischewski, N. Nauyn-Schmiedeberg's Arch. Pharmacol. 1983;322:237.

90. Meszaros, J., Kecskemeti, V. and Szegi, J. Arch. Int. Pharmacyn. Therap. 1982;255:256.

91. Meyler, W.J., Wesseling, H. and Agoston, S. Eur. J Pharmacol. 1976;39:127.

92. Plizga, A. and Holl, J.E. Res. Comm. Chem. Pathol. Pharmacol. 1980;28:13.

93. Roewer, N., Rumberger, E., Bode, H., and Schulteam Esch, J. Anesthesiology 1985;63(Suppl. 3A):274.

94. Salata, J.J., Wasserstrom, J.A. and Jalife, J. J. Mol. Cell. Cardiol. 1983;15:233.

95. Salata, J.J. and Jalife, J. J. Pharmacol. Exp. Ther. 1982;220:157.

96. Roewer, N., Kuck, K-H., Nienaber, Ch., Schroeder, E. and Schulteam Esch, J. Anesthesiology 1984;61(Suppl. A):254.

97. Lascault, G., Asta, J., Cameron, D. and Waxman, M.B. Abstract. Circulation 1982;66 (Suppl):II-294.

98. Durbin, Jr., C.G., Fisher, N.A. and Lynch, C. Abstract. Anesthesiology 1983;59(Suppl):A227.

99. Lynch, C., Durbin, C.G., Fisher, N.A., Veselis, R.A. and Althaus, J.S. Anesth. Analg. 1986;65:252.

100. Bailey, L.E., Shibata, S., Seriguchi, D.G. and Dresel, P.E. Life Sci. 1980;26:1061.

101. Hatae, J., Ohba, M. and Kawata, H. J. Mol. Cell. Cardiol. 1980;12:857.

102. Hollander, P.B. and Ramsey, C.C. Abstract. Fed. Proc. 1977;36:957.

103. Meszaros, J., Kecskemeti, V. and Szegi, J. Abstract. Proceedings of the 28th International Congress of Physiological Sciences, 1980, p. 579.

104. Meszaros, J., Kecskemeti, V. and Szegi, J. Eur. J. Pharmacol. 1981;74:181.

105. Bolling, S.F. and Gardner, T.J. J. Am. Coll. Cardiol. 1985;5:433.

106. Brooks, R.R., Carpenter, J.F., Jones, S.M., Gregory, C.M. and Moore, A.F. Proceedings of the IUPHAR 9th International Congress of Pharmacology, Abstract 455P, 1984.

107. Butterfield, J.L., Carpenter, J.F., Jones, S.M., Brooks, R.R., Ellis, K.O., et al. Abstract. Pharmacologist 1983;25:246.

108. Saltzman, L.S., Kates, R.A., Corke, B.C., Norfleet, E.A. and Heath, K.R. Anesth. Analg. 1984;63:473.

109. Ally, A.I., Horrobin, D.F., Manku, M.S., Morgan, R.O., Karmazyn, M., Karmali, R.A. and Cunnane, S.C. Can. J. Physiol. Pharmacol. 1978;56:520.

110. Conte-Camerino, D., Lograno, M.D., Siro-Brigiani, G. and Megna, G. Eur. J. Pharmacol. 1983;92:291.

111. Bowman, W.C. and Khan, H.H. J. Phar. Pharmacol. 1977;29:628.

112. Graves, S., Dretchen, K.L. and Kruger, G.O. Eur. J. Pharmacol. 1978;47:29.

113. Mahmoudian, M., Dehpour, A.R. and Mofakham, S. Eur. J. Pharmacol. 1981;70:287.

114. Harris, J.D. and Benson, G.S. Urology 1980;16:229.

115. Ayyat, F.M., Lloyd, L.K. and Kuhlemeier, K.V. J. Urol. 1984;132:372.

116. Teague, C.T. and Merrill, D.C. Urology 1978;11:531.

117. Khalaf, I.M., Foley, G. and Elhilali, M.M. Investigative Urology 1979;17:188.

118. Hackler, R.H., Broecker, B.H., Klein, F.A. and Brady, S.M. J. Urol. 1980;124:78.

119. Pederson, E., Harving, H. and Klemar, B. J. Urol. 1978;119:403.

120. Amabile, G., Mercuri, N., Parisi, L., Floris, R. and Pirro, C. In: Benvenuti (ed.). Dantrolene sodium: new theraptutic approach to clinical problems of spasticity. 1982, Turin: Minerva Medica, p. 71

121. Ohta, H., Myojin, K., Sanada, J., Ikeda, H. and Ogawa, M. Hiroshima J. Anesth. 1986;22(Suppl.):1.

122. Colton, C.A. and Colton, J.S. Comp. Biochem. Physiol. 1979;64C:153.

123. Durant, N.N., Lee, C. and Katz, R.L. Eur. J. Pharmacol. 1980;68:403.

124. Roed, A. Eur. J. Pharmacol. 1982;83:83.

125. Ito, F., Fujitsuka, N. and Hanuichi, T. Brain Res. 1984;294:269.

126. Callanan, K. and Keenan, A.K. Irish J. Med. Sci. 1982;151:92.

127. Patti, F., Giammona, G., Reggio, A., Maccagnano, C., Rampello, L. et al. Acta Neurol. 1981;36:384.

128. Ellis, K.O. In: Benvenuti (ed.). Dantrolene sodium: new therapeutic approach to clinical problems of spasticity. 1982. Turin: Minerva Medicine, p. 35.

129. Yamawaki, S., Hotta, I., Tanaka, S., Uchitomi, Y. and Yanagawa, K. Hiroshima J. Anesth. 1986;22(Suppl.):47.

130. Pinder, R.M., Brogden, R.M., Speight, T.M. and Avery, G.S. Drugs 1977;13:2.

131. Farquhar, R, Leslie, G.C. and Part, N.J. Br. J. Pharmacol. 1986;88:79.

132. Janjic, D., Wollheim, C.B. and Sharp, G.W.G. Am. J. Physiol. 1982;243:59.

133. Sharp, G.W.G., Wiedenkeller, D.E. and Pian-Smith, M. Diabetalogia 1984;27:331A.

134. Francis, K.T. and Hamrick, M.E. Biochem. Pharmacol. 1980;29:1669.

135. Kojima, I., Kojima, K., Kreutter, D. and Rasmussen, H. J. Biol. Chem. 1984;259:14448.

136. Kojima, K., Kojima, I. and Rasmussen, H. Am. J. Physiool. 1984;247:645.

137. Kojima, I., Kojima, K. and Rasmussen, H. Am. J. Physiol. 1985;248:36.

138. Bates, M.D. and Conn, P.M. Endocrinology 1984;115:1380.

139. Tanaka, S., Shimonaka, H., Uematsu, H., Yamamoto, M., Shiojiri, H., Kawai, K. and Nozawa, Y. Hiroshima J. Anesth. 1986;22(Suppl.):9.

140. Fukuchi, H., Kuwata, N., Seki, Y. and Morio, M. Hiroshima J. Anaesth. 1979;15:21.

141. Fukuchi, H. and Morio, M. Hiroshima J. Anaesth. 1977;13(Suppl.):119.

142. Tomonaga, F., Kobayashi, T., Yago, K., Kamiya, T., Ishii, M., Tanaka, R., Watanabe, S. and Nomiyama, S. Hiroshima J. Anaesth. 1979;15(Suppl):101.

143. Gronert, G.A., Mansfield, E. and Theye, R.A. In: Aldrete, J.A. and Britt, B.A. (eds.). Second International Symposium on Malignant Hyperthermia. 1978. New York: Grune & Stratton, p. 535.

144. Drechsler, H. Anaesthesist 1981;30:257.

145. Editorial. FDA Drug Bulletin 1979;9:27.

146. Britt, B.A. Can. Anaesth. Soc. J. 1984;31:61.

147. Ryan, J.F. Int. Anesth. Clin. 1979;17:153.

148. Mathieu, A., Gogosian, A.J., Ryan, J.F., Crone, R.K. and Crocker, D. Anesthesiology 1979;51:454.

149. Faust, D.K., Gergis, S.D. and Sokoll, M.D. Anesth. Analg. 1979;58:33.

150. Friesen, C.M., Brodsky, J.B. and Dillingham, M.F. Can. Anaesth. Soc. J. 1979;26:319.

151. Liebenschutz, F., Mai, C. and Pickerodt, V.W.A. Br. J. Anesth. 1979;51:899.

152. Jones, S.E.F. and Thomas, C.H. Anaesthesia 1980;35:987.

153. Hoff, B., Taylor, C. and Kruel, J. (Letter). Critical Care Medicine 1980;8:699.

154. Harrison, G.G. S. Afr. Med. J. 1981;60:909.

155. Hamasaki, T., Isohata, J., Nomiyuama, S., Matoba, M., Nakata, F., Murakami, M. and Tanaka, R. Hiroshima J. Anesth. 1986;22(Suppl.):101.

156. Waite, P.D., Ballard, J.B. and Yonfa, A. J. Oral. Maxillofac. Surg. 1985;43:907.

157. Souliere, Jr., C.R., Weintraub, S.J. and Kirchner, C. Arch. Otolaryngol. Head Neck Surg. 1986;112:564.

158. Baudendistel, L., Goudsouzian, N., Cote, C. and Strafford, M. Anaesthesia 1984;39:1000.

159. Kageshima, K., Yasuda, N., Kageyama, K., Nagano, O., Tanifuji, Y., Kobayashi, K. and Nishijima, H. Hiroshima J. Anesth. 1986;22(Suppl.):111.

160. Yamada, H., Ohwaki, A., Endoh, M., Noguchi, J., Takeda, J. and Kuribayshi, N. Hiroshima J. Anesth. 1986;22(Suppl.):117.

161. Aono, H., Hirakawa, M., Tachibana, T., Takeda, A., Takimoto, E. and Tsuchiya, M. Hiroshima J. Anesth. 1986;22(Suppl.):137.

162. Denborough, M.A. Int. Anesth. Clin. 1979;17:11.

163. Sedwick, L.A. and Romano, P.E. Surv. Ophthalmol. 1981;25:378.

164. Schvaneveldt, J.A. and Stankiewicz, J.A. The Laryngoscope 1981;81:918.

165. Landauer, B. Der Anaesthesist 1980;29:507.

166. Dohlman, L.E. Plastic and Reconstructive Surgery 1982;69:547.

167. Hopkins, S. Nursing Mirror 1980;151:30.

168. Flewellen, E.H. and Nelson, T.E. Anesthesiology 1978;51:1001.

169. Rampertaap, M.P. South. Med. J. 1986;79:331.

170. Burke, R.E., Fahn, S., Mayeux R., Weinberg, H., Louis, K. et al. Neurology 1981;31:1022.

171. Caroff, S.N. J. Clin. Psychiat. 1980;41:79.

172. Khan, A., Jaffe, J.H., Nelson, W.H. and Morrison, B. J. Clin. Psychiatry 1985;46:244.

173. Martin, M.L., Lucid, E.J. and Walker, R.W. Ann. Emerg. Med. 1985;14:354.

174. Kawada, H., Izumi, H., Miyazato, Y., Kazunaga, H., Hayabara, T., Uefuji, T. and Baba, E. Hiroshima J. Anesth. 1986;22(Suppl.):65.

175. In-Nami, H., Okada, K., Kasai, T. and Kobayashi, K. Hiroshima J. Anesth. 1986;22(Suppl.):59.

176. Yamawaki, S., Yanagawa, K., Uchitomi, Y., Yamanaka, T. and Matsuura, Y. Hiroshima J. Anesth. 1986;22(Suppl.):41.

177. Kaplan, R.F., Feinglass, N.G., Webster, W. and Mudra, S. JAMA 1986;255:642.

178. Ueta, I., Takayama, Y., Yoshimoto, Y., Yamamoto, S., Myojin, K. and Tamura, M. Hiroshima J. Anesth. 1986;22(Suppl.):53.

179. Araki, M., Takagi, A. and Sugita, H. Hiroshima J. Anesth. 1986;22(Suppl.):17.

180. Merry, S.N., Werry, J.S., Merry, A.F. and Birchall, N. Aust. N.Z. J. Med. 1983;13:70.

181. Tollefson, G. J. Clin. Psychopharmacology 1982;1:266.

182. Scarlett, J.D., Zimmerman, R. and Berkovic, S.F. Aust. N.Z. J. Med. 1983;13:70.

183. Caroff, S.M., Rosenberg, H., Fletcher, J.E., Heiman-Patterson, T.D. and Mann, S.C. Anesthesiology 1987;(in press).

184. Araki, M., Takagi, A., Kiguchi, I. and Sugita, H. Abstract. Muscle Nerv. 1986;9:221.

185. Editorial. Lancet 1984;1:545.

186. Szabadi, E. Br. Med. J. 1984;288:1399.

187. Katsube, T., Kobayashi, S., Yamaguchi, S., Yamashita, K. and Tsunematsu, T. Hiroshima J. Anesth. 1986;22(Suppl.):71.

188. Denborough, M.A. Med. J. Australia 1982;1:204.

189. Lydiatt, J.S. and Hill, G.E. JAMA 1981;246:41.

190. Gronert, G.A., Thompson, R.L. and Onofrio, B.M. Anesth. Analg. 1980;59:377.

191. Sakai, T., Hosokawa, S., Shibasaki, H., Goto, I., Kuroiwa, Y., Sonoda, H. and Murai, Y. Neurology 1983;33:495.

192. Chyatte, S.B. and Birdsong, J.H. Southern Med. J. 1971;64:830.

193. Gelenberg, A.J. and Poskanzer, D.C. Neurology 1973;23:1313.

194. Chyatte, S.B., Birdsong, J.H. and Roberson, D.L. Arch. Physical Med. Rehab. 1973;54:365.

195. Chyatte, S.B., Birdsong, J.H. and Bergman, B.A. Southern Med. J. 1971;64:180.

196. Mayer, N., Mecomber, S.A. and Herman, R. Am. J. Physical Med. 1973;52:18.

197. Basmajian, J.V. and Super, G.A. Arch. Physical Med. Rehab. 1973;54:60.

198. Chyatte, S.B. and Basmajian, J.V. Arch. Physical Med. Rehab. 1973;54:311.

199. Wilkinson, S.P., Portmann, B. and Williams, R. Gut 1979;20:33.

200. Molnar, G.E., Evangelista, L.A. and Kathirithamby, R. N.Y. State J. Med. 1978;78;1233.

201. Faling, L.J., Petusevsky, M.L. and Snider, G.L. Ann. Int. Med. 1980;93:151.

202. Mai, J. and Pedersen, E. Acta Neurol. Scand. 1979;59:309.

203. Weiser, R., Terenty, T., Hudgson, P. and Weightman, D. The Practitioner 1978;221:123.

204. Feit, H. Neurology 1979;29:1631.

205. Eberstein, A. and Goodgold, J. Exper. Neurol. 1978;62:804.

206. Glass, A. and Hannah, A. Personal communication, Kaiser Foundation Hospital, Terra Linda, California.

207. Joynt, R.L. and Leonard, Jr., J.A. Develop. Med. Child Neurol. 1980;22:755.

208. Herman, R., Mayer, N. and Mecomber, S.A. Am. J. Physical Med. 1972;51:296.

209. Chipman, M. and Kaul, S. Neurology 1972;22:401.

210. Meyler, W.J., Bakker, H., Kok, J.J., Agoston, S. and Wesseling, H. J. Neurol. Neurosurg. Psychiat. 1981;44:334.

211. Ketel, W.B. and Kolb, M.E. Curr. Med. Res. Opin. 1984;9:161.

212. Aquilar Bernal, O.R., Bender, M.A. and Lacy, M.E. J. Roy. Soc. Med. 1986;79:277.

213. Nogen, A.G. Child's Brain 1979;5:420.

214. Collier, C.B. Anaesthesia 1979;34:152.

215. Plotz, J., Braun, J. and Stallenberger, R. Anaesthesist 1981;30:338.

216. American Hospital Formulary Service. Published by the American Society of Hospital Pharmacists, Bethesda, Md. 20814.

217. Agoston, S. (Letter.) Anaesthesia 1980;35:1012.

218. Cornette, M., Gillard, C. and Borll-Hermans, G. Acta neurol. belg. 1980;80:336.

219. Utili, R., Boitnott, J.K. and Zimmerman, H.J. Gastroenterology 1977;72:610.

220. Abernathy, C.O., Utili, R., Zimmerman, H.J. and Ezekiel, M. Toxicol. Appl. Pharmacol. 1978;44:441.

221. Roy, S., Francis, K.T., Born, C.K. and Hamrick, M.E. Res. Comm. Chem. Pathol. Pharmacol. 1980;27:507.

222. Francis, K.T. and Hamrick, M.E. Res. Comm. Chem. Pathol. Pharmacol. 1979;23:69.

223. Petusevsky, M.L., Faling, L.J., Rocklin, R.E., Snider, G.L., Merliss, A.D., Moses, J.M. and Dorman, S.A. JAMA 1979;242:2772.

224. Miller, D.H. and Haas, L.F. J. Neurol. Neurosurg. Psychiat. 984;47:553.

225. Wan, H.H. and Tucker, J.S. Postgrad. Med. J. 1980;56:261.

226. Andrews, L.G., Muzumdar, A.S. and Pinkerton, A.C. Can. Med. Assn. J. 1975;112:148.

227. Rivera, V.M., Breitback, W.B. and Swanke, L. JAMA 1975;233:863.

228. Greenspun, B. and Pacho, A. Arch. Phys. Med. Rehab. 1981;62:521.

229. Pembroke, A.C., Saxena, S.R., Kataria, M. and Silkha, K.D. Brit. J. Dermatol. 1981;104:465.

230. Chapin, J.W., Lou, C.G. and Wingard, D.W. (Letter.) Anesthesiology 1981;54:527.

15

ANAESTHESIA FOR MALIGNANT HYPERTHERMIA SUSCEPTIBLE PATIENTS

J. LERMAN AND J.E.S. RELTON

Department of Anaesthesia, The Hospital for Sick Children, Toronto, Ontario, Canada M5G 1X8

ABSTRACT

Malignant hyperthermia susceptible (MHS) patients requiring elective surgery present special problems for the anaesthesiologist. This chapter discusses the controversy of masseter muscle spasm, prophylactic preoperative dantrolene, anaesthetic drugs in malignant hyperthermia (MH), the anaesthetic management of MHS patients for elective surgery, and the interaction of several common medical and surgical conditions with malignant hyperthermia.

INTRODUCTION

The dramatic decrease in mortality from malignant hyperthermia (MH) in recent years may be attributed to four factors (1):

1. Improved awareness and understanding of the MH syndrome by patients and their physicians;
2. Better preanaesthetic identification of MHS patients;
3. More accurate identification of the triggering agents and the drugs considered to be "safe";
4. Improved intra-operative and post-operative monitoring;
5. The availability of an intravenous preparation of dantrolene.

As a result of this decrease in mortality, surgery may be undertaken in malignant hyperthermia susceptible (MHS) patients with minimal additional risk.

The term MHS refers to those patients in whom an MH reaction may occur during or after exposure to specific stimuli such as anaesthetic agents. The definitive diagnosis of MHS can be made only by means of a muscle biopsy. In many cases, patients are considered to be MHS because of a previous MH reaction, a family history of MHS, the development of masseter muscle spasm or rigidity (MMR) during anaesthesia, familial myopathy, or an increased plasma creatine kinase (CK) concentration. Although patients in these latter circumstances may not be biopsy-positive, they must be considered MHS until proven otherwise.

Britt and co-workers (2) identified those groups of MHS patients who are at high-risk for developing MH reactions:

1. Patients who have survived an MH reaction or have a positive muscle biopsy;
2. First-degree relatives of such patients;
3. Members of known or suspected MHS families with known muscle abnormalities or increased CK values.

Those patients or their relatives who have exhibited MMR during anaesthesia with halothane and succinylcholine should be added to this list (3-11).

All patients with evidence of MHS may be at risk for developing an MH reaction. Those patients in whom the suspicion of MHS exists should receive appropriate investigations and anaesthetic management for MH. As the number of biopsy-positive adults has increased arithmetically, the number of their progeny who are also MHS has increased geometrically. As a result, approximately 150 MHS children are anaesthetized at The Hospital for Sick Children in Toronto per year (representing 0.5-1.0% of the annual case load). Most of these children have a family history of a positive muscle biopsy or an MH reaction. A similar increase in MHS adult patients has also occurred at the Toronto General Hospital. Although it is unlikely that the majority of these patients will prove to be muscle biopsy-positive for MH, we undertake all precautions for MH for all of these patients.

Recent reports have drawn attention to the close association between the incidence of MMR and the MH syndrome (4-6,8-11). Schwartz and co-workers reported that 1% of all healthy children anaesthetized with halothane and paralyzed with intravenous succinylcholine demonstrated evidence of MMR (9). Muscle biopsies from children with MMR were positive in 50% to 100% of cases (6,7,9-11). This very high incidence of MHS children exceeds the accepted incidence of the syndrome in the paediatric population by approximately 100 times (3). Such a high incidence of MHS in otherwise healthy children undergoing elective surgery may have a serious impact on the clinical practice of anaesthesia. In contrast to the data from Boston, the incidence of both MMR and MH reactions in the last twenty years at The Hospital for Sick Children in Toronto has been very low (approximately 1:5,000 anaesthetics). Based on our experience, the incidence of MMR appears to be related to both the induction agents used and the sequence in which they are administered. The coincidental administration of succinylcholine during halothane anaesthesia is more likely to precipitate MMR and an MH reaction than is succinylcholine followed by halothane. Similarly, the in vitro muscle contracture responses of these patients are positive only to the combination of halothane and caffeine and in most cases, not to the individual triggers (12,13). Despite the discrepancy in the incidence of MMR, it must be stressed that all children who develop MMR at induction of anaesthesia must be considered MHS and managed accordingly (1,9,10).

PREOPERATIVE MANAGEMENT

All MHS patients scheduled for elective surgery should be admitted the day before surgery unless they have been examined by an anaesthetist previously. A complete preoperative history, physical examination, and the appropriate laboratory investigations which include a routine haemoglobin, CK, urinalysis, and electrocardiogram are recommended. There is no place for outpatient surgery in MHS patients.

Each MHS patient must be questioned carefully about previous reactions to anaesthesia and surgery, a family history of MH reactions or unexpected death during or after anaesthesia and surgery. The stigmata of MH such as ptosis, strabismus, hernia, clubfeet, joint hypermobility, scoliosis, mitral valve prolapse, intolerance to heat and exercise, and the presence of muscle cramps should be recorded (3). Particular attention must be paid to the state of dentition since poor dentition is not uncommon in MH. If evidence of heart disease is present, a chest X-ray, two-dimensional echocardiogram, and a cardiology consultation should be obtained.

Prophylactic antibiotics should be administered if mitral regurgitation is present (14). A medic-alert bracelet may be the only clue to MHS in the incoherent, senile, or unconscious patient. Physician awareness and suspicion of MH must be considered in all cases.

PREMEDICATION

All MHS patients scheduled for elective surgery should be premedicated in order to attenuate the stress of the pre-induction period. In adults premedication should include an anxiolytic or a hypnotic-sedative (oral diazepam 0.10-0.15 mg/kg) the evening before surgery and 1-1/2 to 2 hours before surgery, and a narcotic (intramuscular morphine 0.15 mg/kg or demerol 1.0 mg/kg) one hour prior to surgery. In children, premedication may include pentobarbitone (2 mg/kg oral or rectal) administered two hours prior to surgery and morphine (0.1-0.2 mg/kg) with atropine (0.01 mg/kg) intramuscularly one hour prior to surgery. Stress may trigger an MH reaction (15,16). Suitable premedication is therefore mandatory during this most stressful period.

An intravenous should be started four hours prior to surgery. In adults, lactated Ringer's solution or normal saline may be infused at a rate of 1.5-2.0 ml/kg/hr. In children, D5/0.2%S may be infused at a rate appropriate for the child's weight (17).

PROPHYLACTIC DANTROLENE

In order to prevent MH reactions, prophylactic dantrolene has been recommended in MHS patients scheduled for elective surgery (18-20).

When dantrolene was first shown to reverse MH reactions in swine (21-23), the oral preparation was the only form of dantrolene available for humans. In order to ensure adequate blood concentrations of dantrolene during anaesthesia, it was necessary to administer the oral preparation several hours prior to surgery to allow sufficient time to achieve an adequate blood concentration. Although Britt reported that MH reactions did not occur in patients receiving at least 4 mg/kg of oral dantrolene (24), breakthrough reactions have been reported in MHS humans premedicated with a lesser dose of the oral preparation followed by trigger-free anaesthetic (25,26). These breakthrough reactions may have been due to insufficient muscle concentrations of dantrolene or to excessive stress under anaesthesia. An insufficient concentration of dantrolene in muscle in these cases could be attributed in part to the variability in the gastrointestinal absorption of dantrolene (27).

Many of the side effects which occur following preoperative oral dantrolene are dose-dependent. The dose of the oral preparation (4-8 mg/kg) which was approved by the Federal Drug Administration is associated with more frequent side effects than lower doses recommended in the literature (28). Oral dantrolene causes drowsiness, diplopia, dysarthria, dysphonia sialorrhea, weakness, nausea, vomiting and diarrhoea (24,28-31). It also depresses the twitch response in striated muscle (29,32) and prolongs the duration of action of neuromuscular blocking drugs (33). It is reported to delay the time to recover 90% of the twitch response in adults paralyzed with vecuronium (33). This residual neuromuscular blockade may cause post-operative respiratory insufficiency. Despite the problems with oral dantrolene prophylaxis, it is readily available, inexpensive, and is still recommended by many experts (30).

Recently, the intravenous route for dantrolene prophylaxis has been recommended (19,20). This obviates the variability in gastrointestinal absorption which occurs with the oral preparation and ensures predictable blood concentrations. It also avoids some of the side effects of the oral preparation since it is given only at induction of anaesthesia. When given prophylactically in susceptible swine (i.e. before an MH reaction begins), intravenous dantrolene (3.5-5.0 mg/kg) (34-36) effectively prevents MH reactions during exposure to halothane and succinylcholine. Using maximum twitch depression as a measure of the effective dose of dantrolene, Flewellen and co-workers have found that the effective dose in human volunteers is 2.4 mg/kg (19). This is similar to the dose of intravenous dantrolene reported to reverse MH reactions in man (24,37). Based on this data, several investigators have recommended that 2.4 mg/kg dantrolene be administered intravenously immediately before induction of anaesthesia in all MHS patients (19,20). However, this dose of dantrolene depresses the twitch response in human volunteers by 75% with a corresponding reduction in grip strength and a subjective feeling of weakness for up to twenty hours after administration of the drug (19). Twitch depression of this magnitude may exacerbate post-operative weakness, particularly of the respiratory muscles. Dantrolene-induced twitch depression is of particular concern because it cannot be reversed pharmacologically (38). In addition, administration of the intravenous preparation through a small peripheral vein may cause thrombophlebitis. These side effects cannot be ignored.

Is prophylactic dantrolene then necessary at all (1,39)? When dantrolene was first introduced, it was recommended for use in MHS patients because it prevented MH reactions in MHS swine exposed to known triggering agents. The evidence that pre-operative prophylactic dantrolene prevents MH reactions in humans is less convincing. Breakthrough MH reactions have occurred following its use albeit with small doses of dantrolene (25,26) and smouldering MH reactions may subsequently occur. In patients in whom prophylactic intravenous dantrolene is indicated (see below), dantrolene should be administered immediately before induction of anaesthesia. The patient should then be monitored carefully since an MH reaction might still occur.

There is evidence that at least in elective surgery, prophylactic dantrolene is not necessary (40,41). This evidence is based on two studies in which the incidences of MH reactions following a trigger-free stress-free anaesthetic for muscle biopsy in MHS patients who received no prophylactic dantrolene were 0.0% (40) and 0.62% (41). Furthermore, the effectiveness of intravenous dantrolene to completely reverse all of the clinical and biochemical manifestations of MH has obviated the need for prophylactic dantrolene (24,37,42). To date, there have been no cases of MH which failed to respond to intravenous dantrolene providing the diagnosis of a reaction was established early and an adequate dose of intravenous dantrolene was administered (37,42). Therefore, whenever any type of surgery is planned for an MHS patient, 36 vials (10 mg/kg for an adult) of the intravenous preparation of dantrolene must be immediately available (43).

Considering the potential side effects and complications from dantrolene, the potential for masking MH reactions, the high cost of intravenous dantrolene, and the infrequency of MH reactions, we do not recommend either oral or intravenous prophylactic dantrolene for all MHS patients undergoing elective surgery but do strongly recommend that an adequate supply of intravenous dantrolene be immediately available.

Prophylactic dantrolene may, however, be indicated in some MHS patients. These indications include: a high-risk MH patient; severe preanaesthetic apprehension; an MH reaction in evolution; strenuous and prolonged exercise in the immediate preoperative period; surgical emergencies, especially those associated with massive muscle trauma; anaesthesia or surgery which is conducive to hypercarbia, hypoxia and/or impaired perfusion of the muscles; or the absence of a vapour-free machine. In smaller community hospitals, a vapour-free machine may not always be available. In these patients, dantrolene should be administered intravenously immediately before induction of anaesthesia. To minimize the risks of an MH reaction in these circumstances, we recommend prophylactic intravenous dantrolene (2.4 mg/kg) immediately prior to induction of anaesthesia.

ANAESTHETIC TECHNIQUE

The choice of anaesthetic techniques for the MHS patient includes local, regional, or general anaesthesia. The optimal technique must include trigger-free drugs and provide stress-free conditions. Stress-free conditions must be maintained particularly during local and regional anaesthesia because stress can aggravate the triggering effects of anaesthetic agents or trigger an MH reaction directly. Therefore, supplemental parenteral neuroleptic and analgesic drugs should be administered to attenuate the stress response to anaesthesia and surgery.

Each anaesthetic drug may be categorized according to its capability of triggering an MH reaction: 1) proven safe; 2) proven unsafe; or 3) controversial. A drug proven safe for use in MH is one which does not trigger an MH reaction under any conditions. A drug proven unsafe has been shown to trigger MH under certain conditions. A controversial drug is one which does not satisfy either of the two preceding definitions. Drugs in the last category may trigger MH reactions in some circumstances and not in others. These controversial drugs will be categorized according to the prevailing opinion in the literature. Six groups of anaesthetic drugs will be discussed: a) local anaesthetics; b) induction agents; c) inhalational agents; d) analgesics and anxiolytics; e) muscle relaxants; and f) anti- cholinesterases and anticholinergics.

a) **Local Anaesthetics**

Anaesthetic drugs used for local and regional techniques may be divided into two groups according to their structure. Ester local anaesthetics (i.e. chloroprocaine, tetracaine, and procaine) are safe to use in MHS patients. These drugs do not trigger MH reactions in swine or in man (24,44). They may in fact prevent MH reactions by reducing the concentration of calcium in the sarcoplasmic reticulum (SR) (45).

Amide local anaesthetics have been contraindicated for MHS patients. Amide anaesthetics were thought to exacerbate MH in humans because MH crises intensified despite their administration. However, amide local anaesthetics were usually administered at a time when the MH crisis was expected to intensify. This cause and effect relationship erroneously led to the conclusion that amide anaesthetics trigger MH reactions. Furthermore, in contrast to ester local anaesthetics, these drugs did not decrease the concentration of calcium in the SR in studies conducted in vitro (45,46). Recent evidence suggests that amide local anaesthetics may be safe. Neither lidocaine nor bupivacaine trigger MH reactions in swine (48,49). Lidocaine has been used to treat arrhythmias during MH reactions, although it was clearly ineffective compared to procainamide (47). Mepivacaine has also been used (in one study) for femoral nerve blocks in 100 MHS patients undergoing elective muscle biopsies without complications or MH reactions (50). The dose of mepivacaine used was in the lower range of the recommended dosage. We believe that amide local anaesthetics (without epinephrine) may be used safely in MHS patients providing the dose used is in the lower range of the recommended dosage (51). If a MH reaction does occur, it may be prudent to avoid amide local anaesthetics not only because they may potentiate an MH reaction, but also because dantrolene may reduce the threshold for local anaesthetic toxicity (52).

b) **Induction Agents**

All barbiturates are safe to use in MH and may in fact protect against MH reactions or possibly delay the onset of a reaction (53). The use of ketamine, a phencyclidine, is controversial (54); most experts do not recommend its use in MHS patients (47). It has been shown to trigger MH reactions in both MHS swine and humans (47). This, together with its potent central sympathomimetic activity has led to the recommendation to avoid its use (51). Etomidate, a new non-barbiturate induction agent (not currently available in Canada or the U.S.A.), does not trigger MH reactions in MHS swine. However, its use in MH remains to be established because it does not delay the onset of MH reactions to the same extent as thiopentone (55).

c) **Inhalational Agents**

Halothane is the inhalational agent most commonly associated with MH or MMR (3,6). Enflurane (56), isoflurane (57,58), methoxyflurane (3) and others (3,59) have all been reported to trigger MH reactions in humans (48). The relative capabilities of four potent inhalational agents to augment caffeine-induced contractures in MHS muscle in vitro have been established: halothane having the greatest capability followed in decreasing order by enflurane, isoflurane and methoxyflurane (60). Recently, the new inhalational agent sevoflurane was also shown to trigger MH reactions in susceptible swine (61). All inhalational agents are, therefore, absolutely contraindicated.

Nitrous oxide is safe (3). Nitrous oxide has been used in a very large number of patients with MH including 956 who underwent muscle biopsies for MH without evidence of a significant number of reactions (41). It has been suggested, however, that it may be a very minor trigger of MH (62) by virtue of its sympathomimetic activity (63,64). Nevertheless, nitrous oxide is regarded as a safe agent for use in MHS patients.

d) **Analgesics and Anxiolytics**

All narcotics are safe in MHS patients. Droperidol (65), diazepam and other benzodiazepines (66) are likewise safe. Recent in vitro evidence suggests that midazolam is not only safe in MH, but as with other benzodiazepines, it may not interfere with the contracture tests used to diagnose MH (66).

e) **Muscle Relaxants**

Depolarizing muscle relaxants (such as succinylcholine) are absolutely contraindicated in MH. The use of non-depolarizing muscle relaxants remains confusing. D-tubocurare has been incriminated as a trigger of MH reactions, although the evidence against d-tubocurare is weak (67). It is considered unsafe despite the absence of any scientific proof that it triggers MH in swine (68) or humans. Pancuronium, a steroidal non-depolarizing muscle relaxant, is the most commonly used muscle relaxant in MH. In swine, it protects against MH reactions (53,69,70) and it remains the muscle relaxant of choice. Newer intermediate-acting non-depolarizing muscle relaxants such as atracurium (71,72) and vecuronium (33,73) have been used in MHS patients undergoing elective surgery without complications. In swine, these drugs neither attenuate nor exacerbate MH reactions (72,73). Because of their lack of effect on in vitro contracture testing, atracurium and vecuronium may be considered safe.

f) **Anticholinesterases and Anticholinergics**

Neostigmine, pyridostigmine, and edrophonium are considered safe for use in MH (40). Anticholinergic agents (atropine and glycopyrrolate), however, have only been recommended cautiously. This latter group of drugs increases both core temperature and heart rate. Since both of these effects are indicative of an MH reaction in evolution, clinicians have tended to avoid these drugs. In MHS children, however, we routinely administer intravenous atropine (10 mcg/kg) (or the equivalent dose of glycopyrrolate) at induction of anaesthesia to block vagal activity and maintain cardiac output. Furthermore, when reversing neuromuscular blockade in children, atropine or glycopyrrolate should always be administered in the dose appropriate for the particular anticholinesterase to prevent bradycardia (40,74,75). This recommendation for the use of anticholinergics in MHS children is based on the extremely low incidence of reactions in MHS children given anticholinergics compared with the potential complications associated with bradycardia (76). In adults, neuromuscular blockade may be reversed using the standard dose of anticholinesterase and a reduced dose of anticholinergic. However, because of unpublished reports of asystole following reversal with neostigmine and atropine, some clinicians prefer to ventilate the patient until the neuromuscular junction recovers spontaneously.

ANAESTHETIC AND MONITORING EQUIPMENT

Optimal anaesthetic management for MHS patients requires diligent preoperative preparation. An anaesthetic machine free from contamination with potent inhalational agents must be used. A "contaminated" anaesthetic machine may be rendered "vapour-free" by flushing the machine with air or nitrous oxide for approximately 24 hours. A new or disposable anaesthetic circuit, reservoir bag, and vapour-free ventilator should be used. Monitoring equipment ought to include a precordial stethoscope, blood pressure cuff, doppler probe, electrocardiogram, axillary and core thermometer probes, end-tidal carbon dioxide analyzer (77-80) and pulse oximeter. Axillary temperature is particularly important to monitor since one of the earliest signs of an MH reaction is an increase in muscle metabolism (exothermic reactions). An arterial cannula may be inserted at the discretion of the anaesthesiologist. Fifty pounds of crushed ice, ice bags, and hypothermia blankets must be available. Anaesthetic induction agents such as barbiturates, tranquilizers, and narcotics are considered safe for use in these patients. Pancuronium, atracurium, or vecuronium may be used for muscle relaxation. Drugs for emergency use must also be available. These include refrigerated normal saline, 7.5 percent sodium bicarbonate, warmed 20 percent mannitol solution, powdered dantrolene (20 mg per ampoule) for intravenous administration (1 mg/kg up to 10 mg/kg), procainamide (1 mg/kg/min up to 15 mg/kg total dose), hydrocortisone (10 mg/kg), furosemide (0.25-1.0 mg/kg), soluble regular insulin, 50 percent glucose solution, chlorpromazine (0.2 mg/kg), and intravenous propranolol (0.015 mg/kg increments up to 10 mg for an adult) (see Chapter 16 for full details).

INTRAOPERATIVE MANAGEMENT

On arrival in the operating room, the patient is placed on a hypothermia blanket (switched to the normothermia position). A second blanket to cover the patient should be available in the room. Temperature probes are placed in the axilla and rectum. It is crucial that these patients not be allowed to become cold. A precordial stethoscope is positioned on the chest, and a blood pressure cuff and doppler probe are placed on one arm. Electrocardiographic leads are attached at the appropriate sites.

Vital signs are measured and recorded. Each patient is preoxygenated for three to five minutes. An intravenous cannula is inserted using local anaesthetic infiltration and anaesthesia is induced with intravenous thiopentone (3-5 mg/kg), and fentanyl (0.002-0.005 mg/kg). Supplemental fentanyl may be given in repeated doses of 0.001 mg/kg. Intravenous diazepam (0.10-0.20 mg/kg) may be given to ensure amnesia. Intravenous droperidol (0.010-0.075 mg/kg) may be given slowly to prevent nausea and vomiting after surgery. Pancuronium (0.05-0.15 mg/kg) or atracurium (0.40-0.60 mg/kg) may be used to facilitate intubation. The patient is then ventilated with 50-70% nitrous oxide in oxygen for approximately three to five minutes. When the patient is sufficiently relaxed to permit endotracheal intubation, 100% oxygen is administered for one minute and the larynx is sprayed with 5% cocaine (maximum dose of 2 mg/kg). The trachea is then intubated. Supplemental thiopentone may be administered to blunt the autonomic responses to intubation. The patient is then hyperventilated with a mixture of 50-70% nitrous oxide in oxygen to maintain an arterial or end-tidal pCO_2 of 30-35 mm Hg. The timing and dosage of muscle relaxants are determined by the response to a peripheral nerve stimulator. Vital signs including axillary and core body temperatures, blood pressure, heart rate, and end-tidal carbon dioxide tension must be monitored and recorded at five minute intervals. Although mixed venous lactate concentration (indicative of increased muscle metabolism) is an early sign of an MH reaction in evolution, its measurement is not immediately available in most centres. Mixed venous and arterial blood gases (pH, pCO_2, pO_2, and base excess) however, are usually available and may be helpful in diagnosing an MH reaction in evolution. Lactated Ringer's solution or normal saline should be administered at a flow rate of 2 ml/kg/hr for adults or at an age-adjusted flow rate for children (17). At the completion of surgery, neuromuscular blockade may be reversed in children with an anticholinesterase (i.e. neostigmine 0.030-0.040 mg/kg (74) and an anticholinergic (i.e. atropine 0.015-0.020 mg/kg) (see Section f for further explanation).

POST-ANAESTHETIC CARE

After surgery is completed, the patient may be extubated in the operating room. If respirations are depressed because of excessive narcosis, incremental doses of naloxone (0.005-0.001 mg/kg) may be administered. If an overdose of diazepam is suspected, physostigmine (30 mcg/kg) can be given (81). The patient is then transferred to the post-anaesthetic room (PAR) with the hypothermia blanket, temperature monitoring equipment, and an intravenous cannula in place. The PAR staff must be fully aware of the nature of MH; the signs, symptoms, and treatment of a reaction. Vital signs must be recorded at five minute intervals. The patient should not be returned to the ward until all danger of a MH reaction has passed, i.e. until vital signs have been stable and the laboratory tests ordered at the conclusion of anaesthesia are found to be within normal limits.

MH reactions have been reported following a prolonged trigger-free anaesthetic (82) and during the recovery period in the PAR (83-85). The occurrence of MH reactions in the PAR may be attributed to several possible causes including stress during the recovery period (83), trace concentrations of triggering anaesthetic agents in the PAR, and hypercapnia. It is very important for the PAR staff to understand this increased risk during the recovery period and to alert the anaesthesiologist of an MH reaction without delay.

After the patient returns to the ward, vital signs are monitored either continuously or hourly for twenty-four hours. The nurse must be instructed to notify the anaesthesiologist if the temperature increases more than 1°C/hour or 1.5°C in total. Post-anaesthetic dantrolene (intravenous preparation) is indicated only if an MH reaction occurs.

ASSOCIATED MEDICAL PROBLEMS

MHS patients may present with co-existing medical problems which could complicate their anaesthetic management.

Fever is not an uncommon finding during anaesthesia and surgery. A clinically significant increase in body temperature or a rapid increase in body temperature (1-5°C/hr) should raise the

TABLE I. Fever Under Anaesthesia

1. Malignant hyperthermia (MH)
2. Iatrogenic fever (environmental, external heat sources)
3. Endocrinopathies: (a) thyroid storm
   (b) hypothalamic dysfunction
   (c) pheochromocytoma
4. Sepsis: (a) bacterial (i.e. respiratory, urinary, endocardium, abscess)
   (b) viral (i.e. hepatitis)
   (c) fungal
5. Malignancy
6. Drugs (i.e. cocaine overdose, monoamine oxidase inhibitors and demerol, atropine)
7. Allergic reactions (i.e. blood transfusions, drugs)
8. Miscellaneous (i.e. fat embolism)

suspicion of MH. Early recognition is important since the mortality from an MH reaction correlates significantly with the maximum temperature attained (3). In some cases, differentiating MH from the other causes of fever under anaesthesia (FUA) (such as thyroid storm) may be difficult (86). A brief list of the causes of FUA is shown in Table I.

MHS patients with asthma (87) and bronchitis may require medical management of their bronchospasm or wheezing during the peri-operative period. Safe treatment of bronchospasm includes methylxanthines (88), steroids, anti-histamines, and disodium cromoglycate. If bronchospasm develops during anaesthesia, a loading dose of aminophylline should be administered over twenty minutes (if adequate blood levels of theophylline are not already present) followed by an infusion appropriate for the patient's requirements (89,90). $Beta_2$ specific agonist drugs are also safe to use in MH. However, alpha- and $beta_1$-agonists are contraindicated (see below).

Cardiovascular disease is not common in MHS patients. Some, but not all anti-anginal medications are safe. Diltiazem has reversed halothane-induced contractures in vitro (91) but has not been proven safe in MH. Verapamil has been used to control supraventricular tachyarrhythmias. Although verapamil has not been shown to trigger MH reactions, there is conflicting evidence regarding its efficacy in the treatment of MH (92). Of major concern, however, are reports of hypotension and asystole in both swine and humans given dantrolene and either diltiazem (93) or verapamil (94-96). These reactions may be dose related phenomena (97). Nevertheless, verapamil should be used cautiously in patients receiving dantrolene. A list of the drugs considered safe and unsafe for use in MHS patients with medical problems has recently been summarized by Tomarken and Britt (98).

Several reports have suggested an association between some muscle disorders (including Duchenne's muscular dystrophy, King Denborough Syndrome and central core disease) and MH (99-104). Although the incidence of MH reactions in patients with these disorders is small, its occurrence must always be considered (103).

ASSOCIATED SURGICAL PROBLEMS

Some surgical conditions, both emergency and elective, may present practical problems for the anaesthetic management of MHS patients.

Emergency surgery, particularly in smaller hospital centres, may require the use of a "contaminated" anaesthetic machine for an MHS patient. In this circumstance, an alternate gas delivery system may be used (a Laerdal bag with a tank of 100% oxygen or entonox). Prophylactic intravenous dantrolene should be administered at induction of anaesthesia if a contaminated machine must be used. Prophylactic intravenous dantrolene is also indicated for anaesthesia after massive muscle injury and after strenuous exercise in MHS patients.

The anaesthetic management of epiglottitis or the uncertain airway in MHS patients (eg. morbid obesity, hypomobility of the tempero-mandibular joint, etc.) presents a complex combination of problems. In non-MHS patients, the management includes a choice of an awake intubation, inhalational induction or tracheostomy under local anaesthesia. Intravenous inductions are usually contraindicated because patency of the airway cannot be guaranteed. In MHS patients, inhalational inductions with volatile anaesthetic agents are contraindicated. In order to secure the airway, two choices are available: an awake intubation (by direct laryngoscopy, rigid bronchoscopy, or fiberoptic laryngoscopy) or a tracheostomy under local anaesthesia. Once the airway is secured by the oral route, the patient may be anaesthetized with a trigger-free anaesthetic, the trachea intubated via the nasal route and the patient then transferred to the intensive care unit.

Several problems may complicate the anaesthetic management of MHS patients undergoing elective surgery. Short procedures may be managed by simple methods. For example, myringotomy and tube insertion is a brief operation (usually lasting five to ten minutes) which requires the close co-operation of surgeon and anaesthesiologist in order to expedite the surgery. We use the following sequence: with the surgeon poised at the microscope, with speculum and myringotome in hand, anaesthesia is induced with intravenous thiopentone (and atropine in children) and maintained with 70% nitrous oxide and 30% oxygen by spontaneous ventilation with a mask. The first myringotomy should be completed during the first dose of thiopentone (5 mg/kg). The second myringotomy is completed during the second dose of thiopentone (3–5 mg/kg). Fentanyl (1–2 mcg/kg) may be administered to provide supplemental analgesia post-operatively.

The avoidance of vasoconstrictor drugs in MHS patients may introduce hazards as threatening as MH itself. Cleft palate and pharyngoplasty surgery may be associated with massive blood loss and disruption of the airway if a vasoconstrictor drug is not infiltrated into the operative site. These complications may be life-threatening. Based on our experience, we recommend the use of low-dose epinephrine to reduce bleeding in children undergoing cleft palate or pharyngoplasty surgery. Such epinephrine treatment should be preceded and accompanied by prophylactic dantrolene.

The management of pregnant MHS patients for delivery has been investigated at the Toronto General Hospital. Three percent chloroprocaine or more recently 2% lidocaine has been used for epidural anaesthesia in these patients. There have been no cases of neurotoxicity. Dantrolene has been studied in pregnant patients with MH. Twenty-five mg of oral dantrolene given four times a day for a maximum of five days caused no significant adverse effects in mothers of babies (Britt, B.A. Unpublished data). One mother and one infant (7%) developed MH reactions in the non-treated group. Immediately after delivery, the infant blood concentrations of dantrolene were approximately 75% those of the mother.

Although dantrolene has been shown to depress the mechanical twitch response in healthy volunteers, its effect in patients with myasthenia gravis is poorly understood. Data from one patient with myasthenia gravis and MH suggested that dantrolene depresses the mechanical twitch response to a lesser extent than in healthy volunteers (105). We recommend careful monitoring of neuromuscular blockade in patients with myasthenia gravis, particularly when dantrolene is administered.

The management of MHS patients for elective cardiopulmonary bypass has recently been reported (106). Inotropic drugs may be required during the post-bypass period to augment myocardial contractility. Two groups of inotropic drugs have been investigated for use in MHS patients: alpha and beta agonists. In susceptible swine, alpha agonists have been shown to induce MH reactions in three out of three swine (107). Alpha agonists may trigger MH by inducing intense muscle or cutaneous vasoconstriction which in turn leads to tissue hypoxia and an increase in core temperature. In contrast, beta agonist drugs do not trigger MH reactions in swine (107). Beta agonist drugs, particularly the $beta_1$ drugs, cause tachycardia and may obscure an important early sign of MH. Epinephrine, which has both alpha and beta agonist activity have not triggered an MH reaction in one MH swine studied (108). Thus, the use of these inotropic drugs is relatively contraindicated; the use of dopaminergic agents, however, has not been investigated. Digoxin and intravenous calcium do not trigger MH reactions in susceptible swine (109) although they have not been accepted for use in MHS humans. Two modalities are available to augment myocardial contractility in the MH patient: 1) afterload-reducing drugs; and 2) mechanical devices. The drugs which we consider safest in these patients are the afterload-reducing agents (vasodilators) such as nitroglycerin or nitroprusside. If medical management proves to be insufficient, then an intra-aortic balloon pump may be indicated as a temporizing measure until adequate myocardial contractility is restored.

ACKNOWLEDGEMENTS

The authors thank Ms. S.L. Loo for her assistance in preparing this manuscript.

REFERENCES

1. Flewellen, E.H. In: Review Course Lectures, 1985, IARS, Texas, pp. 76-83.

2. Britt, B.A., Kwong, F.H-F. and Endrenyi, L. In: Henschel, E.O. (ed.). Malignant Hyperthermia: Current Concepts. 1977. New York: Appleton-Century-Crofts. pp. 63-77.

3. Britt, B.A. and Kalow, W. Can. Anaesth. Soc. J. 1970;17:293.

4. Donlon, J.V., Newfield, P., Sreter, F.A. and Ryan, J.F. Anesthesiology 1978;19:298.

5. Nelson, T.E. and Flewellyn, E.H. N. Engl. J. Med. 1983;309:416.

6. Flewellen, E.H.D and Nelson, T.E. Can. Anaesth. Soc. J. 1982;29:42.

7. Rosenberg, H. and Reed, S. Anesth. Analg. 1983;62:415.

8. Flewellen, E.H. and Nelson, T.E. Anesth. Analg. 1984;63:793.

9. Schwartz, L., Rockoff, M.A. and Koka, B.V. Anesthesiology 1984;61:772.

10. Ellis, F.R. and Halsall, P.J. Br. J. Anaesth. 1984;56:381.

11. Rosenberg, H. and Fletcher, J.E. Anesthesiology 1985;62:212.

12. Fletcher, J.E. and Rosenberg, H. Anesthesiology 1985;63:190.

13. Kennamer, D.L., Belt, D.W., Winsett, D. and Nelson, T.E. Anesthesiology 1986;65:A237.

14. Shulman, S.T., Amren, D.P. and Bisno, A.L. Circulation 1984;70:1123a.

15. Gronert, G.A., Thompson, R.L. and Onofrio, B.M. Anesth. Analg. 1980;59:377.

16. Wingard, D.W. In: Henschel, E.O. (ed.). Malignant Hyperthermia: Current Concepts. 1977. New York: Appleton-Century-Croft. pp. 79-95.

17. Hinkle, A.J. In: Annual Refresher Course Lectures, American Society of Anesthesiologists. 1985;130;1.

18. Pandit, S.K., Kothary, S.P. and Cohen, P.J. Anesthesiology 1979;50:156.

19. Flewellen, E.H., Nelson, T.E., Jones, W.P., Arens, J.F. and Wagner, D.L. Anesthesiology 1983;59:275.

20. Flewellen, E.H. and Nelson, T.E. Anesthesiology 1984;61:477.

21. Harrison, G.G. Br. J. Anaesth. 1977;49:315.

22. Kerr, D.D., Wingard, D.W. and Gatz, E.E. In: Aldrete, J.A. and Britt, B.A. (eds.). Second International Symposium on Malignant Hyperthermia. 1978. New York: Grune and Stratton. pp. 499-507.

23. Hall, G.M., Lucke, J.N. and Lister, D. Anaesthesia 1977;32:472.

24. Britt, B.A. Can. Anaesth. Soc. J. 1984;31:61.

25. Fitzgibbons, D.C. Anesthesiology 1981;54:73.

26. Ruhland, G. and Hinkle, A.J. Anesthesiology 1984;60:159.

27. Leitman, P.S., Haslam, R.H.A. and Walcher, J.F. Arch. Phys. Med. Rehabil. 1974;55:388.

28. Dykes, M.H. JAMA 1975;231:862.

29. Pinder, R.M., Brogden, R.N., Speight, T.M. and Avery, G.S. Drugs 1977;13:3.

30. Wingard, D.W. Anesthesiology 1983;58:489.

31. Watson, C.B., Reierson, N. and Norfleet, E.A. Anesthesiology 1986;65:312.

32. Herman, R., Mayer, N. and Mecombe, S.A. Amer. J. Phys. Med. 1972;51:296.

33. Driessen, J.J., Wais, E.W. and Giden, M.J. Anesthesiology 1985;62:523.

34. Harrison, G.G. Br. J. Anaesth. 1975;47:62.

35. Flewellen, E.H. and Nelson, T.E. Anesthesiology 1980;52:303.

36. Gronert, G.A., Milde, J.H. and Theye, R.A. Anesthesiology 1976;44:488.

37. Kolb, M.E., Horne, M.L. and Martz, R. Anesthesiology 1982;56:254.

38. Flewellen, E.H., Nelson, T.E. and Bee, D.E. Anesthesiology 1980;52:126.

39. Hall, G.M. Br. J. Anaesth. 1980;52:847.

40. Larach, M.G. and Rosenberg, H. Anesthesiology 1983;59:A228.

41. Cunliffe, M., Lerman, J. and Britt, B.A. Anesth. Analg. 1987;66:S35.

42. Liebenschutz, F., Mai, C. and Pickerodt, V.W.A. Br. J. Anaesth. 1979;51:899.

43. Duncan, P.G. Can. Anaesth. Soc. J. 1985;32:205.

44. Nelson, T.E. and Flewellen, E.H. Anesthesiology 1979;50:118.

45. Bianchi, C.P. and Bolton, T.C. J. Pharm. Exp. Ther. 1967;157:388.

46. Adragna, M.G. Anesthesiology 1985;62:99.

47. Britt, B.A. In: Milton, A.S. (ed.). Pyretics and Antipyretics. Handbook of Experimental Pharmacology. 1982. Berlin: Springer Verlag. Volume 60, Chapter 17.

48. Wingard, D.W. and Bobko, S. Anesth. Analg. 1979;58:99.

49. Harrison, G.G. and Morrell, D.F. Br. J. Anaesth. 1980;52:385.

50. Berkowitz, A. and Rosenberg, H. Anesthesiology 1985;62;651.

51. Gronert, G. Anesthesiology 1980;53:395.

52. Rosenblatt, R., Tallman, R.D. Jr., Weaver, J. and Wang, Y. Anesthesiology 1984;61:A209.

53. Gronert, G.A. and Milde, J.H. Anesth. Analg. 1981;60:499.

54. Denborough, M.A., Forster, J.F.A. and Lovell, R.R.H. Br. J. Anaesth. 1962;34:395.

55. Suresh, M.S. and Nelson, T.E. Anesth. Analg. 1985;64:420.

56. Pan, T.H., Wollack, A.R. and DeMarco, J.A. Anesth. Analg. 1975;54:47.

57. Joseph, M.M., Shah, K. and Viljoen, J.F. Anesth. Analg. 1982;61:711.

58. Boheler, J., Hamrick, JC., McKnight, R.L. and Eger, E.I., II. Anesth. Analg. 1982;61:712.

59. Lips, F.J., Newland, M. and Dutton, G. Anesthesiology 1982;56:144.

60. Britt, B.A., Endrenyi, L., Frodis, W., Scott, E. and Kalow, W. Can. Anaesth. Soc. J. 1980;27:12.

61. Shulman, M., Braverman, B., Ivankovich, A.D. and Gronert, G. Anesthesiology 1981;54:259.

62. Ellis, F.R., Clarke, I.M.C. and Appleyard, T.N. Br. Med. J. 1974;4:270.

63. Williams, C.H., Hoech, G.P. Jr. and Roberts, J.T. Anesthesiology 1978;49:58.

64. Gronert, G.A. Anesthesiology 1978;49:59.

65. Fletcher, J.E., Rosenberg, H. and Hilf, M. Anesthesiology 1985;63:A302.

66. Fletcher, J.E., Rosenberg, H. and Hilf, M. Can. Anaesth. Soc. J. 1984;31:377.

67. Britt, B.A., Webb, G.E. and LeDuc, C. Can. Anaesth. Soc. J. 1974;21:371.

68. Harrison, G.G. In: Gordon, R.A., Britt, B.A. and Kalow, W. (eds.). International Symposium on Malignant Hyperthermia. 1973. Springfield: Charles C. Thomas. pp. 271-286.

69. Jones, D.E., Ryan, J.F., Taylor, B., Papp, L., Lopez, J.R., Alamo, L., Sreter, F.A. and Allen, P.D. Anesthesiology 1985;63:A344.

70. Hall, G.M., Lucke, J.N. and Lister, D. Br. J. Anaesth. 1976;48:1135.

71. Michel, P.A. and Fronefield, H.P. Anesthesiology 1985;62:213.

72. Williams, C.H., Dozier, S.E., Ilias, W.K. and Pulfer, R.T. Anesth. Analg. 1985;64:185.

73. Buzello, W., Williams, C.H., Chandra, P., Watkins, M.L. and Dozier, S.E. Anesth. Analg. 1985;64:515.

74. Fisher, D.M., Cronnelly, R., Sharma, M. and Miller, R.D. Anesthesiology 1984;61:428.

75. Fisher, D.M., Cronnelly, R., Miller, R.D. and Sharma, M. Anesthesiology 1983;59:220.

76. Friedman, W.F. Prog. Cardiovasc. Dis. 1972;15:87.

77. Triner, L. and Sherman, J. Anesthesiology 1981;55:482.

78. Baudendistel, L., Goudsouzian, N., Cote, C. and Strafford, M. Anaesthesia 1984;39:1000.

79. Murphy, A.L., Conlay, L., Ryan, J.F. and Roberts, J.T. Anesthesiology 1984;60:149.

80. Verburg, M.P., Oerlemans, F.T.J.J., van Bennekom, C.A., Gielen, M.J.M., de Bruyn, C.H.M.M. and Crul, J.F. Acta Anaesthesiol. Scand. 1984;28:1.

81. Havasi, G., Gintantas, J., Havasi, I. and Racz, G.B. Can. Anaesth. Soc. J. 1986;28:497.

82. Neubauer, K.R. and Kaufman, R.D. Anesth. Analg. 1985;64:837.

83. Schulte-Sasse, U., Hess, W. and Eberlein, H.J. Can. Anaesth. Soc. J. 1983;30:635.

84. Grinberg, R., Edelist, G. and Gordon, A. Can. Anaesth. Soc. J. 1983;30:273.

85. Mathieu, A., Bogosian, A.J., Ryan, J.F., Crone, R.K. and Crocker, D. Anesthesiology 1979;51:454.

86. Stevens, J.J. Anesthesiology 1983;59:263.

87. Vaughn, J., Casson, H. and Hirshman, C.A. Anesthesiology 1983;58:283.

88. Flewellen, E.H. and Nelson, T.E. Anesth. Analg. 1983;62:115.

89. Bukowskyj, M., Nakatsu, K. and Munt, P.W. Ann. Int. Med. 1984;101:63.

90. Szefler, J.S. In: Symposium on Ped Allergy. Ped. Clinics of N. Am. 1983;30:943.

91. Iwatsuki, N., Koga, Y. and Amaha, K. Anesth. Analg. 1983;62:861.

92. Zukaitis, M.G., Hoech, G.P. Jr., Williams, C.H. and Simpson, S. Anesthesiology 1982;57:A228.

93. Saltzman, L.S., Kates, R.A., Norfleet, E.A., Corke, B.C. and Heath, K.S. Anesthesiology 1984;61:A11.

94. Saltzman, L.S., Kates, R.A., Corke, B.C., Norfleet, E.A. and Heath, K.R. Anesth. Analg. 1984;63:473.

95. Durbin, C.G. Jr., Fisher, N.A. and Lynch, C. III. Anesthesiology 1983;59:A227.

96. Rubin, A.S. and Zablocki, A.D. Anesthesiology 1987;66:246.

97. Gallant, E.M., Foldes, F.F., Rempil, W.E. and Gronert, G.A. Anesth. Analg. 1985;64:601.

98. Tomarken, J.L. and Britt, B.A. Ann. Emerg. Med. 1987;16.

99. Seay, A.R., Fiter, F.A. and Thompson, J.A. J. Pediatr. 1978:93:88.

100. Linter, S.P.K., Thomas, P.R., Withington, P.S. and Hall, M.G. Br. J. Anaesth. 1982;54:1331.

101. Kelfer, H.M., Singer, W.D. and Reynolds, R.N. Pediatrics 1983;71:118.

102. Brownell, A.K.W., Paasuke, R.T., Elash, A., Fowlow, S.B., Seagram, C.G.F., Diewold, R.J. and Friesen, C. Anesthesiology 1983;58:180.

103. Rosenberg, H. and Heiman-Patterson, T. Anesthesiology 1983;59:362.

104. King, J.O., Denborough, M.A. and Zapf, P.W. Lancet 1972;i:237.

105. Mora, C.T., Eisenkraft, J.B. and Papatestas, A.E. Anesthesiology 1986;64:371.

106. Byrick, R.J., Rose, D.K. and Ranganathan, N. Can. Anaesth. Soc. J. 1982;29:50.

107. Hall, G.M., Lucke, J.N. and Lister, D. Br. J. Anaesth. 1977;49:855.

108. Harrison, G.G., Saunders, S.J., Biebuyck, J.F., Hickman, R., Dent, D.M., Weaver, V. and Terblanche, J. Br. J. Anaesth. 1969;41:844.

109. Gronert, G.A., Ahern, C.P., Milde, J.H. and White, R.D. Anesthesiology 1986;64:24.

16

TREATMENT OF ACUTE HYPERTHERMIA CRISES

D.E. JONES[1] AND J.F. RYAN[2]

[1]Department of Anesthesiology, Baylor College of Medicine and University of Texas, Houston, Texas; [2]Department of Anesthesiology, Massachusetts General Hospital, Boston, Massachusetts.

ABSTRACT

Early recognition is vital for the successful treatment of malignant hyperthermia. To this end routine monitoring of all patients undergoing anaesthesia must include: heart rate, blood pressure (preferably by electronic blood pressure cuff or direct arterial line), electrocardiogram, tidal volume, airway pressure, respiratory rate, end tidal carbon dioxide analysis, pulse oximeter, airway oxygen monitor, and body temperature. Once malignant hyperthermia is suspected, even remotely, therapy for this condition must be begun immediately. Appropriate treatment for malignant hyperthermia includes: stop all triggering agents instantly; infuse dantrolene 1.0 mg/kg/minute until temperature, muscle tone and heart decline into the normal range, change to a "clean" gas machine which has not previously been contaminated with lipid soluble potent inhalational agents; hyperventilate the patient with 100% oxygen; correct the metabolic acidosis with sodium bicarbonate; lower an early elevated serum potassium with a mixture of regular insulin in 50% glucose and raise a later low serum potassium with infusions of potassium chloride; prevent acute renal failure by administering furosemide and mannitol; and if the temperature is significantly elevated cool the patient by surface cooling, cold intravenous solutions and lavage of stomach, bladder, rectum and peritoneal and thoracic cavities.

INTRODUCTION

The successful treatment of an acute crisis precipitated by the various manifestations of malignant hyperthermia (MH) depends on the early recognition of its symptoms and their causes because the effectiveness of therapy is contingent upon the promptness of its instigation.

It is widely accepted that the excitation contraction (E-C) coupling mechanism in MH patients is defective. Recently intracellular calcium levels have been directly measured in the skeletal muscle of MH susceptible (MHS) swine (1). Baseline intracellular calcium levels were found to be four times controls while during an MH crisis levels were twenty times normal (1). The result is fulminant hypermetabolism in skeletal muscle which is recognized clinically by the following symptoms.

SIGNS	LABORATORY FINDINGS
<u>Tachycardia</u>	<u>Central venous desaturation</u>
<u>Tachypnoea</u>	<u>Central venous hypercapnia</u>
Arrhythmias	<u>Metabolic acidosis</u>
Cyanosis	<u>Respiratory acidosis</u>
Skin mottling	Hyperkalaemia
Fever	Myoglobinemia
Rigidity	Elevated creatine kinase
Profuse sweating	
Unstable blood pressure	

We consider the signs and laboratory findings which are underlined to be the primary ones in MH; the other signs and laboratory findings may be present depending upon the type of presentation of MH and the speed of its detection.

DIAGNOSIS

Successful management of MH reactions depends on early recognition of the crisis. The clinical and laboratory signs of acute MH reactions are described in detail in Chapter 1.

How is the diagnosis made? In many situations the full-blown, unmistakable picture of MH is not present. A tachycardia, occasionally with arrhythmias, is accompanied by a gradual increase in temperature that, due to the brevity of the surgical procedure, is not impressive. The anaesthesiologist must be alert to these abnormalities and document the syndrome by arterial and venous blood studies. As have others (2), we have found that end expired $C0_2$ measurements are most useful. The primary signs and laboratory findings of hypermetabolism are the sine qua non of this disease state. Since the diagnosis is sometimes considered after the anaesthetic is terminated and the MH crisis is over, serial creatine kinase (CK) levels and blood and urine determinations for myoglobin can be helpful. It must be remembered that one-third of patients with MH will have normal CK on all determinations (3). Patients in whom this diagnosis is still equivocal after these measures should have a muscle biopsy performed under the direction of one of the several investigators in this area to confirm or rule out the diagnosis.

The simple, yet important, treatment for rigidity following succinylcholine is to terminate the anaesthetic. For emergency surgical procedures, precautions can be taken so that anaesthesia can be re-initiated utilizing a neuroleptic technique. Such precautions should include the preparations suggested for the anaesthetic management of the susceptible patient described in Chapter 15.

In addition to ceasing surgery, serial CK and myoglobin determinations and immediate blood gas studies are appropriate. CK elevations in the first twenty-four hours are often present from intramuscular injections, surgery, and emotional stress (4). These increases are 100 to 200 percent of normal laboratory values. Patients with rigidity can have astronomical increases that are very helpful in making a diagnosis.

To put the variety of MH manifestations into perspective, the experience of one community hospital in a six-week period is of assistance. The first patient was a seven year old undergoing a tonsillectomy and adenoidectomy. This girl developed full-blown symptoms of tachycardia, tachypnoea, cyanosis, vasoconstriction, and arrhythmias. She had a marked metabolic and respiratory acidosis with myoglobinemia and a CK of 18,000 i.u. per hour. Prompt, vigorous therapy brought about a happy outcome with no sequelae. Three weeks later a five year old boy having orchidopexy received thiopentone and succinylcholine for anaesthesia induction. He became quite rigid but was still able to be ventilated. The anaesthetic was terminated. No myoglobin was found in the blood or urine, but CK levels at six and twelve hours were 31,000 i.u. Two weeks later the operation was performed. It included a muscle biopsy and a neuroleptic technique was used. The third patient, a nine year old boy requiring tonsillectomy and adenoidectomy, developed a ventricular arrhythmia during halothane-nitrous oxide (50 percent)-oxygen (50 percent) induction by mask. Surgery had not yet begun. Due to lack of response of the ventricular arrhythmia to lidocaine and controlled ventilation, the anaesthetic and surgery were postponed. That evening, a CK determination was ordered. The result was a level of 105,000 i.u. Further studies and a muscle biopsy confirmed the diagnosis. These examples show that it is important for the anaesthesiologist to realize that MH can have a varied pattern of severity and symptomatology.

THERAPY

It is important to define--prior to anaesthesia and surgery--the steps one would take to handle MH. These are:

1. Stop the anaesthetic.
2. Administer 100 percent oxygen.
3. Start sodium dantrolene therapy 2.5 mg/kg.
4. Hyperventilate the patient.
5. Correct acidosis.
6. Treat hyperkalaemia.
7. Initiate cooling.
8. Maintain urinary output (1 to 2 ml/kg/hr)
9. After the syndrome is arrested, administer IV or PO dantrolene for 72 hours in an intensive care setting.
10. Prepare the treatment protocol and keep it accessible.

The first thing to do is to turn off the inhalation agents. Fear has been expressed by many that the patient will awaken if the anaesthetic is stopped, but this does not occur.

The next step is to administer 100 percent oxygen to the patient. In experimental animals, and most likely in humans, the first sign of the syndrome is marked venous desaturation.

If possible, the anaesthesia machine should be changed for MH cases to one free of contamination with inhalational anaesthetic agents. Alternatively, hoses, bag and soda lime should be changed.

Hyperventilation should be instituted since respiratory acidosis is a component of MH. PCO_2 recordings have been in the range of 100 mm Hg or greater during the syndrome, and thus there is little fear of decreasing cerebral blood flow owing to hyperventilation.

Once these initial steps have been introduced, monitoring should be instituted for the following:

1. Electrocardiogram
2. Temperature
3. Urine output (Foley catheter)
4. Arterial pressure
5. Central venous pressure

After the diagnosis of MH has been made and a report of the initial blood pH and PCO_2 is received, an appropriate dose of sodium bicarbonate should be administered. If blood gas determinations are not available but the anaesthesiologist has made a clinical diagnosis of MH, an initial dose of 132.9 mEq of sodium bicarbonate (i.e. three ampoules) given to a 70 kg patient would seem reasonable. For a child, 2 mEq per kilogram is appropriate. Further administration of sodium bicarbonate wil depend upon how long the fever continues to rise, and whether arrhythmias develop or cardiovascular status deteriorates. Trihydroxymethylamino methane has inexplicably but almost invariably been associated with a fatal outcome, and it is therefore suggested that sodium bicarbonate be used in preference to this agent.

Dantrolene sodium, 1-[[5-(p-nitrophenyl)furfurylidine]amino]-hydantoid sodium salt hydrate, was first shown by Snyder and associates (5) to be representative of a new class of muscle relaxants that apparently act directly on the skeletal muscle. It dissociates E-C coupling in skeletal muscle (6) and is useful in controlling clinical spasticity resulting from serious chronic disorders such as spinal cord injury, stroke, cerebral palsy, and multiple sclerosis (7). Dantrolene is the drug of choice for both MH prophylaxis and treatment. Dantrolene has been used to treat MH in animals (8-10) and humans (11,12). It is the only medication that will terminate MH in pigs when administered after the onset of the syndrome (13). Further details of the mode of action and use of dantrolene can be read in Chapter 14.

The following case report illustrates our early experience with a patient to whom dantrolene sodium was administered after the failure of conventional therapy in an episode of MH.

Case Report

A 68-year old Caucasian male was scheduled for surgical decompressive laminectomy to relieve intractable lumbar pain of unknown cause. No personal or family anaesthetic history could be correlated with MH. Preoperative medication consisted of 10 mg of Valium (diazepam) orally, 75 mg of Demerol (meperidine hydrochloride with acetaminophen) intramuscularly, and 0.6 mg of atropine administered one hour before surgery. The patient arrived in the operating room with a blood pressure of 120/80 and in atrial fibrillation with a ventricular rate of 84 beats per minute. He was given 2 mg of d-tubocurarine followed by 200 mg of thiopentone and 80 mg of succinylcholine, all intravenously. No trismus was noted on intubation, but immediately thereafter the patient became increasingly difficult to ventilate. The extremities became rigid. The patient's skin also felt hot. Blood pressure rose from 140/80 to 220/110 mm Hg and the pulse rate was 120 beats per minute. Nitrous oxide which had been administered in 50 percent oxygen, was discontinued, and the inspired oxygen concentration was increased to 100 percent. A rectal temperature probe was immediately inserted, giving an initial reading of 39.4^{o}C (103°F).

The patient was treated with 100 mg of Lasix (furosemide), 1 gram of Solumedrol (methylprednisolone), 500 mg of Pronestyl (procainamide) in 500 cc of 5 percent dextrose in water (D5W), and 2 ampoules of mannitol.

Arterial and venous blood gas determinations were obtained, revealing excellent arterial oxygenation with extreme venous desaturation. Initially, no bicarbonate was administered, although later, as the patient developed a persistent and severe acidemia, he required six injectables of bicarbonate (of 50 mEq each) to maintain pH in the 7.20 range. Arterial pH was 7.24, pCO_2 was 48, and pO_2 was 316 (FIO_2 100 percent) 40 minutes after the onset of the syndrome. Base excess was calculated to be -6.8.

In spite of all therapeutic measures, the temperature still remained at 38.9°C (102°F) with a persistent metabolic acidosis. Gastric lavage was begun with iced saline, and the CO_2 absorber was packed in ice. These measures still produced no success, with the temperature still 38.6°C after one hour's treatment.

Out of desperation, 0.5 gm of dantrolene was dissolved in 500 ml of D5W by breaking open the capsules. To enhance its solubility, 500 mEq of bicarbonate and 25 grams of mannitol were added to the D5W. The mixture was then infused through a Sorensen micropore blood filter. An additional 0.5 gram of dantrolene was prepared the same way and administered. After the dantrolene infusions, the temperature fell rapidly to 37.2°C within 20 minutes. Blood gases stabilized with a pH of 7.38, PCO_2 of 36, and PO_2 of 275 (FIO_2 100 percent). The patient remained normothermic and after four hours was transferred to the intensive care unit, where he was later extubated. Throughout the entire episode, serum potassium levels were within normal limits without therapy. No disseminated intravascular coagulopathy appeared clinically or on serial clotting studies. Urine output was high (more than 2 litres in three hours) with no myoglobinuria. The remainder of his hospital course was complicated by a low-grade fever which developed three days later and pneumonia with staphylococcus aureus, although blood cultures were sterile.

The final concentration of the dantrolene suspension prior to administration was 3 ug per mililiter. Harrison arrived at a similar concentration using mannitol and NaOH to dissolve this drug in a saline solution for experimental use in swine. Saline, in retrospect, was a poor choice of solvent since increasing the sodium concentration decreased the solubility of dantrolene. However, assuming that the syndrome did not suddenly subside spontaneously and that the dantrolene was therapeutic, this final concentration in saline was nevertheless effective in our patient.

ADDITIONAL THERAPEUTIC MEASURES

Arrhythmias that are seen in MH frequently result from high plasma potassium levels. It is important to remember that plasma potassium levels are pH dependent and that the most rapid method of dealing with elevated plasma potassium concentrations is during the administration of sodium bicarbonate. Sensible administration of sodium bicarbonate requires blood pH, PCO_2 and base deficit determinations. In a fulminant crisis, it is very difficult to correct pH above 7.20; often it cannot even be raised to 7.10. Vigorous treatment of acidosis is necessary as there may be rapid plasma potassium changes, and ventricular fibrillation and cardiac standstill may occur. Since MH affects other membranes besides skeletal muscle, arrhythmias may be due to direct cardiac effects of the syndrome. Huckell and his co-workers (14) at Toronto General Hospital reported cardiovascular abnormalities among MH patients diagnosed by caffeine contracture tests, and have been previously described by Britt (15). Huckell and associates warn that any young patient with a family history of sudden death, even if not related to anaesthesia, or one with unexplained cardiomyopathy, unexplained elevated CK, an abnormal EKG, or life-threatening arrhythmias may, although normothermic, be susceptible to MH.

In children, surface cooling with ice has been shown by Digby Leigh and his group (16) to be uniformly successful. In adults, a more rapid means of cooling is necessary. Intravenous cooling with iced saline solution is most effective. Other routes are:

1. gastric

2. rectal
3. vesicular
4. incisional.

Cooling is stopped when the patient's temperature falls to 38.3°C (101°F) and restarted if it rises again above this temperature. It is vital that the operating room always have a supply of iced saline solution present and available for immediate use. Rapid administration of iced saline solution (1000 ml every ten minutes for thirty minutes) has been shown to be helpful in lowering core temperature. Saline helps cool the aortic blood perfusing the whole body. Lavage of the stomach, bladder, rectum, peritoneal cavity, and thorax with iced saline is also useful. Cardiopulmonary bypass and gastric hypothermia have been utilized to exchange heat. These methods, however, can rarely be used in most hospitals, and unless they are part of a planned hospital protocol, they should not be attempted as a first line of defense in cooling the patient. Furthermore with the advent of dantrolene therapy, such treatment is now of secondary importance.

It is important to attain and sustain a high urine flow because myoglobinemia may precipitate into casts, yielding renal failure. Renal failure has resulted despite what appeared to be successful therapy. Myoglobinemia is a problem only if urinary output falls--myoglobinemia and myoglobinuria are not harmful if flow is maintained at a high level. Mannitol and Lasix should be administered to provide a urine flow of at least 2 ml per kilogram per hour. The generous use of iced intravenous fluids, with proper attention to the response in the urinary catheter and the level of the central venous pressure (CVP) line, will help prevent the possibility of postsyndrome renal failure.

Our policy is to administer a 20 percent dextrose solution in a bolus, accompanied by 10 units of regular insulin. Dextrose provides an energy source, and insulin affords the means of transferring sugar into the cells where it is utilized. The efficacy of this practice has not been established, but in view of the tremendous metabolic strain placed upon the whole organism, it seems to be a reasonable procedure.

LATE COMPLICATIONS

The late complications of MH have been outlined and their management discussed (17,18). One other problem has emerged in the past three years as a late complication of this syndrome: a recurrence of symptoms of MH following successful resolution of the initial episode. In reality, the initial onslaught continues, and renewed therapy is necessary to terminate the syndrome completely and prevent its recrudescence. The following two patients exemplify this complication.

Case Reports

The first was a muscular, 20 year old, 104.3 kg, 190.5 cm male with a closed comminuted fracture of the left tibia and fibula. There was no significant past medical history and no significant family anaesthetic history. After surviving an actively treated initial episode of MH, the patient was successfully treated for diffuse intravascular coagulation. During the next 20 hours, he was stable except for continued mild muscle rigidity, a markedly increased volume requirement (14.5 kg weight gain in 18 hours) to maintain central venous pressure, and a poor urine output that had increased to 0.48 ml per kilogram per hour after appropriate fluid and mannitol therapy. Leg abduction and flexion were met by moderate resistance throughout the postoperative period. The initial episode passed, but thirty hours later the patient suddenly became nervous and cyanotic, and his masseter muscles tightened. He became dyspneic and tachypneic. Despite ventilation with 100 percent oxygen, within twenty minutes the patient had a cardiac arrest. The presumptive diagnosis of potassium-induced arrest and recrudescence of MH was made. He was placed on cardiopulmonary bypass and dialysis. The first serum potassium level on bypass was 14.2 mEq/L. Arterial blood gas measurements showed a pH of 6.94, pCO_2 of 54, and PO_2 of 400. A treatment regimen of 15 mg/kg of intravenous dantrolene, 1.5 gram of mannitol, glucose, sodium bicarbonate, insulin, and cooling was instituted. When the serum potassium fell below 7.8, the heart spontaneously defibrillated and its output increased while on partial bypass. Despite vigorous dialysis, the potassium could not be reduced below 6.6 mEq/L.

Under pressor support, an unsuccessful attempt was made to wean the patient from bypass, but he succumbed to pulmonary edema and heart failure.

The second patient was a 31 year old, 177 cm, 102 kg man who was extremely muscular in appearance. He was admitted for bilateral inguinal herniorrhaphies. There was no prior history of surgery, and except for allergy to tetracyclines, he had no significant medical history. There was no family history of an abnormal response to anaesthesia. With cardiopulmonary bypass, the patient survived the initial episode of MH. He remained intubated for sixteen hours and demonstrated increasing rigidity of all extremities in the recovery room. Also, despite increased fluid administration, CVP remained low, urine output fell, and weight increased 24 kg in thirty hours. Serum potassium continued to rise, from 4.4 to 8.3 mEq/L, despite Kayexalate (sodium polystyrene sulfonate) enemas, peritoneal dialysis, mannitol infusion, and glucose and insulin therapy. A cardiac arrest occurred when the serum potassium reached 8.8 mEq/L. The patient was resuscitated, but despite pressor infusion, initiation of hemodialysis was not tolerated and he had a cardiac arrest and died.

PROTOCOL FOR TREATMENT OF MALIGNANT HYPERTHERMIA

Since one of the most important components of the treatment of this catastrophic complication is a planned regimen of response by attending medical personnel, we have prepared a card for each anaesthesia machine which the Massachusetts Society of Anesthesiologists, under the instigation of Dr. Pierce, has distributed to all its members. The value of this precaution has been verified in its first year of use. It is suggested that each hospital group of anaesthesiologists secure a similar written protocol.

The following guidelines for therapy have been printed on an index card:

1. **Definition:** A fulminant hypermetabolic crisis triggered by anaesthetic agents.

2. **Incidence:** Usually estimated at 1 per 20,000 to 1 per 40,000; men more than women; familial genetic transmission; 50 percent of offspring affected.

3. **Causative Agents:** Most frequently implicated are halothane and succinylcholine. Cases have been reported after methoxyflurane, enflurane, nitrous oxide-meperidine, spinal, ether, trichloroethylene, cyclopropane, gallamine, d-tubocurarine, isoflurane, lidocaine, and Carbocaine.

4. **Choice of Anaesthetic for Elective Surgery in Susceptible Individuals:** Neuroleptanalgesia (no cases of MH reported in over 800 anaesthetics with this technique); conduction anaesthesia using ester type local anaesthetic drugs. Especially AVOID local anaesthetics of the amide type, halogenated agents, depolarizing neuromuscular blockers.

5. **Clinical Findings:** Variable course; tachycardia and tachypnoea, pyrexia usually not the presenting sign.

A. **Early:** Sometimes muscle rigidity following succinylcholine; TACHYCARDIA; also tachypnoea, unstable blood pressure, arrhythmias, cyanosis, profuse sweating, rapid increase in temperature.

B. **Late:** Skeletal muscle swelling, left heart failure, disseminated intravascular coagulopathy.

C. **Laboratory:** Respiratory and metabolic acidosis, increased levels of K, Mg, myoglobin, serum creatine phosphokinase, hypoxaemia, hypercapnia ($PC0_2$ to 200 mm Hg) (pH 7.20 to 6.80), myoglobinuria.

6. **Mandatory Monitoring:** ECG, temperature, pH, $PC0_2$, $p0_2$, K, Na, CVP, arterial line for BP, blood gases, electrolytes, urinary output (Foley catheter).

7. **Supplies:** Have immediately available: sodium bicarbonate (12 amps.); iced IV saline (12 x 1000 ml bottles); furosemide (4 x 100 mg amps); mannitol (4 x 12.5 gram amps); dantrolene (36 x 20 mg amps); regular insulin (100 U x 1 amp); 50 percent D/W (x 50 ml); ice chips, extracorporeal perfusion apparatus in hospitals with experience in its use.

8. **Treatment:**

A. STOP ANAESTHESIA AND SURGERY IMMEDIATELY. Change rubber goods and anaesthesia machine.

B. HYPERVENTILATE with 100% O_2 (1 to 10 liter O_2 flows).

C. ADMINISTER DANTROLENE SODIUM IV 2.5 mg/kg IV or until syndrome is reversed.

D. ADMINISTER SODIUM BICARBONATE, 100 mEq STAT and up to 600 mEq over time, guided by arterial pH and PCO_2. Since increasing the blood pH drives potassium into cells, sodium bicarbonate is highly effective in reversing the severe hyperkalaemia.

E. ACTIVELY COOL PATIENT.

a) IV iced saline (NOT Ringer's lactate), 1000 ml every ten minutes for 30 minutes.

b) Lavage of stomach, bladder, rectum, and peritoneal and thoracic cavities with iced saline (3 to 6 liters).

c) Extracorporeal circulation and heat exchanger (femoral to femoral).

d) Surface cooling with ice and hypothermia blanket.

F. MAINTAIN URINE OUTPUT. Mannitol 12.5 gm IV; furosemide 50 mg IV (up to 4 doses each). Urine output greater than 2 ml/kg/hr will help prevent myoglobin casts in kidney leading to renal failure.

G. GIVE INSULIN. Add 10 units of regular insulin to 10 ml of 50 percent D/W and give as IV bolus.

REFERENCES

1. Lopez, J.R., Jones, D.E., Alamo, B.A., Allen, P.D., Papp, L., Gergely, J. and Sreter, F.A. Muscle and Nerve, in press 1986.

2. Triner, L. Anesthesiology 1981;55:482.

3. Britt, B.A. New Engl. J. Med. 1974;290:1140.

4. Nevins, M.A., Saran, M. and Bright, M. JAMA 1973;224:1382.

5. Snyder, H.R., Davis, C.S. and Bickerton, R.K. J. Med. Chem. 1967;10:807.

6. Ellis, K.O., Castellion, A.W. and Honkomp, L.J. J. Pharm. Sci. 1973;62:948.

7. Chyatte, S.B. and Birdsong, J.H. South. Med. J. 1971;64:830.

8. Anderson, I.L. and Jones, E.W. Anesthesiology 1976;44:57.

9. Gronert, G.A., Milde, J.H. and Theye, R.A. Anesthesiology 1976;44:488.

10. Harrison, G.G. Br. J. Anaesth. 1975;47:62.

11. Britt, B.A. In: Milton, A.S. (ed.). Handbook of Experimental Pharmacology, Pyretics and Antipyretics. 1982. Heidelberg: Springer-Verlag, pp. 547-615.

12. Wingard, D. Personal communication.

13. Harrison, G.G. Br. J. Anaesth. 1977;49:315.

14. Huckell, V.F., Staniloff, H.M. and McLaughlin, P.R. In: Aldrete, J.A. and Britt, B.A. (eds.). Second International Symposium on Malignant Hyperthermia. 1978. New York: Grune and Stratton.

15. Britt, B.A. In: Britt, B.A. (ed.). Malignant Hyperthermia. Int. Anesth. Clin. 1979;17:63.

16. Digby Leigh, M., Lewis, G.B. and Scott, E.B. Anesth. Analg. 1971;50:39.

17. Ryan, J.F. In: Henschel, E.O. (ed.) Malignant Hyperthermia--Current Concepts. 1977. New York: Appleton-Century-Crofts, pp. 47-56.

18. Wade, J.G. In: Gordon, R.A., Britt, B.A. and Kalow, W. (eds.). International Symposium on Malignant Hyperthermia. 1973. Springfield: Thomas.

Index